Conservation of Biodiversity for Sustainable Development

Edited by
O. T. Sandlund, K. Hindar and A. H. D. Brown

Scandinavian
University Press

Scandinavian University Press (Universitetsforlaget AS), 0608 Oslo 6, Norway
Distributed world-wide excluding Scandinavia by
Oxford University Press, Walton Street, Oxford OX2 6DP

London New York Toronto
Delhi Bombay Calcutta Madras Karachi
Kuala Lumpur Singapore Hong Kong Tokyo
Nairobi Dar es Salaam Cape Town
Melbourne Auckland

and associated companies in
Beirut Berlin Ibadan Mexico City Nicosia

Cover design: Knut Kringstad
Published with a grant from the Norwegian Institute for Nature Research (NINA)

British Library Cataloguing in Publication Data

Sandlund, Odd Terje
Conservation of Biodiversity for
Sustainable Development
I. Title
333.7

ISBN 82-00-21508-3

Printed in Norway by PDC, Aurskog

SERIES PREFACE

We are pleased to present "Conservation of Biodiversity for Sustainable Development" as the first volume in the series entitled "Studies in Conservation Biology". The series will be edited under the auspices of the Norwegian Institute for Nature Research (NINA) and published by the Scandinavian University Press. The series will publish high quality monographs and conference proceedings dealing with conservation biology in a wide sense.

Progress in conservation biology is a prerequisite for understanding the role of biodiversity in the functioning of the biosphere, and represents important steps towards a sustainable management of the world's natural resources. Scientists working in this field are invited to suggest topics for future volumes.

Karl Baadsvik
Director

Nils Chr. Stenseth and Bernt-Erik Sæther
Series' editors

CONTENTS

2, 6, 9, 10, 17, 19, 20

FOREWORD

In September 1990, 110 international and Norwegian experts in conservation biology and related fields of research and management met in Røros, Norway, to participate at the International Conference on the Conservation of Genetic Resources for Sustainable Development. The Conference, which was organized by the Norwegian Institute for Nature Research (NINA), summarized the state of the art of conservation biology, and analyzed the problem of biodiversity conservation from a wide array of perspectives.

Genetic and ecological research forms the basis for our understanding of conservation. However, to achieve any long-term conservation of Earth's biodiversity, the scientific results must be implemented within a wider social, economic and political perspective. This was clearly demonstrated at the Conference in a number of lectures and in the discussions leading up to the set of Recommendations from the Conference.

This volume makes the information presented at the Conference available to a wider audience. I feel that the topics discussed here are of utmost relevance to all scientists, managers and politicians who are engaged in the struggle to conserve and use sustainably the living natural resources of this planet. The issues discussed are of crucial importance to achieve a sustainable existence of humankind, and I hope that the publication of this volume contributes to the understanding of biodiversity and its conservation.

I want to thank the invited lecturers for their excellent contributions to the Conference, and for their additional efforts to process the lectures into the papers presented in this volume. Thanks are also due to the Organizing Committee, Kjetil Hindar, Bror Jonsson, Bernt-Erik Sæther, Odd Terje Sandlund and Nils Chr. Stenseth and to all Conference participants for their contribution to the successful event.

The editors of this volume, Tony Brown, Kjetil Hindar and Odd Terje Sandlund are acknowledged for their expedient work in turning out this volume within one year after the Conference.

The Conference at Røros and the publication of this volume were sponsored by the Norwegian Research Council for Science and the Humanities (NAVF), the Norwegian Ministry of Environment, the Norwegian Ministry of Foreign Affairs, the Directorate for Nature Management, the Norwegian Research Council for Agricultural Sciences, and the Norwegian Fisheries Research Council, which are all gratefully acknowledged.

Karl Baadsvik
Director,
Norwegian Institute for Nature Research

INTRODUCTION

A dominating problem in today's world is how to provide a decent standard of living to the Earth's human population without destroying the life systems on which the very existence of humanity depends. The solution to this problem has been termed "sustainable development", which may be defined as the use of natural resources without destroying the possibility for future generations of using the resources in the same way, if they wanted. This may seem a utopian goal, embracing two factors that cannot be reconciled: economic development, and conservation of biological resources. However utopian: there is no alternative to facing the challenge.

This volume is based on the International Conference on the Conservation of Genetic Resources for Sustainable Development, which took place at Røros, Norway, in September 1990, as one part of the effort by the Norwegian scientific community to follow up the conclusions of the World Commission on Environment and Development. In their report "Our common future", the Commission states that "Conservation of living natural resources – plants, animals, and microorganisms, and the non-living elements of the environment on which they depend, is crucial for development" (p. 147), and "..the effort to ensure food security requires .... a holistic approach focused on ecosystems at national, regional, and global levels..." (p. 144). The Conference was initiated by the Norwegian Research Council for Science and the Humanities (NAVF), and convened by the Norwegian Institute for Nature Research (NINA).

The aims of the Conference were: (1) To describe the state of the art of conservation biology, (2) to clarify central conservation problems and suggest research priorities, and (3) to outline how genetic and ecological knowledge can be used in international efforts for a sustainable development. This volume, based on the lectures and discussions at the Conference, maintains these aims.

In addition to the important updated review of the status of conservation biology, concerning both wild and domesticated organisms, this volume stresses that the scientific know-how must be placed in a social and political framework. Successful management of living resources depends on a proper basis of information from a wide array of scientific disciplines. While humankind through most of its history has manipulated nature for its own benefit, modern technology has provided means of more profound manipulation than ever before. This raises important philosophical and ethical questions of humankind's role in nature, and stresses that management of living resources also contains an ethical aspect.

Resources for survival The future of humankind depends in large measure on the genetic resources of wild and domesticated organisms. This is obvious when these organisms produce food or shelter, but it is equally true for the less conspicuous organisms performing their function in the ecosystems that comprise life on Earth. Even from a strictly utilitarian point of view, any organism might in the future turn out to be of the greatest direct benefit for man. Suffice it here to mention the fungus which provides penicillin.

Concurrently with an increasing acceptance of the essential importance of biodiversity, the present activities of humans are causing the highest rate of reduction in biodiversity ever. In this we are all to blame. The industrialized countries pursue an economic policy that exploits natural ecosystems globally for short-term economic gain. Through their overconsumption, the industrialized countries also cause more than a fair share of the world's environmental change. In developing countries, the ever increasing demand for agricultural land leads to large scale deforestation and erosion.

Conservation and management of living natural resources All over the world, natural ecosystems are replaced by low-diversity agricultural systems, or by roads, industry, mining, and cities.

The encroachment of human activities on all natural ecosystems means that an important element in the efforts to save natural biodiversity is the establishment and maintenance of conservation areas (national parks). Biological research related to this concerns such questions as the size of the conserved area necessary for saving the various species and ecosystems involved. One important subject for debate is whether the focal point of conservation biology should be the ecosystem or the population, and whether keystone species may be used to monitor the survival of whole ecosystems. The preservation of rare or near-extinct species is another important subject. Many conservation areas have been established to save remnant populations of spectacular species, and some species are only surviving in very small numbers in captivity. Related to this, research on small populations has focused on concepts like minimum viable population size, and whether genetic or demographic processes are the more crucial for the long-term viability of small populations.

In this context, the political, cultural and social setting of the conservation area may pose a more immediate problem than the resolution of the scientific issues. The lesson to be learnt from a number of cases all over the world is that to ensure long-term conservation of any natural area, neighbouring communities must be involved in its management. The benefits provided by the conservation area, in terms of income opportunities, water supply, and protection from erosion, must be realized by the local community. The alternative, at least in many poor communities, is encroachment on the conservation area through poaching, slash and burn agriculture, or clear-cutting.

As the additional land areas available for cultivation dwindles, the only possibility to feed the increasing human population lies in the increased productivity of agriculture. The cultivated plants and animals that produce most of the food consumed by humans have been selected through hundreds or thousands of generations for traits beneficial to humans. The selection performed by the local community or individual farmer has created a diverse array of local varieties of many species. Thus, traditional agriculture has a wide genetic basis. Modern industrial agriculture, however, bases its production on a few genetically homogeneous high-yield varieties, so that genetic variation in domestic organisms may be very low over large areas. The high-yield varieties possess little resilience to new biotypes of pests and diseases, or environmental changes, and depend on high inputs of water, fertilizers and pesticides to produce the high yield.

To maintain the necessary genetic variation in high-yield varieties, the different traditional varieties and the wild relatives of domestic plants and animals must be available to the agricultural scientists. For this purpose, a well developed system of gene banks for seeds and sperm has been established for a number of domestic organisms.

Gene banks for wild organisms also exist in the form of botanical and zoological gardens. Although originally established for other purposes, these institutions now play a crucial role in securing continued existence of numerous wild species deprived of their natural habitats.

The development of gene technology has radically improved our ability to manipulate organisms for our own ends. The potential to "tamper with life" is perceived as ethically disturbing by scientists and laymen alike. However, the potential benefits gained from gene technology are so manifold that our only solution is to learn to live with it, and the practical use of gene technology must be based on sound ethical codes of conduct. In addition, the ecological and genetic consequences of releasing genetically engineered organisms to the environment must be carefully evaluated prior to any such release.

A challenge for science and management

The extinction of habitats and species, the reduction of genetic variation in cultivated organisms, and the manipulation with genomes, all pose a threat to the genetic resources on which sustainable development depends.

The basis for a sound management of living natural resources is information produced by biological and ecological research. The many aspects involved in conservation and sustainable management demand holistic approaches and collaboration between professions, without diminishing the demand for quality. The output from scientific work may also be more easily applicable in practical management if scientists evaluate their results in a wider social and cultural context.

It is our intention that this volume will contribute to the progress of the various aspects of research dealing with the conservation of biodiversity. Sustainable management of living natural resources demands that the biological and ecological information is combined with profound knowledge of local socio-economic and cultural structures. The slogan "think globally; act locally" illustrates pointedly the guideline for actions to safeguard Earth's genetic resources for the future.

We wish to extend our gratitude to Liv Bjørnaas, Knut Kringstad, Hilde Meland, Eva M. Schjetne, and Kari Sivertsen at NINA for their help with figures and typing.

Odd Terje Sandlund, Kjetil Hindar
Norwegian Institute for Nature Research,
Tungasletta 2, N-7005 Trondheim, Norway

and

Anthony H. D. Brown
CSIRO Division of Plant Industry,
Canberra, A.C.T. 2601,
Australia

SECTION 1

Biodiversity

Today the world's biodiversity – the structural and functional variety of life forms at genetic, population, species, community, and ecosystem levels – faces the most serious threat since humans appeared.

The biodiversity crisis is a global as well as a local problem. Consequently, its solution must be found on the global, regional, national, as well as the local level. In this section the actions needed on the global level are outlined, and an example of a well developed national plan for long term biodiversity conservation is presented.

Jeffrey A. McNeely (Chapter 1) argues that the primary cause for the loss of biodiversity in tropical countries is the inequity between rich and poor countries in distribution of power, information and resources. Global trade undervalues natural resources, and poor countries that only produce biological raw materials are kept in poverty. Although every country in the long run benefits from biodiversity conservation, the poor countries bear a disproportionate share of the costs. There is an urgent need for a global strategy for the conservation and sustainable use of biodiversity. The strategy must define the options and opportunities for action capable of achieving global goals while addressing local priorities. The International Union for the Conservation of Nature, the World Resources Institute and United Nation's Environmental Program are presently preparing such a strategy in collaboration with a number of other agencies. The strategy deals with all aspects of biodiversity, including marine, freshwater and terrestrial ecosystems at all latitudes, and addresses themes like international cooperation in biodiversity conservation, institutional and policy reform, and financing mechanisms for conserving biological resources.

An innovative national plan to come to grips with many of these problems on the local level is presented by Daniel H. Janzen (Chapter 2). Costa Rica's National Biodiversity Institute (INBio) is the focal point of one of the first examples worldwide of concrete action to fulfill a national plan for sustainable use of biodiversity. The biodiversity plan also includes a reorganization of the country's national parks to ensure that as much as

possible of the benefits and revenues created by the national parks are realized by the local communities. INBio's plan is to create an inventory of Costa Rica's 500,000 species within a decade, while disseminating the collected biodiversity information to all parts of society, including the educational system, the tourist industry, and the biotechnological industry. The mapping of biodiversity is done by the development of a set of new professions like parataxonomists and apprentice curators. The parataxonomists are "laypersons" from a variety of backgrounds. They receive a short course and subsequent on-the-job training in the technical skills and basic taxonomy needed for collection and preparation of biological samples. The apprentice curators have a 3–5 year biology education from a local university, and receive on-the-job training in curatorial skills by collaborating with international taxonomists staying at INBio for varying periods of time. INBio is establishing itself as a professional partner to biotechnological and pharmaceutical industries in the trade with biodiversity information. This generates foreign currency incomes and job opportunities without destroying the conserved ecosystems. The Costa Rican model is currently being imitated by several other countries such as Mexico, Chile, and Taiwan. The present, detailed description of the experiences of INBio in Costa Rica provides a host of ideas as to how to go about conservation and sustainable use of natural biodiversity in developing as well as developed countries.

1 The biodiversity crisis: challenges for research and management

Jeffrey A. McNeely
IUCN – International Union for Conservation of Nature and Natural Resources
CH-1196 Gland,
Switzerland

ABSTRACT

Biodiversity must be conserved as a matter of principle; as a matter of survival; and as a matter of economic benefit. Despite its importance, biodiversity is more threatened than at any time since the extinction of the dinosaurs. The immediate causes of the biodiversity crisis are habitat destruction and degradation, but the ultimate causes include: inequities in the distribution of power, information and resources; the effects of global market forces and market failures that misvalue natural resources; separation of environmental conservation and economic development, often involving the decline of indigenous systems of resource management; unsustainable levels of resource demand at a global level; and lack of an ethical commitment to sustainability. Inequities within and between countries are especially pronounced with respect to biodiversity; everyone benefits from conservation of biodiversity, but lower-income countries bear a vastly disproportionate share of the costs. Meeting the management challenge of the biodiversity crisis requires transforming attitudes and practices concerning biodiversity; building a global alliance to conserve biodiversity; empowering communities to conserve biodiversity and use it sustainably; integrating conservation of biodiversity with economic development; and reducing and stabilizing human impacts on biodiversity.

INTRODUCTION

Our species entered the industrial age with a population of one billion and with biological diversity – the total of genes, species, and ecosystems on Earth – possibly near an all-time high. Biological resources – the portion of diversity of actual or potential use to people – were freely available for exploitation to support development. In the late 20th century, we are coming to realize that biological resources have limits, and that we are exceeding those limits and thereby reducing biological diversity. This is a time of extraordinary change in the relationship between people and the biological resources upon which their welfare depends. More people are added to the human population each year than ever before, spe-

Conservation of Biodiversity for Sustainable Development.
Edited by O. T. Sandlund, K. Hindar and A. H. D. Brown. 1992. Scandinavian University Press, Oslo.

cies are becoming extinct at the fastest rate in the past 65 million years, and climate appears to be changing at its fastest rate ever.

In times like these, maintaining maximum biological and cultural diversity assumes far greater urgency because diversity provides the raw materials with which different human communities will adapt to change. The loss of each additional species reduces the options for nature – and people – to respond to changing conditions.

But instead of conserving the rich resources of forest, wetland, and sea, current processes of development are depleting many biological resources at such a rate that they are rendered essentially non-renewable, leading to forms of development that are not sustainable. Instead of developing, the world is being impoverished by the loss and degradation of its most fundamental capital stock – its genes, species, habitats, and ecosystems. Some scientists predict that if present trends continue, some 25 percent of the world's species will be lost in the next 25 to 50 years; deforestation, desertification and destruction of wetlands are widespread phenomena; and even in economically important species, genetic diversity is being lost as systems become more simplified.

The "biodiversity crisis" is a tragedy, in the classic sense of an inevitable conflict between the protagonist and a superior force, which reaches a sorrowful or disastrous conclusion that excites pity or terror *(Webster's 3rd New International Dictionary)*. Our planet's diversity is the protagonist, and I'm afraid that the growing consumption of resources by our species is the superior force. The outcome is likely to be pitiful and terrible, unless we are able to mobilize much stronger forces for conservation.

This paper will suggest why we should care about losing biodiversity, why diversity is being lost, and what can be done about it.

WHY CONSERVE BIODIVERSITY?

Many people feel that biodiversity should be conserved as a matter of principle. And indeed governments have agreed with the ethical position of conserving biodiversity, accepting in the United Nations Charter for Nature that all forms of life warrant respect regardless of worth to people. The Charter states that the genetic viability on the earth shall not be compromised; the population levels of all life forms, wild and domesticated, must be at least sufficient for their survival, and to this end necessary habitats shall be safeguarded.

The loss of the living richness of the planet also has profound implications for development as the World Commission on Environment and Development recognized in its influential 1987 report (WCED 1987). Biological resources are renewable: forests, fisheries, wildlife, and crops reproduce themselves and can even increase when managed appropriately. Natural habitats – those which maintain their productivity without significant management inputs from people – have long provided local people with the means for survival, supplying food (meat, nuts, fruits, vegetables), fodder, firewood, construction materials, medicinal plants, wild genes for domestic plants and animals, and so forth. In more recent years, forests and fisheries have been major contributors to export-oriented economic development. Further, the highly diverse natural ecosystems which support this wealth of species also provide important ecological services. These include maintaining hydrological cycles, regulating climate, contributing to the processes of soil formation and maturation, storing and cycling essential nutrients, absorbing and breaking down pollutants, and providing sites for inspiration, tourism, recreation, and research (see McNeely et al. 1990, for further discussion of the above).

THE DISTRIBUTION OF COSTS AND BENEFITS: THE MAJOR PROBLEM

Given the importance of biodiversity, why is society permitting so much of it to be lost? At the risk of over-generalization, it seems to me that the major problems of conservation lie not in the biology of the species concerned – the topic that most interests ecologists – but rather in the social, economic, and political arenas within which people operate.

Biological resources are being abused because some people – typically those with the greatest economic stake in development – are making money out of depleting wildlife and habitat. And they are making money because they are harvesting the benefits of nature without paying the environmental costs. Instead, the costs are passed on to society as a whole, to be paid by all of us, now or in the future. The unscrupulous have always needed to be constrained by society. Unfortunately, it has taken society a long time to recognize that quick-buck exploiters and messy polluters both externalize costs and damage society's wider interests.

Who benefits from over-exploitation?

A considerable body of information demonstrates that living natural resources are being over-exploited (that is, harvested at rates far in excess of what can be sustained, leading to reduced productivity), and that many species are being lost as a result. For example, the most recent estimates suggest that more than 142,200 km^2 of tropical forests were lost in 1989, a rate of about 27 ha per minute (Myers 1989); only a small portion of this land was converted to permanent agriculture, with most being degraded to low-productivity pasture, scrub, or secondary forest. Since tropical forests hold well over half the world's species, and many of them are confined to relatively small areas, the number of species being lost is certain to be very high indeed; some experts suggest that at least several hundred – or even thousand – species are being lost every year (mostly invertebrates). Where these species are lost as a result of over-exploitation – that is, non-sustainable use – the public has great cause for concern.

Many forces are blamed for this over-exploitation, but it is useful to ask who has benefitted from it. The short answer is that the lion's share of the benefits have flowed into relatively few pockets, and that most of the profits are earned by the wealthier sectors of the population. In the case of tropical forests, most governments have nationalized their forests and let harvesting rights to timber concessionaires while collecting only a small portion of the economic rents available (Repetto and Gillis 1988). In Brazil, World Bank studies have shown that much of the deforestation has been stimulated by government subsidies of a small handful of relatively wealthy ranchers who have cleared the forest for cattle pastures (Binswanger 1987). Numerous similar examples are now available in the economic development literature.

In general, wild nature (which contains most non-domestic biological diversity) has been treated as a public good, benefitting everybody now and in the future, without individuals being charged for these benefits (or even realizing that they are benefitting). When the natural system being harvested appears to be limitless, or has a very low price placed upon it – which is often the situation of tropical forests – then the system is easily subject to over-exploitation. Such over-exploitation is often in the interest of the individual seeking to maximize profits (Clark 1976).

Another, much less comfortable, answer is also available. The beneficiaries of over-exploitation are all consumers, including us. Over-exploitation keeps prices down, which we like; we ignore the fact that these prices are kept artificially low because we don't pay the full costs. But – like national debts – this painful fact is easy to ignore.

from those systems are inevitably reduced as well (Hodgson and Dixon 1989). The timber concessionaire typically externalizes such environmental costs, if indeed he is even conscious of them.

Who pays the costs of over-exploitation?

Some conservationists would answer this question by saying "everybody", to the extent that everybody benefits from biological diversity and therefore suffers when it is reduced. This is not a totally satisfactory answer. Nature certainly has some built-in redundancy, and some species could disappear (indeed are disappearing) without anybody missing them. But few data are available on which species are particularly important in the functioning of ecosystems, so we can't really determine the extent to which the general public is suffering from the loss of biological diversity.

But in the case of tropical forests, the people who pay the greatest costs of over-exploitation are very often the people who live closest to the forest. These local people had long used the forests as the source of their irrigation water, construction materials, medicinal plants, and game animals.

It is also informative to consider who does **not** pay the costs. Very often, the individuals who are earning excessive rents from resource exploitation do so because they do not need to pay the environmental costs (in economic jargon, they "externalize" the costs). For when logging activities affect downstream agricultural systems, as by reducing the quality or supply of irrigation water or reducing the storage capacity of a reservoir through increased siltation rates, the affected farmers pay. When harvesting timber disrupts functioning ecosystems, as by killing coral reefs by run-off sedimentation, the benefits derived

Who earns the benefits from conserving biological diversity?

Two broad kinds of benefits are derived from conservation. Direct benefits are obvious from sustainable harvests of biological resources, contributions to ecological functions, tourism, and other measurable goods and services. It is less easy to define or measure the benefits from the more abstract existence, option, and bequest values, though many economists are attempting to do so.

It is clear that conserving biological diversity – that is, using biological resources in a sustainable manner which does not reduce the overall diversity of the ecosystem involved – can bring very considerable benefits to those who directly harvest the biological resources. Such conservation – what many of us would call effective management for sustainable yields – has been a common approach of rural people since time immemorial; and indeed, a precondition for the survival of societies throughout history has been the capacity to manage their resources for a sustainable yield. Those societies which were unable to use their resources sustainably have been forced either to change, or to disappear; the archeological record is littered with examples. The modern consumer-based society based on international trade has forced traditional cultures to adapt to the needs of the dominant society. The new governments established in the tropics in the post-colonial period have established government resource management agencies which have tended to fragment responsibility along sectoral lines and have made insufficient investments to be very successful.

When protected areas – an invention of the Industrial Age – are managed for tourism or game production, insufficient benefits have gone to the local people. In Zambia's Luangwa Valley, for example, both national parks and hunting areas have been managed primarily for the benefit of outsiders (tourists and safari hunters), with the earnings going to the central government and the private sector in Lusaka. Not surprisingly, the local residents, denied access to resources previously under their control, have become increasingly impoverished and resentful (Leader-Williams 1990; Marks 1984). While there are a few bright spots, the generalization is that relatively few direct benefits have flowed to the people who live closest to the over 3,000 protected areas that have been established in tropical countries over the past 50 years. They may, however, have earned benefits from the conservation of ecological functions such as watershed protection; some of India's well-protected tiger reserves, for example, are now providing year-round water instead of seasonal flows.

Many of the benefits earned from existence value are being earned by "free riders," who obtain them without having to pay the costs involved. Conserving the tiger in India is of much greater importance to comfortable urban-dwellers than to the local people who lose cattle or family to tigers; school children in northern cities may care far more about orangutans and Javan rhinos than do farmers in Java. Farmers (and consumers) in Europe benefit when Peru conserves wild relatives of potatoes, or when Mexico conserves wild relatives of maize. When tropical countries conserve a natural area containing a high degree of biological diversity, the wealthy nations of the north earn considerable benefits while paying few costs (enjoying what economists call a "consumer surplus"). But the tropical countries also earn option and bequest values, which presumably contribute to their conservation decisions.

Who pays the opportunity costs of conserving biological diversity?

At both the international and local levels, the opportunity costs of conserving biological diversity – that is, the benefit foregone by using a scarce resource for one purpose instead of for its best alternative use – are paid disproportionately by the people who live closest to the greatest biological diversity. In one of conservation's greatest ironies, the nations which have the most species tend to be the poorest developing countries, and within those countries, the individuals who live in the most diverse ecosystems – the tropical forests – are often the poorest of the poor.

In another of conservation's ironies, when a central government establishes a national park because of its outstanding value to the nation and to humanity, it simultaneously denies access to any marketable value of the area to the local people who once benefitted from the goods and services that the area formerly provided, and which the national park was established to protect. For example, in mid-1990, 6,000 Bolivian amerindians began a 640 km protest march about the disruption to their lives caused by a scheme set up by conservationists in the USA to preserve part of the Amazonian rainforest where they live, since they had been denied their traditional rights to the resources in the area.

So the opportunity costs of modern conservation programmes which restrict access to resources are falling disproportionately upon the very nations and communities that development assistance agencies are most interested in assisting. If conservation programmes are to be socially accepted, then new and more appropriate means of apportioning opportunity costs, or providing compensation for them, need to be sought. Economic incentives can provide one mechanism for doing so (McNeely 1988).

If the mal-distribution of costs and benefits of over-exploitation and conservation is a major problem facing humanity today, then solutions should be sought through the better distribution of these costs and benefits.

CHALLENGES FOR RESEARCH AND MANAGEMENT

As suggested above, the major problems of conserving biological diversity lie not in the biology of the species concerned but rather in the social, economic, and political arenas within which people operate. Finding solutions to the biodiversity crisis in this quagmire is our challenge.

How can we reverse the steady trend toward depletion? The simple answer is that policy-makers need to define the actions that are both necessary **and sufficient** to save the biological diversity upon which society depends, and present these actions in terms that are likely to be implemented. The actions need to be broadly conceived, and be endorsed at a high level by the institutions whose cooperation is essential to the conservation of biological diversity.

In short, we need to awaken social conscience and reaction. This has already begun to work for pollution, which is now a crime in some countries; some courts are beginning to assess heavy damages on the irresponsible. We now need to generate a similar response for those who deplete biological resources.

But this simple solution is exceedingly difficult to implement. Biological diversity has global, national, and local dimensions, but actions at these various levels are not very well coordinated, and indeed often work against each other. While one country may wish to save its tropical forests, for example, other countries are promoting trade in commodities which replace tropical forests; all too often, the different sectors are each working on their own priorities, with little attention paid to the impacts of this work on "external" sectors.

Therefore, a strategy is required to define the options and opportunities for action capable of achieving global goals while addressing local priorities. The strategy needs to be supported by regional, national, and local institutions, and backed up by sectoral strategies and action designed to meet specific needs. The International Union for Conservation of Nature and Natural Resources (IUCN), the World Resources Institute (WRI), and the United Nations Environment Programme (UNEP) have just begun to prepare such a strategy in collaboration with a number of other agencies.

Dealing with all aspects of biodiversity, including both marine and terrestrial ecosystems at all latitudes, the strategy is designed to:

- Stimulate vastly increased international cooperation on the conservation of biological diversity;
- Spur institutional and policy reform to address the root causes of the loss of biological diversity and to encourage sustainable use of species and ecosystems;
- Foster agreement on a long-term work plan to defend, understand and use wisely biological resources;
- Promote the development and application of new financing mechanisms for conserving biological resources; and
- Encourage a substantially increased emphasis on research that must accompany conservation action, including the inventory of the world's biological diversity, the study of ecological pro-

cesses, and the study of conservation biology.

The Strategy will convert these broad goals into reality through developing new action in seven major interconnected areas:

1. Building a broader constituency for conservation

Conservation has often suffered from being seen as a highly specialized sector of only limited relevance to the real needs of society. But consider some of the other sectors which depend on nature's bounty for their welfare: agriculture depends on wild genetic resources and biological controls; medicine draws on nature for pharmaceuticals; industry requires raw materials, all of which ultimately come from nature and many of which are biological in origin; tourism increasingly bases its attractions on natural amenities; indigenous peoples' groups have a deep vested interest in ensuring the conservation of the biological resources upon which their cultures are built; water resources often require intact forests to protect watersheds and prevent siltation; the military uses natural forests along international boundaries to maintain national security; fisheries depend on swamp forests and mangroves as nursery grounds; disaster prevention needs natural vegetation to prevent landslides and floods. Many others could be mentioned. Each of these sectors needs to be reminded of their reliance on biological diversity, and provided with opportunities to invest in ensuring the continued survival of the biological basis of their prosperity.

At a more personal level, conservation action needs to be promoted and implemented by development practitioners, engineers, politicians, rural sociologists, agronomists, and economists, in addition to the more usual (though still regrettably thin on the ground!) biologists, protected area managers, and ecologists.

A broader constituency also implies broader responsibility. While support for government conservation agencies needs to be increased, we also need to look for opportunities for conservation where the effectiveness of government action is necessarily limited. Even the most prosperous governments are not able to conserve their nation's biological diversity by themselves; instead, incentives and disincentives are required to promote conservation-oriented human behaviour. This will require that a wide range of resource management structures be established in the private sector, involving exlusive use rights, zoning, easements, tax benefits, private reserves, game ranches, and so forth. These "carrots" need to be backed with with appropriate "sticks," such as legal constraints, financial penalties, taxes, and public opinion.

2. Bringing local people into the conservation movement

One of the best ways to build stronger support for conservation is to support indigenous conservation efforts by local people. Part of the richness of human life around the globe is due to the inter-relations between people and their local environments. The loss of cultures, or of traditional knowledge within cultures undergoing rapid change, is a problem which is at least as serious for humanity as is the loss of species. All who follow will share the loss of traditional knowledge about the local environment. Crucial knowledge about how that environment might be used to provide benefits on a sustainable basis may be lost forever, along with the species that have supported human welfare for thousands of years.

The consumer culture will not last forever, but the earth abides. The numerous local cultures which are adapted to sustainable use of locally-available resources may well provide the resilience which will enable our species to adapt to whatever changes the future may bring.

Ironically, conservationists have often considered local people as the enemy rather than as potential allies. True enough, local people are often poachers or encroachers on protected areas. But they are often also the ones who have the best knowledge about the local species, and who have a real stake in ensuring that local biodiversity can continue to support their communities. Problems arise when outside forces – such as foreign markets for ivory or rhino horn, large numbers of tourists, or timber concessions on lands traditionally owned by the community – start to compete for local resources, leading to a free-for-all where the biggest losers are the wildlife, and ultimately the local people. But where local people are given greater responsibility for resource management – when they are given a real stake in conservation – they have shown both a willingness and a capacity to conserve. Extractive reserves in Brazil and the Kuna Yala in Panama are just two of numerous examples that could be quoted. What kinds of local involvement can help conservation, and under what conditions?

If biological resources are to be used to generate productive employment, better health care, and improved nutrition, and make other major contributions to sustaining society, we need to help build local capacity to study and manage biological resources. This capacity needs to include the resource management agencies, agricultural research institutions, universities, local people, and others. Such local capacity is essential for building management structures which are consistent with the local cultures, and may often build upon the traditional management approaches which have proven successful for thousands of years (such as the practices of forest-dwelling peoples in Amazonia); but more modern approaches are also required.

3. Expanding the utility of biological resources

It may seem a bit strange to attempt to conserve biological resources by finding new ways to use them. But on the other hand, people tend to value the parts of the environment that they use; and we all "use" nature, even in a wilderness – it renews the purity of air and water, and maintains wildlife resources of value to society. Developing new approaches to using wild biological resources in a sustainable way could help ensure their survival.

Plant breeders fight a continuous battle against evolving pests and changing soil and climate conditions; plant breeding, often drawing on the wild relatives of domestic plants, is essential to the future of human civilization. Further, agricultural development to date has focussed on relatively few crops that respond well to market forces, mechanization, irrigation, and agricultural chemicals. Further development may well require much more site-specific adaptations, drawing in the first instance on plants that are already well adapted to local conditions. On the other hand, breakthroughs may also come from exotics, or at least from genes from outside the system; this again calls for conserving the diversity of crops wherever they may occur. Many more plants can provide benefits to humanity than are currently recognized by international agencies; these plants are often well recognized by local people, and their wisdom needs to be better harnessed in the development process.

Finally, the moral justification for a system of protected areas or any other programme to conserve biological diversity is strengthened when it generates a flow of benefits to people. The more people benefit directly from the proper conservation and use of wild resources, the greater the incentive for them to protect the resource and the lower the cost to government of doing so. If a government decides that an area is important to be conserved, and this involves opportuntiy costs to local communities, then it would seem only appropriate that these local communities be compensated for their opportunity costs through a series of economic incentives. In fact, one could argue that the cost-benefit ratio of conservation must ultimately be positive for the local people if the resources are to survive.

4. Ensuring that sufficient information is available

Our ignorance about biodiversity is vast. While only about 1.4 million species have been described, most biologists believe that 10 million is a reasonable minimum figure, with the maximum figure perhaps reaching 80 or 100 million (May 1989; Stork 1988). And we are very poorly placed to fill in the blanks. For example, the distinguished Harvard entomologist E. O. Wilson has pointed out that while ants and termites make up about a third of the animal biomass in tropical forests and cycle a large part of the energy in all terrestrial habitats, only eight entomologists worldwide have the general competence to identify tropical ants and termites (Wilson 1985).

It seems self-evident that increasing knowledge about the kind and variety of organisms that inhabit the earth – and the ways that these organisms relate to each other and to humans – must be a foundation of conservation action.

But ignoring our ignorance for a moment, we don't even utilize the information that we do hold; most decisions which affect biodiversity today are taken without anything remotely resembling a full consideration of the available information. The Biodiversity Programme is working with the World Conservation Monitoring Centre to prepare a compendium of current knowledge on the status and distribution of global biological resources, the efficiency of resource management, the uses and economic values of biological resources, and the legal, financial and institutional basis for conservation. It is hoped that this will help ensure that data are available in the proper form for those who need it, when they need it. We can no longer allow ignorance to be an excuse for inaction, or for inappropriate action.

5. Applying science to management

Very considerable field work has been carried out on biological resources, and the science of conservation biology has made important contributions to the theory of species relationships, population genetics, minimum viable population sizes, energy flow, protected area design, and many others (see, for example, Soulé 1986, 1987). But far greater effort is required to ensure that the results of this science hit the ground, especially in the tropics. New ways need to be found for interpreting science, and transforming it into practical action which can be implemented by resource managers, including farmers, pastoralists, foresters, and park wardens. An essential element in this process is building a scientific infrastructure within the tropical countries, an area where far greater effort is required, both in promoting the necessary research and in applying research to management.

Many threatened species find their last stronghold in protected areas, but as protected areas come under various social, economic, and political pressures, the survival of these species and ecosystems becomes increasingly problematic. The survival of other threatened species require that additional protected areas be established to ensure their survival. For still others, no viable population can be expected to be contained in a protected area; migratory species and large carnivores such as African wild dogs are good cases in point. What are the management options available to governments? What ecological information is required to monitor endangered species, and how can it be generated and presented in a form helpful to management agencies? How can wildlife managers in developing countries determine what management procedures are required and possible, and how can they obtain the expertise required for the more high-tech management approaches?

Another area where science can make a very major contribution is in restoring degraded ecosystems. Vast parts of the globe have been so abused by past human activity that they are no longer productive. With some effort, many such areas can be brought back to productive levels where they can make a contribution to society, either by supporting important biological resources or by taking pressure off of more natural areas.

6. Stimulating far greater investments in conservation

Once we demonstrate that conservation is a mainstream concern, as essential to human welfare as health, security, and education, then governments and development agencies will see new opportunities for investment. This is already beginning to happen, and the report of the World Commission on Environment and Development gave solid support to the trend of greater public support for environmental issues (WCED 1987). A new international convention on biodiversity is being negotiated, with the intention of making significant additional funds available for conservation; and the World Bank is exploring new ways to fund environmental projects (World Bank 1990). But to capitalize on this trend, conservationists need to package their ideas in terms that will be palatable to those controlling the very significant funds which can either deplete or conserve natural habitats. This will require demonstrating broad public support for conservation, and demonstrating in economic terms the contribution biological resources make to a country's social and economic development.

And then what if somebody agreed to provide, say, US$ 500 million a year for conserving biodiversity. Where would we invest first? And even this amount would be insufficient for doing everything – especially when tens of billions are spent in exploiting biodiversity – so efforts spent in deciding what to do first are usually well repaid in savings of time, finances, and personnel.

Determining priorities is a complex task. The world contains more variability than can be expected to be protected by explicit conservation programs; further, the capacity of governments or private organizations to deal with environmental problems is limited and many urgent demands compete for their attention. So those seeking to conserve biological diversity must be selective, and ask which species and habitats most merit a public involvement in protective measures. No generally accepted scheme exists for establishing priorities for the conservation of biological diversity, nor is it either possible or advisable for such a scheme to be devised. Different organizations and institutions can be expected to have different ways of establishing priorities because of their differing goals, for example focussing on selected species, centres of endemism or diversity, areas of highest biomass, or particular habitats. Each system has its own strengths and weaknesses, with the major point of difference being the objective for which the system was devised. But we need to specify the mechanisms that are available for determining priorities, and identify those specific areas which contain the greatest biological wealth, and which are under the greatest threat.

But we would be foolish to expect massive funds to continue flowing from afar in the long run. Instead, we need to identify mechanisms that will enable resource management systems to become eventually self-financing, ensuring that sufficient income earned from biological resources is returned to managing those resources in ways that will sustain their productivity.

7. Devising new institutional approaches to conserving biological diversity
Since biological resources provide the basis for sustainable forms of development in a wide variety of sectors, it is apparent that many institutions need to be involved in their conservation. A major constraint is the fundamental mismatch between the way natural systems function and the way governmental organizations are designed to manage these systems; natural systems are highly inter-related structures with many complementary functions, while government agencies for managing these systems are characterized by fragmentation, conflicts in use of resources and jurisdiction over that use, polarization of interests, and poor coordination and communication.

Certainly the direct resource management agencies need to be strengthened; but additional institutional mechanisms are required, expecially in sectors which use biodiversity as an attraction (e.g. tourism and recreation, zoos and botanic gardens), which are directly concerned with production of yields from biological resources (e.g. agriculture, forestry, and fisheries), sectors whose impacts on biodiversity may be indirect (e.g. highways, finance, education, the military), and sectors which draw on biological resources for direct consumption (e.g. traditional medicine, rural development), or to support production (e.g. manufacturing and handicrafts). Universities must be enabled to work more closely with the management agencies, and long-term research and monitoring efforts must become part of the management enterprise. If we are successful in expanding the constituency for conservation, what sorts of coordinating mechanisms are likely to be effective under what conditions? What needs to be done to further encourage governments and the private sector to work to the same objectives? How can government institutions separate those activities which they are uniquely qualifed to carry out from those which would be more efficiently or effectively done by other means, for example, non-govermental organizations or local communities?

CONCLUSION

Our most challenging times are still ahead, as the human population continues to increase, agricultural pests continue to breed resistance to our chemical warfare, our abuse of the atmosphere begins to change the climate, and our desire for material comfort further encourages us to ever higher levels of consumption of limited resources. Further, the easiest economic development activities have already been implemented, so acceptable solutions to social and economic problems are likely to become increasingly difficult in the coming years.

In the times of ecological, social and economic instability that are just around the corner, if not upon us already, the renewable and locally available biological resources – and the knowledge of how to use these resources sustainably – will be more important than ever. But it is time to go on the offensive, to stop fighting the biodiversity crisis as if it were a series of discrete battles over species extinctions. We need to get in front of the problems, and attack them at their social, economic, and political source.

LITERATURE CITED

Binswanger, H. P. 1987. Fiscal and legal incentives with environmental effects on the Brazilian Amazon. World Bank Report ARU 69: 1–48.

Clark, C. W. 1976. *Mathematical Bioeconomics: The Optimal Management of Renewable Resources.* Wiley, New York.

Hodgson, G. and J. A. Dixon. 1989. Logging versus fisheries in the Philippines. The Ecologist 19(4): 139–143.

Leader-Williams, N. 1990. Black rhinos and African elephants: Lessons for conservation funding. Oryx 24: 23–29.

Marks, S. A. 1984. *The Imperial Lion: Human Dimensions of Wildlife Management in Central Africa.* Westview Press, Boulder, CO.

May, R. M. 1989. How many species? pp. 61–81. In: L. Friday and R. Laskey (eds) *The Fragile Environment.* Cambridge University Press, Cambridge.

McNeely, J. A. 1988. *Economics and Biological Diversity: Developing and Using Economic Incentives to Conserve Biological Resources.* IUCN, Gland, Switzerland.

McNeely, J. A., K. R. Miller, W. Reid, R. Mittermeier and T. Werner. 1990. *Conserving the World's Biological Diversity.* IUCN, World Resources Institute, World Bank, WWF-US, and Conservation International, Washington, D.C.

Myers, N. 1989. *Deforestation Rates in Tropical Forests and Their Climatic Implications.* A Friends of the Earth Report, London, UK.

Repetto, R. and M. Gillis (eds) 1988. *Public Policies and the Misuse of Forest Resources.* Cambridge University Press, New York.

Soulé, M. E. (ed) 1986. *Conservation Biology: The Science of Scarcity and Diversity.* Sinauer, Sunderland, MA.

Soulé, M. E. (ed) 1987. *Viable Populations for Conservation.* Cambridge University Press, Cambridge, UK.

Stork, N. E. 1988. Insect diversity: facts, fiction and speculation. Biol. J. Linn. Soc. 35: 321–337.

World Commission on Environment and Development. 1987. *Our Common Future.* WCED, Oxford University Press, Oxford, UK.

Wilson, E. O. 1985. The biological diversity crisis: A challenge to science. Issues Sci. Technol. (Fall 1985): 20–29.

World Bank. 1990. Funding for the Global Environment. IBRD Discussion Paper, Washington, DC.

2 A south-north perspective on science in the management, use, and economic development of biodiversity

Daniel H. Janzen
Department of Biology, University of Pennsylvania
Philadelphia, Pennsylvania 19104,
USA

and

Instituto Nacional de Biodiversidad (INBio)
3100 Santo Domingo, Heredia,
Costa Rica

ABSTRACT

The only certain way to save tropical biodiversity is to use it sustainably. To use it requires saving it as conserved wildlands and determining what is in them. Both these steps must be carried out by local human resources and a home-grown scientific and administrative system, but the development of both of these requires facilitation by the international scientific community. This facilitation requires that the international scientific community abandons its contemporary view of tropical biodiversity as largely an intellectual resource to be harvested by the extra-tropical academic community. Sustainable use of tropical biodiversity requires goal-directed search for intellectual and economic uses, realistic evaluation of impact by users, recognition that the wildland administration itself is a user, and the establishment of a mechanism by which the user community truly pays for use. Income ranges from on-site entrance and use fees to direct payment for off-site environmental services to royalties derived from the direct commercialization of biodiversity information such as phytochemicals and genes. A mechanism must be found whereby these fees come to pay the management costs for tropical wildlands. A National Biodiversity Institute is a mechanism to promote and develop all of these processes in coordination with the institutional management of the conserved wildlands themselves. Examples from Costa Rica are used to illustrate all of these problems and processes.

INTRODUCTION

The only certain way to save tropical biodiversity is to use it sustainably. The users must range from its immediate neighbors to the most global corners of society. Under this premise, the tropical conservation challenge is converted from "how to select, decree and guard what tropical national park", to "how does one manage a biodiverse wildland area in such a manner that

Conservation of Biodiversity for Sustainable Development.
Edited by O. T. Sandlund, K. Hindar and A. H. D. Brown. 1992. Scandinavian University Press, Oslo.

its biodiversity remains relatively intact yet society's intellect and purse benefit from its presence"?

The process of biodiversity conservation through use has three overlapping steps, each of which is necessary but not sufficient for biodiversity conservation:

1. Save it.
2. Know what it is.
3. Use it sustainably.

Such a view moves the newly-labeled academic domain of "conservation biology" or "conservation science" from being a self-perpetuating process to being one of many means to achieve a goal. However, if conservation science is viewed as a means to a goal other than self-perpetuation, then its activity needs to be defined and structured with respect to that goal, rather than with respect to the traditions in which nearly all of its practitioners received their initial academic training.

If we scientists expect laymen and the commercial world to change in favor of biodiversity conservation and sane management of the environment – and surely the current demand for a 'sustained yield economy' is such a change – then we must also be willing to submit our own academic indoctrination to an intellectual and pragmatic goal-directed revision. No one said it would be easy. But all of us can see that it is necessary.

BACKGROUND

This essay touches on the interplay of systematics, conservation science, phytochemical ecology, and other kinds of biological science in the emerging area of the conservation of tropical biodiversity through sustainable use. I attempt to adopt the viewpoint of the tropical user society, the society that receives a product from its own tropical biodiverse conserved wildland – whether it is called a national park, game refuge, forest reserve or other similar legal designation. This viewpoint will appear relatively unsympathetic to many of the academic traditions in which I and many other biologists have spent, and still are spending, substantial parts of our lives. However, it is time to seek a new kind of fine-tuned and responsive contract between tropical academic science, local human problems, and local environmental processes.

Scientific research, and especially academic scientific research, is only a small – even if critically important – portion of the array of forces, processes and institutions that demand attention and fine-tuning if there is to be sustainable use and development of wildland biodiversity. The specific kinds of research required for real sustainable uses of tropical biodiversity are relatively straightforward and already practiced throughout the tropics by many people. However, if the questions and the researchers are isolated from society, the results generally only very slowly find their way into society. A yet more difficult problem is that research truly directed at the sustainable uses of biodiversity rarely involve the megaquestions and sweeping generalities that are so prized by the academic community.

I urge the reader to not confound scientific and academic traditions. The scientific way of thinking (double-checking facts, describing things as the way they are rather than as is politically convenient, posing clear questions, encouraging public discussion of hypotheses and conclusions, conducting controlled experiments, viewing as public the information used to arrive at a decision, placing a philosophy ahead of pecuniary benefit, non-acceptance of contradictions, the separation of a process or idea from the person who happened to think of it, the unwillingness to appeal to higher authority as a justification for a conclusion, etc.) is an extremely important methodology to bring to bear on the problems of the conservation of tropical biodiversity. This is particularly true in the tropics, where the distance between the scientific way of thinking and the lay processes of social life are often substantially greater than is the casc in much of the developed world. However, academia and quasi-academia (museums, government research institutes, etc.) have quite reasonably developed their own more specific traditions as to how the tools of science are to be applied and to what. It is these traditions – ranging from what is viewed as an accomplishment to who is a legitimate practitioner of an activity – that require minor to major revision in an effort to put tropical biodiversity to work sustainably.

Economics also enters into the south-north contrast in an unexpected way. For at least several decades, the application of science to tropical problems will be largely funded by the developed countries, either by direct subsidy, or indirectly by the in-country tax base generated by export goods (just as it has been for past decades). In this situation it is especially hard for tropical in-country scientists to be responsive to the needs and desires of their own society. However, it is responsiveness to these needs and desires that will eventually make that society want to fund the in-country science establishment, thereby allowing major portions of the scientific community to operate with the social freedom so characteristic of science in developed countries. Funds spent on 'nationalization' of a tropical country's science community may do as much for the conservation of tropical biodiversity as funds spent on commercialization of tropical science in response to an international market. The latter produces an increase in national GNP, the former an increase in votes for biodiversity conservation.

My goal is to use thoughts provoked by current experiences in Costa Rica to broaden the contemporary discussion of conservation of tropical biodiversity.

THE BASIC FRAMEWORK

There are three co-occurring and overlapping contrasts that strongly affect any effort to conserve tropical biodiversity through use.

1. We of the developed world are accustomed to having massive resources available to apply to a particular conservation situation. We are also accustomed to a populace that rarely threatens a conserved wildland by stating that if it is not immediately turned into material production, it will lose its conservation status. By way of contrast, in most of the tropics all resources are scarce, and conserved wildlands are frequently threatened by direct economic interests. In effect, extra-tropical conservation takes place today in a relatively friendlier social milieu than does tropical conservation, not because tropical peoples are inherently more malicious but because there are fewer resources per person.
2. Scientific research, and especially research in academia, has a major ritualistic and display component; there is ample evidence that academic scientists as a group will be very human and dance to this tune if left to themselves. Most academic scientists have long ago come to the conclusion that they pay adequate dues to society by being solid citizens within their universities and research institutes. However, there is enormous potential for academic scientists to participate in the upcoming conservation of tropical biodiversity through use by society at large (e.g. phytochemical prospecting, Eisner 1990). The realization of this potential requires significant deviation from academic goals, geographic focus, partnership mores, etc.
3. A major body of scientific research applies to the act of actually managing

and using a conserved wildland. This is where most, if not all, of conservation biology, wildland management, etc. has been directed for the past decade. Meanwhile, we lack a major body of scientific research on the use of the biodiversity that is present in a tropical conserved wildland. Research on biodiversity's non-destructive uses is essential for sustainable use of conserved wildlands by tropical societies. These two quite different areas of research overlap at many points, and conflicts over conservation research priorities are often due to failure to distinguish between them. For example, a species inventory is of paramount importance to the development of a national park as a library of phytochemical resources, but a species inventory of that park may be of very low priority to the park manager protecting the park from man-made fires (what species would be incinerated by a fire is far less important than keeping them from being incinerated in the first place).

All three of these sets of contrasts come together and underlie the realization of the following description of a specific country:

> "Costa Rica is a corporation with 50,000 km^2 of land, on which are 12,000 km^2 of greenhouses. In those greenhouses live 500,000 species. This corporation has 3,000,000 stockholders. At present, there is $1,500 worth of GNP per stockholder. Costa Ricans aspire to a standard of living that is normally associated with about $10,000-$15,000 worth of GNP."

THE FIRST STEP: SAVE IT

In the past 1–3 decades many tropical countries have set aside 5–20 % of their national territory as wildlands ostensibly conserved for their biodiversity (national parks or similar). This step has been costly and has required much energy and dedication by individuals, non-government organizations and governments. The community of academic tropical field biologists has mainly contributed by indicating where there are patches of vegetation, species or populations of particular interest. They have done this especially at the time when these units are approaching major perturbation or extinction. The more conspicuous contribution by the bulk of the overall academic community has been largely aimed at theoretical considerations of what sizes and configurations of reserves are optimal to support a maximum number of species. Most recently, the theoretical importance of inter-reserve corridors, edge effects, potential impact of climate shifts, keystone species, reintroductions, and introduction of exotics has been touched on by the same academic circles.

However, there has been little serious effort at hands-on integration of conservation science findings and understandings with the remainder of the process of selecting, establishing, and managing actual specific conserved wildlands. Almost all of these decisions have been made in isolation by professional conservationists, (usually) well-meaning legislators, wildlife managers, and enthusiastic amateurs, with minimal participation by the academic biological science community. The academic community has yet to provide high quality human resources, time and energy for on-site participation in serious planning and designation of tropical conserved wildlands – the "save it" part of the process.

It costs approximately four months of serious full-time mental investment to generate a quality manuscript for Annual Review of Ecology and Systematics, for example. The outcome is a widely read paper, a "scientific advancement", and a healthy jump in the value of the author's CV on the academic job market. Those same four months dedicated to the "save it" stage of a 50,000 ha national park may result in tens of thousands of saved species and subsequent tens of thousands of pages of science about their biology over upcoming decades. However, the academic scientist who engages in the latter career choice for a tropical national park is almost certain to go unheralded, receive no credit from peers, receive no financial approval from his or her department chairperson or research institute head, and jeopardize an "academic" career.

The blame for this sorry state of affairs is due to characteristics of both conservation projects and academia. Very few tropical conservation systems reach out eagerly for quantity and quality input from academia. This is partly because the academic offerings that appear haphazardly on the managers' doorsteps have often been ephemeral, insensitive, self-centered, fickle and mediocre. However, it is also because the management administrations are often defensive, petty, unreceptive, resentful and quite unclear as to overall goals. This conflict is not restricted to tropical regions. Two recent cases are the on-going struggle to determine the best system of management and goals for Yellowstone National Park (e.g. Chase 1987, 1989), and the recent exemplary revision of the US Government management plans for Luquillo National Forest in Puerto Rico following scientific and public protest (e.g. Alcock 1990).

Unfortunately, the first steps of the "save it" stage often create a major obstacle to conservation-through-use. Tropical conserved wildlands are often created when they are under immediate threat from direct harvest. The traditional method for confronting a physical threat is physical guarding backed up by laws. This method attracts a certain kind of person as administrator and guard (ranger, guardaparque), and leads to legislation and institutional attitudes to support this person in his role as guard. The park or reserve effectively becomes a "castle", with the "enemy" out there and the "good guys" inside. By definition, a state of siege is generated. And castles virtually never survive a siege. The longer the state of siege lasts, the more firmly it becomes institutionalized in salaries and jobs, budgets, regulations, legislation and neighborhood traditions. Moving from this state to the "use it sustainably" part of tropical conservation becomes difficult because such a change conflicts very strongly with a mindset and social institution that is sometimes necessary in the first step of conservation. The situation is not unlike the transition from a frontier to a stable rural agroecosystem.

The amount of conflict generated by the exact location of a park's boundaries is a major determinant of future siege intensity. This conflict in turn affects the later evolution of the park into a more civilized state of integration with the surrounding society. The academic biologist, with his or her fascination with reserve size, location of corridors, attempts at saving absolutely every species, etc. has proven to be a particularly difficult member of the conservation team at this point in the evolution of a given conserved wildland. It is appalling to hear an academic tropical biologist state flatly that he does not know anything about sociology and therefore will only consider biological factors. It is as though a doctor were to state that seat belts are an engineering problem unrelated to medicine.

The principle of "paying 5 % of biodiversity to save the other 95 %" is as important in the early "save it" stages as it is when later determining the intensity of permissible park use by visitors, yet this principle is commonly ignored in the discussions between academic biologists and park planners/administrators. For example, a chronic complaint of rural societies is that people are relocated to establish a national park. While some of this relocation or expropriation is in fact necessary, much of it may not be if the park establishment is viewed as (by necessity) a negotiation with society rather than a regal imperative. Just as the establishment of a new highway system is negotiated with land owners, traditional land uses, and other social barriers, so must be national parks.

Consider Costa Rica. The country has set aside approximately 25 % of its land area (Figure 1) for conservation of biodiversity and/or wildlands that are conserved for sustainable direct production with minimal loss of biodiversity. This "save it" stage was carried out over the past two decades, which means that an entire generation of employees and regulations have grown up with it. Fortunately, Costa Rica is only now reaching an intensity of land use where a siege mentality is a severe threat to park

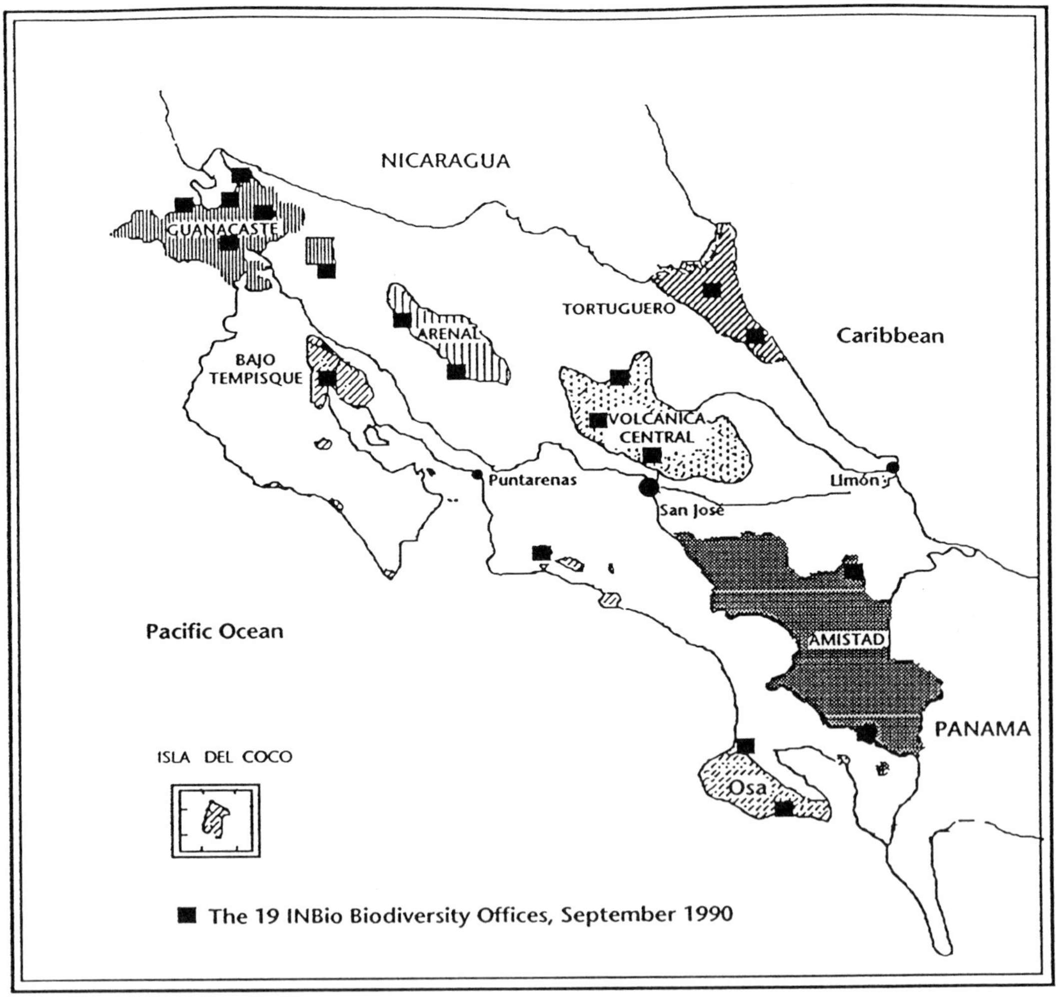

Figure 1. The location of Costa Rica's eight Conservation Areas (Areas de Concervacion) (various shadings) and 19 INBio Biodiversity Offices, September 1990.

existence. Equally fortunately, the wave of interest in moving into the "use it sustainably" stage has appeared simultaneously (and partly driven by the impending crush).

It is striking that academic professional biologists, and what is commonly viewed as academic conservation biology, have played virtually no role in the choice, siting, size or boundary details of Costa Rica's national parks. Rather, the decisions were based on: (1) a strong appreciation of seemingly undisturbed nature by select Costa Ricans, (2) accidents of donation or land ownership, (3) absence of occupants and land viewed as of low agricultural value, (4) timber reserves and water source conservation, and (5) the advice of some internationals with little to some academic exposure to professional biology.

A single example will underline this view. Costa Rica has a very fine conservation system, yet not one of the Conservation Areas (Figure 1) is established on rich agricultural soils. Academic consideration of Costa Rica's conservation planning would have avoided this error, an error that has occurred throughout the tropics.

In Costa Rica, the traditional conservation biologist's concern for such things as reserve size, reserve configuration, corridors, endangered species, species reintroductions, etc. has appeared almost as an afterthought. Ironically, these discussions can be quite disruptive in Costa Rica at this stage, through disrupting emergent activities of integration of a conserved wildland with its neighbors. Expansions of established reserves are particularly disruptive in a siege situation. Even a castle-like reserve can come to an uneasy peace with its neighbors over the years. However, if the boundary **then** changes (commonly in response to a well-meaning academic biologist suggesting that "if it were bigger it would protect more species") – thereby threatening immediately adjacent agroindustry investments, family farm plans, rural development schemes, etc. – there is an immediate conflict. The social reaction is "hey, if we have to cut off a finger, lets do it and get on with living with the consequences. Don't keep threatening, or worrying at the wound."

This situation has proven to be especially difficult when administrators and legislative decision-makers cannot make up their minds about park expansion proposals, or when their plans are influenced by vague possibilities of raising more funds for later land purchase. It is far better to rapidly come to a peaceful settlement with society. Determine the site and shape needed for conservation of biodiversity, negotiate its boundaries once and for all, and get on with making it the best possible park on the one hand, and the surrounding area the best possible agroscape on the other. Conserved wildlands are not sandboxes for academic conservation biologists to pursue their personal research agendas, irrespective of how regal, elegant and "important" the research is deemed to be by academia or by those who borrow its "respectability" to add to their tool box.

Academic considerations of reserve size have one easily identifiable negative effect. The nature of the discussion is oriented towards "what is best", with size being the independent variable. Obviously bigger is better in the biological sense. However, this automatically converts small into "worse". Quite understandably, the strong arguments from academics result in the logical conclusion that small areas are going to gradually decompose, and therefore are not "worth conserving". This can be like discarding a unique 14th Century painting because it has a large hole burned in its lower left corner. Equally, it would be like failing to conserve an island wildland because islands are generally poor in species as compared with adjacent mainland areas of equal size. It is evident that a small and damaged habitat will retain a very large number of species of organisms in a stable manner, as is the case throughout much of the tropics sprinkled with semi-wildland habitats that have been under heavy human use for millenia. From the viewpoint of the neighborhood user, such a small imperfect area is much better than no conserved wildland at all. Yes, we need a Library of Congress, but it would be ridiculous for a country with a Library of Congress to discard the concept of public libraries in small towns (Janzen 1986a, 1988a).

Large reserves can also cause problems. A bigger castle may well contain more species than a small castle. However, if the bigger the castle the greater the social disruption and resentment, then large size is a cost to balance against the gain in social approval derived from being less demanding. A large reserve may maintain everything except a few species of very large raptors, which may be only two out of the 200,000 species in the reserve. Increasing the size of the reserve by a half again on behalf of those raptors may easily create so much antagonism that the entire reserve is assaulted by a populace that would have been willing to accept it at a smaller size.

THE SECOND STEP: KNOW WHAT IT IS

Traditionally, the decision to establish a conserved tropical wildland has been based in part on some very general understanding of what biodiversity occurs at a site, rather than specific information. Specific information is often presented as lists of species and habitats, with some maps of habitat types as well. However, such lists and maps, while attractive in annual reports and on posters, are generally of little use to the within-park manager or to the neighborhood society. As implied earlier, most tropical wildlands have been conserved for their general appearance and the pristineness of the site, rather than because of specific knowledge of what is there. If viewed as a "castle", almost all tropical conserved wildlands can be fully protected and managed with virtually no knowledge of the biodiversity that they contain (see the fire example in the introduction). In this context, a list of the five species of megamammals in a national park containing 100,000 species of organisms does not count as a "biodiversity inventory".

The academic community has sometimes participated enthusiastically in preparing lists of what biodiversity is present in a national park. However, such activity is rarely driven by actual needs of the administration of the conservation unit. Instead, it is almost always done to further a personal or academic institutional agenda. Worse, most tropical conservation area administrators/managers (guards) understand so little of wildland management that they do not know what to request from the academic biologist when assistance is volunteered. Administrators generally respond to such queries with requests for what they see the academic world doing, if for no other reason than to receive approval or credibility from the academic world.

Esoteric scientists and commercial collectors have a long history of neotropical biodiversity inventory. In the 19th Century, numerous European and North American biologists came to neotropical (huge and unconserved) wildlands and sent home specimens of tens of thousands of species. For example, of the Costa Rican macromoths that are currently being inventoried by the Instituto Nacional de Biodiversidad (INBio, see below), better than 50 % of the 7,000-plus species were described from specimens collected somewhere in the Neotropics, decades before Costa Rica even thought about having a national park system or a national inventory.

Following the adventuresome explorer-scientists, there have been several recent decades of academic biologists – such as myself and many others in Costa Rica – pursuing taxonomic research as it suited their own specific interests, cavorting in the forest quite unconcerned with whether it would still be present decades hence. For example, when I came to Costa Rica in 1963, the forest was viewed as essentially a large and diverse supermarket where one collected according to inclination. While many of these researchers were internationals, a significant number were also citizens of the countries in which they studied. Finally, only in the past decade, this effort is beginning to shift toward having a major component of in-country participation. However, the agendas that they follow with respect to conserved wildlands are still largely an extension of the academic agendas set by the (usually international) institutions where they received their academic training. The recently formed Costa Rican Instituto Nacional de Biodiversidad (INBio) is attempting to depart from this behavior (see below).

The gathering of biodiversity information from conserved wildlands in order to further sustainable use of that wildland is an infant science (e.g. Eisner 1990). A major hurdle to be overcome is that of determining the future uses of the information, a determination that will guide which items of information to focus on gathering. A guiding principle for a general-purpose biodiversity inventory is to view the effort as making a huge library functional. The inventory puts call numbers on the new books, confirms existing call numbers and their synonyms, develops a card catalogue, establishes a data base about the general content of the books, sets up a retrieval and loan system for the books, advertises the library's content to potential users, and helps to determine how much to charge the user public for services.

In Costa Rica, INBio has embarked on this exercise (Lewin 1988; Tangley 1990). The INBio has set the ten-year goal of inventorying the half million species of organisms conserved in Costa Rica's 12,000 km^2 of conserved wildlands, and getting that information into a usable format (Janzen in press; Janzen and Hallwachs in press). In the initiation of this activity, the INBio has encountered several conflicts with the world of academic taxonomy and systematics. Here, I mention a few of these conflicts, so as to give the reader the flavor of the situation.

The parataxonomist

The international taxonomic community will not mobilize sufficient human and economic resources to inventory Costa Rica's biodiversity within a decade. This short time frame is necessary if the inventory is to be the precursor to a sustainable use that will in turn facilitate and cause Costa Rican society to preserve her biodiversity. The obvious solution to this impasse is to train a large number of Costa Rican adult working-class rural residents in the technology and philosophy necessary such that they can conduct much of the inventory process, and get on with the task. Such persons are termed "parataxonomists" by the INBio, a term borrowed from the "paramedic" of the medical community. At the time of this writing, the INBio has trained 31 parataxonomists (with funding by USAID, National Science Foundation and USDA, The Natural History Museum (UK), the Pew Charitable Trust and the J. D. and C. T. MacArthur Foundation) and they are taking the first steps for the national inventory. They are generating hundreds of thousands of properly prepared specimens per year, rapidly gaining natural history information and experience, and spreading biodiversity information to the neighbors of the 19 Biodiversity Offices scattered through Costa Rica's conserved wildlands (Figure 1; Janzen in press).

Such a highly pragmatic step, which removes much of the basic work of a national inventory from the domain of the university-educated academic taxonomist, appears on first glance to be ideal. It involves the national community in its own biodiversity management, it increases the human resources that are experienced in biodiversity matters, it expands employment opportunities based on brain rather than brawn, it relieves the academic taxonomist of much time-consuming field work, it is cheap in dollars and time per specimen prepared, and it can be easily and cheaply expanded until the desired inventory intensity is attained.

However, such a scheme has met significant resistance from the Latin American academic sector (and some similar resistance among some older members of the North American taxonomic community). It is viewed as demeaning to the PhD taxonomist to publicly demonstrate that much of what was considered to be his or her professional domain can be carried out by a person with enthusiasm, grade-school education and six months of training. It deprives the same academic of the opportunity to state that the slow pace of his taxonomic research is caused by the difficulty of getting to the field and a shortage of funds. It likewise eviscerates the comment that we must have more (very expensive) resources and the time to train more PhDs before we can get on with an expansion of taxonomic services and activities to encompass such a task as large as a national inventory. Finally, it is the first step in popularizing an activity that, like the white lab coat, has maintained the academic at a socially prestigious level, with residence in a university or government research laboratory setting in or near a capital city at a comfortable salary. Ironically, a field task of a scientific nature is viewed as a step upward by the rural resident while a scientific task in the field is viewed as a step downward by the urban resident (who may well have spent several generations struggling to move from a rural to an urban background).

There is also an interaction between international universities, in-country tropical academics, and parataxonomists. If serious science can be done by the relatively uneducated in the field, then perhaps there is less justification for extended periods of postgraduate study on comfortable fellowships in the universities of developed countries. It is quite amazing to hear an academic lament the "shortage of taxonomists" and then criticize a parataxonomist because "he could not possibly learn or know the detailed specialized collecting methods used by a specialist in his or her group of organisms".

The parataxonomists also do not mesh well with the natural desire, or at least the fashionable feelings, of developed country academia to offer advanced degree programs to international students as their contribution to the development of human resources in undeveloped countries. A parataxonomist neither has the other academic preparation nor language skills to enter into a graduate program, nor is interested in leaving family and friends behind for years while "studying" something in Wisconsin that will be at best of marginal use to his or her daily work as a parataxonomist.

The apprentice curator

Now that the parataxonomist program is generating a very large amount of high-quality biological material, a large bulk of mounted and labeled specimens is accumulating in a central depot. In the case of Costa Rica, this depot is the buildings of the INBio, housing the National Insect Collection, National Plant Collection (National Herbarium), etc. Traditionally, in such a circumstance, various international taxonomists would then be deluged with boxes of specimens to identify and study. This would simply add to their already unbearable backlog of specimens. This would create yet more resentment amongst taxonomists that "society" is not willing to pay the maintenance costs of the taxonomic community in the developed countries (to say nothing of the undeveloped countries).

Enter the apprentice curator. Many Latin American countries generate a moderate number of university graduates with a 4- or 5-year degree in biology. Sending them off to international centers for a PhD in systematics is costly in dollars and time, and generates people with some specialization in a group or a method, but no practical experience as to how to make taxonomy a functional activity within a society (since that is not a part of academic training anywhere). For example, there is not a single advanced degree curriculum in systematics in the US that emphasizes the skills and abilities necessary to, for example, develop a layman's field guide to the identification of a major and common group of organisms. One conspicuous solution is to give tropical university graduates on-the-job training in basic curational skills and uses of taxonomy, thereby gradually training them up to a level where they are, according to their individual abilities, actually conducting the same processes expected of a PhD taxonomist when curating a major collection.

The name apprentice curator is applied to such a person because a major way to speed and facilitate the on-the-job training process is to apprentice him or her to a practicing PhD taxonomist (local or international). However, such a collaborating specialist requires two distinctive traits. These traits are rare within the systematics community but examples do exist (e.g. Janzen in press). The specialist needs to be willing to spend a serious block of time (generally several months per year at a minimum) working in the field and INBio with the apprentice curator in Costa Rica. Second, some serious part of that time has to be dedicated to the process of teaching the apprentice curator as much as is possible about the group of organisms (as well as benefiting from the apprentice curator's labors). On a short time scale, this is not the most efficient use of the PhD specialist's time, as measured in production of traditional products (monographs and/or specialized collections of focal groups). However, on a slightly longer time scale, the rewards can be huge in terms of specialized collections, companionship in productivity, and eventual production of taxonomic works by the tutor and apprentice. Fortunately, within the past several years the National Science Foundation of the US, the USDA, the Smithsonian Institution, and The Natural History Museum (UK), have come to view such a time investment by a master taxonomist as legitimate investment of time (e.g. Janzen in press).

Apprentice curators have, to date, seemed to be less depreciated by the Latin American and North American science community than have been parataxonomists. The apprentice curator is, after all, just a close subspecies of the tutor taxonomist, and many tutor taxonomists began their professional careers as informal apprentices in their youth.

However, there is still one major barrier. The funds that are generally available for a Latin American recent university graduate to go to the US and study for an extended period so as to obtain a higher degree are not generally available to support an apprentice curator in his or her on-the-job training within the developing country. One of the major reasons, though generally unspoken, is that the "student" studying and curating beetles in INBio, for example, does not conform to the traditions and reward systems of that same student studying and curating beetles in a US university. The US university funds to "produce degrees", not to accomplish a conservation objective.

There has been a second reason given as to why US university funds are not available for on-the-job in-country training. It has been noted that the apprentice curator is actually working, thereby allowing the excuse that "he or she should be supported by the INBio". Unfortunately, the INBio is not able to provide a salary that is comparable to that in the developed world for the ability and responsibility that he or she is expressing as an apprentice curator. This is not because there is particular discrimination against taxonomists or curators in Costa Rica, but because salaries are generally very low (the Minister of Natural Resources receives about the same salary as my secretary/assistant at the University of Pennsylvania). Low salaries are commonly compensated for by having several jobs and by putting substantially less than eight hours of work into a "full time" job (in Costa Rica there are even two kinds of "full time" jobs: "full time" and "exclusive dedication"). However, the INBio apprentice curator is held to standards of performance of the developed world, and therefore should be compensated accordingly.

As with the parataxonomist, the invention of the job category "apprentice curator" allows a redefinition of curatorial and systematic responsibilities in accordance with the needs of the social system of biodiversity use in Costa Rica, as opposed to the social needs of the US academic system. For example, from the beginning, the INBio curational task requires an ability to both identify specimens for various public sectors, and generate field guides, natural history synopses, and other informational services. An apprentice curator is also encouraged to invest some time in evaluating environmental impacts, teaching parataxonomists, assisting master systematists, advising the government on biodiversity questions, fund-raising, etc. In the traditional academic world, analogous activities are seen as an imposition rather than as an integral and legitimate use of a staff-member's time. Even though INBio started from bare ground in early 1989, in fact it incorporated various staff members from other more traditional institutions, staff members already somewhat steeped in traditions brought from academia both within and without the tropics. The readjustment of a mindset for these people has not always been smooth.

The inventory

An inventory of the biodiversity of a country or region has traditionally been expressed as a list of species. Many taxonomists feel that this is the end point for their participation in conservation and environmental management. However, an "inventory" as conducted by the lay community normally contains information of value about the species as well as a list of the species themselves. For example, a warehouse inventory may discuss what kinds of bolts are in the warehouse, how many there are of each kind, what they cost, how long for a backorder, where they were obtained, etc. In other words, the inventory is not an end but rather a means to the end of efficient use of a repository of diverse objects too complex to be censused at a glance.

When the INBio speaks of an inventory of the biodiversity of the half million species of organisms occurring in Costa Rica, it is not speaking of just a species list. True, the species list can be generated at any moment by the flick of a computer key, and is something that will grow continually toward an asymptote for the next decade. However, while such a list may often be cited on the international scene when comparing Costa Rica with this or that other area, it alone is virtually useless within Costa Rica. What really matters is the information stacked up in the computer base behind each of the entries in the species list.

The parataxonomist activity is initially focused on collecting samples of the species present, so as to generate the raw material that the apprentice curators and their international connections can use to put the Costa Rican species' nomenclature in (relatively) stable order. At this point, a species' presence in the inventory changes from a number to a Latin binomial (and higher categories) that allows access to the rest of the world literature about that organism (if any be), and allows taxonomic inference to play its powerful role in the perpetual guessing game about natural histories as yet unknown. But in getting the first specimens, the parataxonomist also generates the first data about the geographic and temporal location of the species within the country. As further collection occurs at the nodes in the widespread net of Biodiversity Offices (Figure 1), a better understanding of the ecological distribution of species will appear.

Some groups of species reach taxonomic stability and eco-geographic understanding quite rapidly. Virtually all vertebrates (except for a few small snakes, salamanders and obscure frogs), and about 70 % of the plant species attained this stage before the INBio was even a dream (e.g. Burger 1977; Janzen 1983). The same may be said for butterflies (DeVries 1987), saturniid and sphingid moths (D. H. Janzen and W. Hallwachs, unpublished), ophiine ichneumonid wasps (Gauld 1988), scarab beetles (A. Solis, unpublished), and dragonflies (C. Esquivel and S. Brooks, unpublished). Many other groups are still very much at their beginning, but being steadily accumulated by the parataxonomists and organized by apprentice curators and international taxonomists.

But, the INBio biodiversity database very quickly becomes a repository for Costa Rican natural history and ecological information as well as a source of locations, taxonomic certainty, literature connections, and taxonomic inference for future users.

As the taxonomic status of a group moves beyond the early "inventory" stage, the parataxonomist and even to some degree the (no longer apprentice) curator have the option of concentrating more heavily on as-yet uninventoried groups (such as a parataxonomist with years of experience with insects turning to diatoms or fungi), or focusing his or her efforts on the accumulation of information on natural history, physiology, behavior, morphology, etc. Thus the "paraecologist" is born. As the years pass, the INBio and similar organizations elsewhere will be continually confronted with the question of how best to divide their human and support resources among the activities of parataxonomists, paraecologists, curators, and unforeseen categories of specialists. This allocation will be heavily influenced by the societal involvement with biodiversity, just as the agricultural research of a major land-grant university in the US is in part guided by the crops and society of the region.

This view of an inventory meets with mixed responses by the academic (and museum) science community. In a variety of subtle ways, it departs from well-established – but recent – traditions. It is very easy to forget that the first major inventories of the neotropics by European society during the past century were, for example, largely driven by commercial drug, food and ornamental interests rather than esoteric academia. First, and perhaps foremost, such an INBio-type inventory casts the taxonomist and his or her actions as integral partners in a larger effort, rather than controllers of a self-contained and self-responsible body of endeavor (largely beholden to a dimly remembered academic exposure in a developed country). Taxonomists are, as a group, among the most independent of all kinds of biologists (witness the paucity of coauthored papers in taxonomy). However, the person who is content to spend major blocks of his or her life cloistered in a small room thinking individually about long-dead organisms tends to be at the more introverted end of the social spectrum. The inventory workplace postulated here is substantially extroverted.

Second, the products of an inventory are considered substantially less elegant, in an academic context, than are massive monographs that at least in theory "clean up" a major taxonomic group. However, traditional monographs may require decades of preparatory work, are not organized for lay readers, and are often impenetrable except to other specialized taxonomists.

Third, such an inventory is regional or even more local in scope and solutions. However, many if not most tropical organisms have geographic ranges, phylogenetic relationships, and scientific paper trails that extend far beyond this area (e.g. Janzen 1986b). To focus on putting a regional flora or fauna into a user-friendly format robs the taxonomist of the time and energy needed to ferret out these additional extra-regional connections, some of which will be important in the taxonomic designations that are assigned within the region. It cannot be forgotten that for many taxonomists, much of the original attraction to the endeavor was the great puzzle of understanding how things are related to each other. However, the clues to a taxonomic puzzle often lie geographically far outside of the region of pragmatic interest.

Fourth, an academic taxonomist commonly receives a salary appropriate to his or her level of academic training rather than to the level of productivity as measured in products that can be directly used by others. Associated with this, many taxonomists work for extremely low salaries, if their annual salaries are divided by the very large number of hours they put into their profession. In return, they are generally not asked to account for their efforts in anything but the most general manner. In the "inventory environment" discussed here, salaries and other rewards are unavoidably tied more directly to the amount and nature of directly usable output. The taxonomist is at risk of feeling much more like a cipher in the marketplace.

Finally, in an inventory such as that described here, a taxonomist's output is in fact used, and used within a short time. Taxonomists have long suffered, or enjoyed – depending on your view – nearly complete freedom from feedback from disgruntled or euphoric users. Keys, species descriptions, curated specimens in museums, monographic descriptions of ranges and morphologies, etc. commonly languish for many decades before being glanced at by, perhaps, one or two other specialists. Such specialists are adept at wending their way among the errors, oversights, lack of technology, etc. characteristic of someone working decades to centuries before them. On the other hand, in the more heavily funded areas of biology it is commonplace for quality control to be promoted by repetition or use of any significant discovery within a few days to years. This is the kind of double-checking that is simultaneously nervewracking and exhilarating for the scientist. It also throws the taxonomist into the turmoil of the marketplace, taking away a freedom that has always been one of the pleasures of most taxonomic research.

THE THIRD STEP: USE IT OR LOSE IT

Once the "save it" and "know it" steps are in motion, the "use it sustainably" stage can be entered into with vigor. In effect, the save it and know it steps are a giant production process ending in information dissemination to a major group of users (Figure 2).

However, the products from the sustainable use of a wildland conserved for its biodiversity are much more than just information. The products fall into three large and somewhat overlapping areas:

1. Physical production for the area outside of the conserved wildland (light or no human presence required): carbon storage, water generation, water flow control, erosion control, wind and temper-

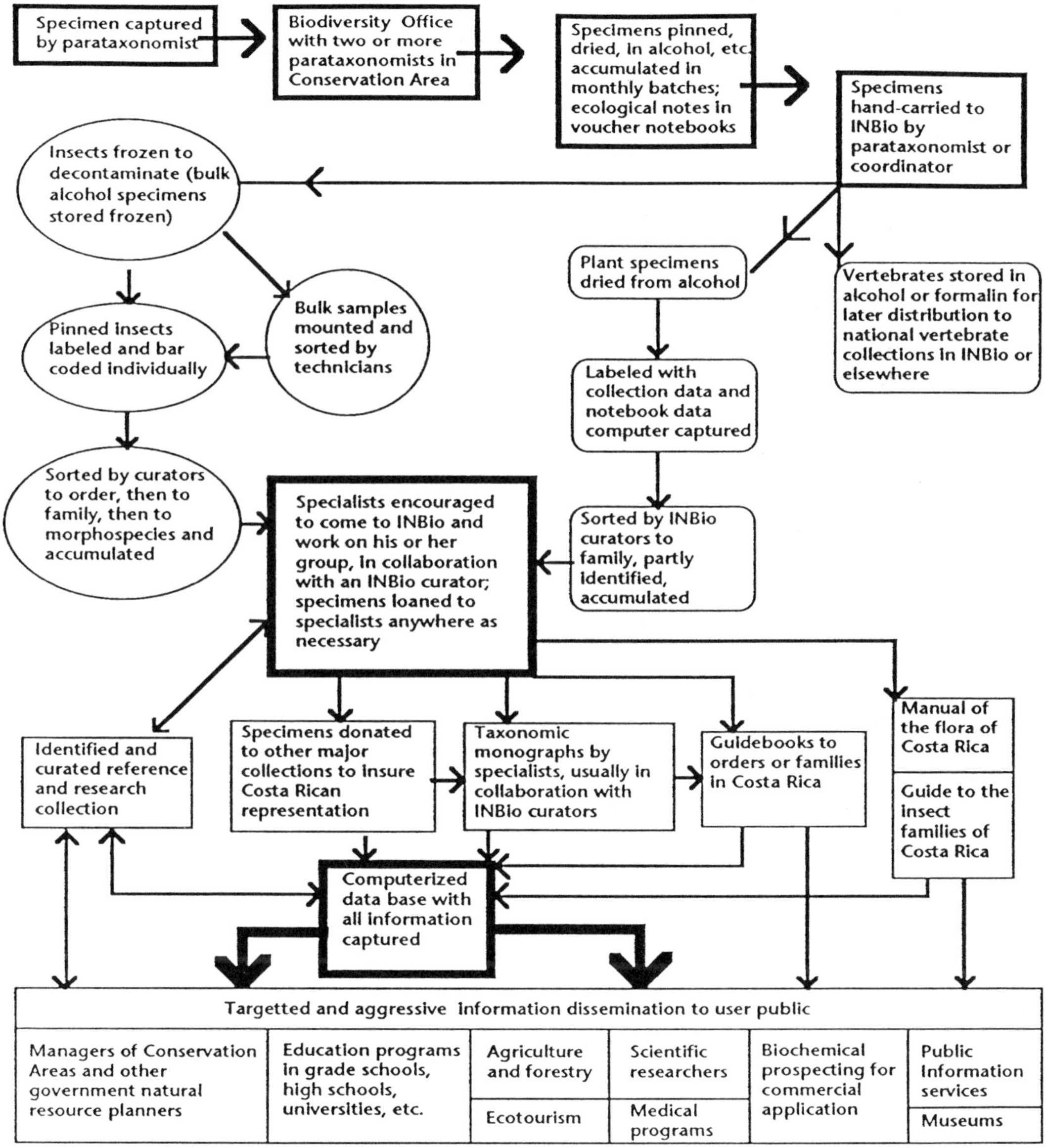

Figure 2. National Biodiversity Institute of Costa Rica (INBio): the inventory flow from field to public use.

ature modification, game refuging, fish nurseries, visual recreation (landscape beauty), gene and species banking, etc.

2. Information extraction for the outside world (very light human use): gene capture, intellectual information, research results, organisms for reintroduction elsewhere, consultation on environmental management, etc., and
3. On-site consumption (light to heavy human use): biology education programs for all ages from gradeschools to adult ecotourists, hands-on research, recreation from hiking to populous sporting events, living centers for administrative personnel and visitors, roads and trails, etc.

The phenomenon of direct use: history and background

The direct human use of the conserved wildland (items 2 and 3 above) will cause some loss of biodiversity if the naturally occurring (i.e., non-human affected) patterns of species interactions and demographies are viewed (correctly) as part of an area's biodiversity. However, this "loss" is qualitatively different from the loss of biodiversity associated with the extinction of species or populations, since naturally occurring interactions and demographies generally reappear when the perturbing influence is removed. Massive habitat restoration in some tropical national parks is based on this concept (Janzen 1988b), and the process permits tolerance of many kinds of wear and tear. For example, rotating the impact of tourists among different parts of a park over the years depends entirely on the ability of animal and plant arrays to reconstitute themselves.

When we speak of paying a few percent of the biodiversity in order to generate the products that will insure conservation of the area into perpetuity, it matters whether we are speaking of irretrievable biodiversity loss (local extinction) or restorable biodiversity loss (biotic restoration following wear and tear). It is of interest that virtually all sustained-yield agroecosystems are the latter type. It is also of interest that restoration processes are temperature-dependent. For example, the effects of fine-scale forest clearing may require millenia to disappear in Canada but disappear within decades in the hot lowland tropics, yet be extremely slow to disappear from the top of a tropical mountain (e.g. Janzen 1973).

Any consideration of use of a wildland conserved for its biodiversity products quickly encounters the question of what is "natural". In the New World tropics, about 10,000 years ago a hunting society extinguished a major set of widespread new world large vertebrates, with subsequent dramatic changes in vegetation (Janzen and Martin 1982). This means that there is no vegetation in the neotropics that is undisturbed by man. To think otherwise is to label Amboseli National Park "undisturbed" after someone has shot all the large mammals and many others (small bovids, vultures, mongooses, civets, canids, cats) have consequently starved to death. It clarifies one's vision to recall that the horse evolved in the New World, went to the Old World on its own, was subsequently extinguished in the New World by humans, and then given back to us by the Spanish. The horse is a native member of the neotropical megafauna and it certainly belongs in some conserved wildlands – though to replace it requires simultaneous replacement of some predation in the form of selective castration and the occasional rifle shot. In the Old World, humans have obviously been part and parcel of natural biodiversity for millions of years.

Overall, one is forced to the subjective and arbitrary conclusion that we should attempt to save at least the amount of biodiversity that persists when human interference is roughly that characteristic of hunter/gatherer humans (or less). To be satisfied with less is simply to accept the omnipresence of the agricultural landscape that humans have become so adept at producing as their special way of surviving (Janzen 1984, 1990) and depriving society of most biodiversity products.

These considerations invariably lead to considering the conservation of tropical wildlands as an act of conserving the garden of genetic evolution and the basket of forces thought to keep a species what it is. However, this view does not match well with what we understand of evolutionary biology.

First, all indications are that, at least for large continental areas, the species present in a given conserved wildland arrived by migration and then underwent little genetic differentiation in response to their new environment. In effect, they arrived and they "ecologically fit" (Janzen 1985, 1986b), just as we have recently witnessed in the case of the African honey bee population introduced into Brazil. This unconventional view is demonstrated by the enormous number of species that retain their characteristic life forms, colors, behaviors, morphologies, etc. over a very wide variety of habitats and selection regimes throughout many degrees of latitude in the tropics. Evolutionary biology has quite understandably long been fascinated with genetic change and its causes, and has left us rather unprepared to deal with the obvious fact that there is an enormous amount of genetic stasis in nature, tropical and otherwise.

Second, the quickest way to promote speciation – otherwise known as the generation of biodiversity – is to produce small populations, each subject to a different set of directional selective forces. It is no accident that practically all the textbook examples of observable natural selection are on real or man-made islands. If we really wanted to promote neotropical tree biodiversity, for example, we would take a tree that is distributed from south Mexican semi-desert to Brazilian rainforest and chop the distribution up into many little bits, otherwise known as national parks, and let natural selection go to work. Examples such as this force us to consider which kinds and how much of biodiversity we really want. And that depends on the user community.

Consideration of what kind of biodiversity to save leads to yet more harrowing considerations. Yes, we simply have to live with whatever kind of species and selective regimes we generate by saving a patch here and a patch there of the once-expansive areas of tropical vegetation, and then subjecting these patches to global climate changes. However, there is another kind of biodiversity conservation over which we have more control. To return for a moment to the neotropical extinct megafauna, it is no secret that neotropical vertebrate biodiversity could be substantially raised by the introduction of a large array of species of wild vertebrates from the Old World tropics. The same undoubtedly applies to trees, insects, snakes, etc.

There is absolutely no *a priori* reason to assume that any given neotropical habitat is "full". What a given park contains today is in great part the result of which species have migrated there. While there are ecological processes that set the upper limits to species richness in any given habitat, immigration rate is one of the major parameters in these processes. The fact that a southeast Asian forest has ten species of squirrels and its ecological analogue in Costa Rica has three is not determined by some magically different ceiling to squirrel species diversity in each of the two geographic areas. The paucity of plant and parrot species in African rainforest as compared with species-rich neotropical rainforests is clearly historical rather than directly ecological.

Artificially elevating an immigration rate does not automatically result in the adjustment of other processes to where the total species richness stays constant. It is an easy prediction that the introduction of the neotropical agouti (*Dasyprocta punctata*, large seed-dispersing and seed-killing rodent; Hallwachs 1986) to Ugandan rainforest would probably not result in the extinction of other animals (or plants), just as there is no evidence or *a priori* reason to suspect that the introduction of free-living *Apis melifera* automatically removed a species of insect from each of the multitude of neotropical habitats it now occupies. But both introductions would increase biodiversity. In fact, it is easy to predict from solid ecological theory that the recent introduction of the screwworm fly into North Africa (Palca 1990) from the neotropics could substantially increase both plant and vertebrate biodiversity per habitat. On the other hand, the current widespread and wholesale introduction of neotropical trees into Africa for firewood, erosion control, timber, etc. has enormous potential for decreasing plant biodiversity in conserved wildlands.

Society has to decide very rapidly for what products its biodiversity is being conserved, because that in turn determines the actual on-site management decisions. Such a decision by society requires serious effort by the science sector as well as other sectors to identify, develop and publicize those potential products, production costs, feedback mechanisms from consumer to producer, etc.

The direct impact of users

The managers of tropical conserved wildlands desperately need field biology studies of the true impact, and associated resiliency processes, of various kinds of use of those wildlands. The information available today is mostly a serendipitous byproduct of esoteric research that was conducted by academics in search of behavioral, ecological, physiological, etc. understanding for its own sake. Explicit tropical management research such as what kinds of rat poison to use to protect Galapagos birds (Cruz and Cruz 1987), how to save Sri Lanka wild fish by translocation (Wikramanayake 1990) and what are the patterns of invasion of tropical abandoned pastures (Janzen 1988c), is poorly funded and scarce.

However, the byproducts of esoteric research do tell us that certain aspects of, for example, animal behavior can be extremely sensitive to humans. For example, ordinary footpaths used as rarely as one time per day may dramatically alter movement patterns and social behavior of medium-sized terrestrial mammals living within tens of meters of the paths (W. Hallwachs, personal communication). On the other hand, a large and healthy jaguar population may co-occur with stable and intense cattle ranching and many people, if the jaguars are not hunted or chased. Such species-specific idiosyncracies are commonplace. All of the adult mahogany trees may be logged out of a tropical forest, yet a century later there be a large and robust population of juveniles and new adults. On the other hand, the removal of a single species of inconspicuous seed-dispersing large rodent from a habitat may lead to the rapid extinction of a tree species that was dependent on the rodent for its seed dispersal (e.g. Hallwachs 1986; Putz et al. 1990) or for suppressing the density of its primary competitors through seed predation (Janzen 1970).

If a wildland manager is to rotate users through different habitats, so as to allow maximum human visitation yet minimal permanent or long-term damage to biodiversity, then much understanding of restoration and reconstitution of plant and animal arrays is needed. It may well be that controlled harvest of deer from a large area of a tropical national park that is undergoing restoration might be best for both conservation of biodiversity and for the neighboring society. The studies associated with such management decisions are long-term and iterative by nature, and are thus best built into the normal research program of the wildland's administration, and/or conducted by local research expertise. This is yet one more reason why the development of local science capacity is so critical for sustainable use of tropical wildlands. On the other hand, such management decisions are also often aided by the natural history and ecology studies of shorter-term duration conducted by academic researchers using the park as a large biological station.

The impact of use has two major components. First, there is the actual impact on organisms, as demonstrated by their reactions. Second, there is the impact on society's image of what is to be conserved. A beer can on the bottom of a pristine stream is far more offensive to the visitor than is the gallon of kerosene dumped out by the same visitor, but the impact on aquatic biodiversity is the opposite. A well designed cement or asphalt trail through a muddy rainforest may both lower the impact of herds of visitors on biodiversity, and yield visitors who are more willing to support that rainforest by payments and voting; however, such a man-made trail is anathema to some kinds of ecotourists. A carefully managed five-story hotel in the middle of a conserved wildland can be a horrendous eyesore, but have no more real impact on biodiversity than a carefully landscaped and inconspicuous parking lot. In fact, if the right kind of people come to enjoy the hotel, the park may suffer far less fire damage, boundary conflict, poaching, and pesticide drift than if the hotel had never been built in the first place. The art, of course, is to plan the location of the hotel so that the use benefits acrue to the park **and** there is no eyesore. Once again, we find ourself stressing that a conserved wildland must be designed as a whole organism, not as a collection of structures and concepts, each reflecting a battle lost or won with a given social threat over the past several decades. However, in order to avoid inventing a giraffe with an elephant's trunk, we have to have a clear vision of the ecological role of any given conserved wildland vis à vis its community of users.

Social conflict among the direct users of a conserved wildland is a very complex area of management. Researchers fight over who gets to do what to this or that plot; it is striking how the politeness and reserve necessary for survival in their home university departments are abandoned when they get to the frontier of the conserved wildland. Researchers disagree with ecotourists as to who gets to enjoy what; researchers have the habit of forgetting that they are just a subspecies of ecotourist. Ecotourists go to a conserved wildland to see wild biodiversity, not other ecotourists (sometimes). Administrators can easily forget that a transportation system designed around the desires and incomes of the park's neighbors can be easily used by the international tourist, but not vice versa. Children in a biology course may not even notice the (adult) ecotourists but be horrified if they see a researcher kill an experimental animal. A Costa Rican gold miner standing in the middle of a flowing stream of mud, that he had made from a once crystal clear river, remarked to me that he much preferred the national park over noisy and congested San Jose, and that he had been certain to never cut a tree or hunt a single animal. One is reminded of the president's staff making certain that the ambassadors of two warring countries never get invited to the same dinner party. I am left with the feeling that the most fertile ground for the recruitment of park administrators should be industrial psychologists.

The park administration as a user

The administration of a conserved wildland is in a contradictory situation. Almost by definition, the staff live in those parts of the nation that are most distant from "civilization" and often viewed as the least threatened by humans. But the staff collect firewood, divert streams for drinking water, wash in streams, clear areas to maintain livestock, run noisy generators, clear roads and trails for access, have dogs and cats, etc. These things are all done with the sense of abandon that is characteristic of frontiers – where wild biodiversity is a threat, not a sustainable resource. Furthermore, the miniature police state so characteristic of most parks during their early years adds to the impression that any impact by outsiders and anything done "in here" is justified by making life more pleasant for the guards in their backwoods situation.

It takes a gentle breed of highly political scientist to point out these things to an administration and gradually bring it around to abandoning what seemed to be such reasonable behavior. What has been, in my experience, most successful is to inform about the kinds of natural history that are disrupted by collecting firewood, dumping soap into streams, etc.

As a park administration gradually evolves into maintaining the wildland for its biodiversity products, it enters into the odd situation of doing many things that seem illogical for the protection of nature. It is critical that the staff have a good scientific understanding of why they do these things, or else they will get caught in a credibility gap between their own tasks and the users' view of biodiversity conservation. The public does not take lightly to the park administration arresting and penalizing the setter of a fire that burns into the park, while at the same time the park burns large areas to maintain a particular vegetation type or create large "blackline" fire breaks. It takes a special effort to educate the neighboring ranching public that cattle can be used as biotic mowing machines and seed dispersers (under certain circumstances) as part of the forest restoration process, while that very land was expropriated from a rancher for its conservation.

A conservation area should be quite capable of harvesting certain kinds of biodiversity products (seeds, animal parts, water, animals for reintroduction elsewhere, etc.). However, this harvest can easily create national and international conflicts when the same products are being protected elsewhere by absolute bans on their sale, transport, use, purchase, etc. Repeatedly the difficulties in wildland management are not the technology of the science and biology within the conserved area, but the complexity of interactions betweens the wildland and the surrounding society. There are many occasions when the only successful park will be one that spends more of the budget outside than inside the park (K. Miller, personal communication).

The payment for use

Humans pay for their use of an object in many kinds of currency. In the tropics, as elsewhere, the currency of concern ranges from a simple park-neighbor agreement not to use those pesticides that will later be washed into the park by rain, to direct payment to the park management budget for use of the water that flows from the park, to royalties paid to the nationwide park system by an international corporation in return for phytochemical information from park plants. The corporation called Costa Rica is paying its greenhouse maintenance bill with input from all levels of society. Simultaneously, society often asks what contribution the park makes to the overall health and development of the corporation.

I earlier suggested a list of products that are a conserved wildland's contribution. My concern here is to focus on the role of the scientist in improving the quality of these products to the user, with commentary on how this may be reflected in the willingness of the user to compensate the conserved wildland for its products.

The introduction of the conserved wildland into the marketplace implies the right on the part of the consumer to have a say in the nature of the products. Automobile-users do not passively receive Detroit's products, but rather generate an ocean of regulations about the behavior of these products. What does this imply for the community of scientists and administrators of conserved wildlands? Consider that a tropical mountain reclothed with ten species of trees, all introduced, may be just as effective in modulating a watershed as was that mountain with its original 300 species of biodiverse native trees. If that does not make one pensive, then perhaps this can be achieved by the realization that agricultural inviability is the best friend that tropical conservation has ever had (Janzen 1987). If there was a corn plant that grew well on tropical rainforest soils, we would not have any tropical rainforest to argue about today. And modern science is working overtime to produce such crop plants. We had better very soon come to recognize a conserved tropical wildland as simply another kind of crop, albeit with certain special needs for its cultivation.

Fees paid by on-site users

Many uses of a conserved wildland involve the direct and frequent presence of human visitors. They, like any visitor, arrive in search of something – photographs, peace, intellectual stretching, escape, exploration, novelty, self, companionship, etc. Almost all of these activities are enhanced by an increase in the accuracy, timeliness, appropriateness, etc. of on-site information flow. This requires more and better signs, trails, guidebooks, public lectures, teaching experiments, field trials, etc. For example, an enormous contribution to tropical conservation would be made by the invention of a method to make an indestructable field sign, quickly and cheaply with formal letters of many different sizes (and options for hand-drawn patterns). Even better would be if it was made from materials that can be purchased in a local hardware store.

The source of the park's information offered to visitors is the background of the scientist, the literature, on-site research, local knowledge, and the knowledge base that field scientists carry in their heads. For example, if a national park is going to offer a major program in biological literacy to neighborhood schoolchildren, and thus be paid back by the votes of those children in later life, the park has to have access to serious biological content to teach. Slogans and platitudes will not suffice. For many tropical national parks, providing detailed biological information to be packaged and offered to the voting visitor is a much more important function of in-park research than is "management" or "contribution to the world".

As a conserved wildland becomes part of the market economy, its users do become willing to pay the opportunity costs, legislation and direct cash for a conserved wildland. However, willingness to pay is quite directly related to the quality of the product that is encountered during a visit. However, a conserved wildland differs from the commercial world in one major respect. Traditionally, the salaries (and other rewards) for a park's staff are not linked in any direct way to the quality of the products that are generated by the park. A basic human motivation for quality performance has been removed. It is imperative that each conserved wildland, in the context of its own society's mores, find a way that individual quality initiative is rewarded, just as it is in the business world. There are diverse methods for direct and indirect financial compensation (merit pay, preferential access to houses and vehicles, help with dependents' schooling, allowing the profits to return to the park endowment and be used for improvement of **that** park's staff, etc.) but the responsibilities associated are also diverse. The price of merit pay is eternal evaluation.

A scientist can play a major role in the non-financial compensation process. The staff of a conserved wildland is constantly being trained upward in technical and philosophical ability. In most tropical rural societies, such training is associated with translocation to progressively more urbanized ("civilized") parts of society. However, the conserved wildland asks its well-trained employees to stay "out in the bush". Such a fate may be pleasant to the jaded international scientist in search of a few months tranquility in a national park, but it can be rural hell to the park staff. One solution is to bring civilization to the park. A major part of this is accomplished by better housing, better schools, better health, etc. provided by the overall administrative system. However, the scientist can also be a major contributor by participating in continuous adult education. Many scientists are walking universities. In a tropical national park these persons are surrounded by a populace not yet jaded by thousands of hours of TV specials, school science laboratories, music summer camps, language jubilees, etc. The scientist pays in time, energy and the effort required to become multilingual. The scientist is repaid through the conservation of the biodiversity on which his or her scientific career is based, through the conservation of the raw materials of biological science.

This is not the place to discuss the complexity of how to determine a fee scale for different users – the school child, the adult ecotourist, the researcher – or how those fees are best guided back into the park's maintenance costs. What should be stressed, however, is that field research by the international scientist can easily be big business. Costa Rica's system of conserved wildlands is, for example, really a network of enormous biological research stations. Today they have minimal logistic facilties, but they are growing fast. When an international researcher obtains an NSF grant for a half million dollars of field research over three years in those parks, that is a half million dollars worth of incoming foreign currency. Furthermore, a very substantial portion of that income will be spent in the neighborhood of the park. The overall park system of Costa Rica could easily accommodate $100 million/year worth of basic biology field research with no damage to its biodiversity. I am not capable of determining here what Costa Rica would actually have to pay in additional goods and services to bring in that amount of foreign currency, but much of the infrastructure is largely in place except for more within-park living and working buildings, some slight modification of existing legislation, and some time-consuming thought by INBio and the Sistema Nacional de Areas de Conservación (SINAC) as to how to best facilitate the process.

Payment for off-site environmental services

Throughout the tropics, environmental services provided deliberately or serendipitously by conserved wildlands are not explicitly charged to the consumer. Yes, some tiny fraction of a rice-grower's taxes may be used to help finance the montane national park that generates his irrigation water, but such a contribution to the park maintenance budget does not begin to approximate the market value of that water. The same applies to erosion control, local climate modification, wind protection, etc. The tropical conserved wildland would be well-advised to formalize its position in the market economy by moving as fast as possible into a social position whereby its environmental services are paid for in one form or another.

However, my focus here is not direct

environmental economics but rather to underline that the scientist can play a very central role in this marketplace. The scientists doing research within conserved wildlands are capable of being a large body of local experts on the potential impact of this or that management action on this or that environmental service that emanates from the park. Most, if not all, tropical national parks and their zone of influence are not serviced by a large university, government research branch, consulting firm, etc. that can be counted on to provide evaluations, be expert witness, draft management plans, etc. for the area neighboring the park. If the situation is correctly structured administratively, the scientist working within the park can even make a serious contribution with a minimum loss of time.

At this point an interesting conflict between the scientist, and especially the academic scientist, and the park administration appears. The administration that figures out how to resolve this apparent conflict will tap a major source of resources for the park administration and for the neighboring society. For a very large number of academic scientists, time is substantially more valuable than money, and success is measured by results rather than a paycheck. For many members of park staffs, time spent is money earned and success is measured by the size of the paycheck. For a scientist, a day in court is a day lost, while for park staff a day in court is a good meal in a good restaurant and perhaps a break in the monotony of park tasks. Somehow, scientists have to learn how to be less contemptuous of the park administration's need for basic needs, and the park administration less contemptuous of the scientist's tight time budget.

Payment for biodiversity information

Nearing the end of this essay, we open the largest can of worms. Without doubt the actual information in biodiverse tropical wildlands has the highest market value of any of its products, and the parataxonomist – apprentice curator – scientist – INBio chain is a major mechanism in putting it on the market (e.g. Eisner 1990). If every cup of coffee in the world had a two cent (US) tax on it, the income would pay all the world's basic tropical wildland conservation management bills forever. As it is, the only income for tropical conservation from that hot soup made from the ground-up seeds of an African rainforest understory shrub is that which is allocated to parks from national budgets of countries that grow coffee. Let me put it in a different format.

I know a tropical forest tree fruit that is extremely rich in antibiotics. However, this information cannot move into the processing system of a pharmaceutical company until Costa Rica's INBio has been able to arrange a royalty arrangement guaranteeing a signficant financial return to the greenhouse maintenance bill. The days of tropical green oil exploration and exploitation without payment of royalties to the host country are over.

The academic scientist, with associated collaborators, is both the solution and the problem. For nearly a century, the academic scientist has been supported in developed country universities on the premise that basic and applied research eventually repays the investment. The academic scientist in today's tropical universities are cut from the same cloth. The academic gains approval, benefits, salary, prestige, awards, etc. for the free, rapid and open flow of information into the greater social community – be it academic or business – through conversation and publication. Note that privately financed research institutes and laboratories have long increased a researcher's financial incentives while reducing the freedom to build personal prestige/value through free dissemination of research results. The files of major drug companies are replete with private information on plant analyses that academic researchers are unknowingly struggling to replicate. For two decades of my academic life, I never thought about the potential commercial use of any tropical biodiversity finding, and published as rapidly and as freely as possible. But discovering that the same receiving society had to be cajoled, wheedled, browbeaten and courted to extract even a tiny financial resource to save a large block of tropical biodiversity led to a quite rapid rethinking of the situation.

We could discuss whether the academic in the developed world is adequately reimbursed for the labor and love of science, but that is not where I feel redress is in order. There is a much greater imbalance. The developed world is just beginning to contribute to the budget for the maintenance of the enormous pools of tropical biodiversity information from which it has extracted information for centuries, and to which it is today turning even more strongly. Academic scientists are very much a key part of the extraction process. A developed-world scientist quite innocently submits a research proposal to the administration of a large tropical national park to study the location and species composition of nickel-accumulating plants throughout the park, with the stated intent of publication of the results in scientific journals. Quite innocently a conservation-minded ethnobotanist describes with glee how he "fooled" a rainforest shaman into divulging which plants are used to treat which ailments, with subsequent immediate glowing free distribution to the developed world's medical research community. Would the same enthusiasm greet a non-US electrical engineer describing to a US customs agent how he fooled a Silicon Valley technician into revealing the design of a new computer chip for free distribution to his country's computer industry? Do those who argue that the developed world's botanical gardens should be major facilitators of tropical phytochemical prospecting devote as much resource and mental energy to insuring that the source country and its wildlands receive a truly significant fraction of the profits from the development of the chemicals that are located?

Often, the academic researcher and backers respond to such questions by noting that the source country "was not doing anything with the biodiversity information" and therefore the finder can reasonably harvest at will. Well, let's see how far someone gets with that principle when he discovers oil in your backyard, oil that you neither knew was there nor were doing anything with. It is no secret that hundreds of thousands of valuable organisms are growing in a tropical country's greenhouses. The fact that we cannot (yet) read the books in these libraries does not lessen their value. Tropical biodiversity information indisputably has commercial value, but there is no system in place for the payment and fair distribution of the royalties. INBio is a pilot project with the intent to facilitate such a system, but there is much yet to be done.

How much does one charge for biodiversity information?

Biodiversity information is no different from all the other kinds of information in the marketplace – there is some that is very cheap and some that is very expensive. The answers center on "what the market will pay". Opportunity costs, research and development costs, inter-institutional espionage, company takeovers, price wars, stockholder pressure, interest income, risk evaluation on invested information, and all the other market processes apply. In other social sectors, information management and pricing is becoming every day a larger part of life. The pricing of tropical biodiversity information will undoubtedly borrow ideas and methodologies from other sectors, and offer some insights of its own. Pricing (and/or patenting) country-specific biodiversity information about a particular plant perfume from a particular rainforest will be quite different from the equitable distribution of new orange tree genes, a decision that will affect the health of literally billions of tropical rural residents.

But however biodiversity information is priced, the tropical academic scientist studying biodiversity walks a fine line between being, on the one hand, unexpectedly and unconsciously an industrial spy and on the other hand, being the source of life-giving economic and social support for the conservation of tropical wildland biodiversity.

ACKNOWLEDGMENTS

This essay was supported by NSF grant BSR 86–10149; the Ministerio de Recursos Naturales, Energia y Minas, Costa Rica; the Swedish International Development Authority, Stockholm, Sweden; the Swedish Society for the Conservation of Nature; the C. S. Fund; the Pew Charitable Trust; the Noyes Foundation and the J. D. and C. T. MacArthur Foundation. I am deeply indebted to W. Hallwachs, C. Fowler, I. Gauld, P. Waterman, T. Eisner, S. Viederman, J. Rosales, R. Garcia, S. Marin, A. Ugalde, P. León, A. Umaña and R. Gámez for inspiration during its development. W. Hallwachs aided greatly with editorial commentary.

LITERATURE CITED

Alcock, J. E. 1990. Summary of the draft supplemental environmental impact statement and draft amended forest land and resource management plan, Caribbean National Forest/Luquillo Experimental Forest. USDA Forest Service, Southern Region, Atlanta, GA: 20 pp.

Burger, W. (ed) 1977. Flora Costaricensis. Fieldiana Botany 40: 1–291.

Chase, A. 1987. *Playing God in Yellowstone*. Harcourt Brace Jovanovich, New York.

Chase, A. 1989. Greater Yellowstone and the death and rebirth of the national parks ideal. Orion 8(3): 44–55.

Cruz, F. and J. B. Cruz 1987. Control of black rats (*Rattus rattus*) and its effect on nesting dark-rumped petrels in the Galapagos Islands. Vida Silvestre Neotropica 1: 3–13.

DeVries, P. J. 1987. *The Butterflies of Costa Rica and their Natural Nistory*. Princeton University Press, Princeton, NJ.

Eisner, T. 1990. Prospecting for nature's chemical riches. Issues Sci. Technol. (Winter 1989–1990): 31–34.

Gauld, I. D. 1988. A survey of the Ophioninae (Hymenoptera: Ichneumonidae) of tropical Mesoamerica with special reference to the fauna of Costa Rica. Bull. Brit. Mus. (Nat. Hist.) 57: 1–309.

Hallwachs, W. 1986. Agoutis (*Dasyprocta punctata*): the inheritors of guapinol (*Hymenaea courbaril*: Leguminosae), pp. 285–305. In: A. Estrada and T. H. Fleming (eds) *Frugivores and Seed Dispersal*. Dr. W. Junk Publishers, Dordrecht, The Netherlands.

Janzen, D. H. 1970. Herbivores and the number of tree species in tropical forests. Amer. Nat. 104: 501–528.

Janzen, D. H. 1973. Rate of regeneration after a tropical high elevation fire. Biotropica 5: 117–122.

Janzen, D. H. (ed) 1983. *Costa Rican Natu-*

ral History. University of Chicago Press, Chicago.

Janzen, D. H. 1984. The most coevolutionary animal of them all. Crafoord Lectures, Royal Swedish Academy of Sciences, Stockholm, Sweden, pp. 2–20.

Janzen, D. H. 1985. On ecological fitting. Oikos 45: 308–310.

Janzen, D. H. 1986a. The future of tropical ecology. Ann. Rev. Ecol. Syst. 17: 305–324.

Janzen, D. H. 1986b. Biogeography of an unexceptional place: what determines the saturniid and sphingid moth fauna of Santa Rosa National Park, Costa Rica, and what does it mean to conservation biology. Brenesia 25/26: 51–87.

Janzen, D. H. 1987. Conservation and agricultural economics. Science 236: 1159.

Janzen, D. H. 1988a. Tropical ecological and biocultural restoration. Science 239: 243–244.

Janzen, D. H. 1988b. Guanacaste National Park: tropical ecological and biocultural restoration, pp. 143–192. In: J. Cairns, Jr. (ed) *Rehabilitating Damaged Ecosystems*, Vol. 2. CRC Press, Boca Raton, FA.

Janzen, D. H. 1988c. Management of habitat fragments in a tropical dry forest: growth. Ann. Miss. Bot. Gard. 75: 105–116.

Janzen, D. H. 1990. Sustainable society through applied ecology: the reinvention of the village, Foreword, pp. xi-xiv. In: R. Goodland (ed) *Race to Save the Tropics*. Island Press, Washington, DC.

Janzen, D. H. (in press). How to save tropical biodiversity: the National Biodiversity Institute of Costa Rica. Amer. Entomol.

Janzen, D. H. and W. Hallwachs (in press). Ethical aspects of the impact of humans on biodiversity. Pontificum Academiae Scientiarum Scripta Varia.

Janzen, D. H. and P. Martin 1982. Neotropical anachronisms: the fruits the gomphotheres ate. Science 215: 19–27.

Lewin, R. 1988. Costa Rican biodiversity. Science 242: 1637.

Palca, J. 1990. Libya gets unwelcome visitor from the West. Science 249: 117–118.

Putz, F. E., E. G. Leigh and S. J. Wright 1990. Solitary confinement in Panama. Garden (March – April 1990): 18–21.

Tangley, L. 1990. Cataloging Costa Rica's diversity. BioScience 40: 633–636.

Wikramanayake, E. D. 1990. Conservation of endemic rain forest fishes of Sri Lanka: results of a translocation experiment. Conserv. Biol. 4: 32–37.

SECTION 2

Genetic resources and conservation biology

In the biodiversity crisis, our focus naturally ranges from the species level, on up through habitats and communities to whole ecosystems. The lowest unit of conservation in this hierarchy is the species. The term "genetic resources" refers to even lower levels of the biological hierarchy, to the variation within the species right down to the DNA level.

Genetic resources will play at least two fundamental roles in meeting the biodiversity crisis. First, if species populations are to endure in the long term, they must evolve to meet future environments. In principle, the more genetically variable a species is, the better equipped it will be to undergo such changes. Second, the genetic resources in the plants and animals that humans directly use for their own existence play a key role in maintaining and improving the productivity of those species. If our farms, plantations and ranching systems are made more efficient and less polluting through genetic improvement, the pressure on further destruction and alienation of the remaining natural areas will be reduced.

The introduction of genetics into nature conservation is only two decades old. Soulé and Mills (Chapter 3) recount the high points of the interaction between genetics and conservation biology; from its origin in concerns for the genetic resources of crop plants to present criticism of its relevance to ecosystem approaches to conservation. They conclude that the single-species approach (itself based on ecological and genetic criteria) and the ecosystem approach are not a true dichotomy but two essential elements of the one task.

The genetic resources that enable species populations to survive and evolve essentially comprise allelic diversity, heterozygosity and divergence between populations. Hedrick and Miller (Chapter 4) turn explicitly to the avoidance of inbreeding depression or maintenance of heterozygosity and the maintenance of genetic variation in captive populations of endangered species. Pedigree analysis is an important tool for those involved in managing such populations for identifying important individuals and matings, and maximizing founder allele survival. They also discuss ways to maintain a specified amount of additive genetic variance in a metapopulation.

These notions have already influenced programmes of *ex situ* propagation in zoos (de Boer, Chapter 13).

Finally Brown and Schoen (Chapter 5) consider the extent of genetic divergence among the natural populations of a plant species. In outcrossing species, the evidence from isozyme surveys has typically been presented as suggesting that any single population harbours the majority of the total species genetic resources. However when variation at the morphological level, or alternatively at the DNA level is examined, the authors reach a contrary conclusion. Populations of plant species show a rich and varied structure, differing in the kinds and amounts of genetic diversity they contain. The challenge is to devise strategies to conserve an adequate sample of these resources.

3 Conservation genetics and conservation biology: a troubled marriage

Michael E. Soulé and **L. Scott Mills**
Environmental Studies,
University of California – Santa Cruz,
Santa Cruz, California 95064,
USA

ABSTRACT

In two prophetic papers, Otto Frankel presented the conceptual and moral agenda for the discipline of conservation genetics. Within the last decade the penetration of genetics into species-level conservation programs has been accomplished, completing the first phase of the field's development. Simultaneously, population genetics, demography, and population ecology have presented new approaches to the estimation of minimum viable populations; these estimates have increased by orders of magnitude, presenting conceptual and political problems to conservationists. In addition, we are seeing a growing hostility to both genetical and species-level programs in favor of so-called ecosystem approaches. We argue that the species – ecosystem dichotomy is a false one, though it will lead to a more integrated synthesis of conservation theory and practice.

A GLANCE BACKWARD

The chapters in this section constitute an authoritative status report on conservation genetics. Instead of presenting a summary of these summaries, we first tip our hats, as it were, to the people and events that helped shape the present, and second, peer ahead at some looming controversies. Some of these controversies could jeopardize our mission to protect genetic and biological diversity.

Until the middle 1970s, most of the people in charge of conservation ignored genetics, and most of the people in charge of genetics ignored conservation. But beginning around 1970, plant geneticists started to become alarmed about the disappearance of primitive or traditional crop varieties and their replacement by modern, genetically uniform, cultivars. Geneticists suspected then as they do today that the seeds of the green revolution contained the agents of their own ultimate collapse, namely, genetic uniformity. They began to question the wisdom of discarding, in a single human generation, the products of thousands of

Conservation of Biodiversity for Sustainable Development.
Edited by O. T. Sandlund, K. Hindar and A. H. D. Brown. 1992. Scandinavian University Press, Oslo.

years of artificial selection for adaptation to microhabitat diversity, and for resistance to stress and drought, disease, and attacks by pests.

Several publications appeared in the early seventies that called for or contained protocols for the sampling and maintenance of germ plasm collections (Bennett 1965; Frankel and Bennett 1970; Marshall and Brown 1975; Frankel and Hawkes 1975; see Frankel and Soulé 1981 for further citations). An international system of gene banks was endorsed by the United Nations Conference on the Human Environment in Stockholm, 1972. The International Board of Plant Genetic Resources (IBPGR) was established within the Food and Agriculture Organization of the United Nations in the same year in order to further the collection, conservation, documentation, and use of germplasm for crop species (Williams 1988). Plant geneticists, we can now see with hindsight, knew that sustainable agroecosystems were impossible without genetic, as well as species diversity. Today, a fairly sophisticated, global system of economic plant germ plasm collections is assembled.

In 1970 and 1974, Otto Frankel, the prophet of conservation genetics, published two landmark papers (Frankel 1970, 1974). In the second paper entitled "Genetic conservation: our evolutionary responsibility" he issued a challenge to the field of genetics, a rather unusual one too, in that it was basically ethical. To quote him (1974), ". . . genetics has social responsibilities in two directions: first to collaborate in planning the biological system of conservation so as to establish the highest possible evolutionary potential; second, to help in establishing an evolutionary ethic, as part of our social ethics, which will make it acceptable and indeed inevitable for civilized man to regard the continuing existence of other species as an integral part of his own existence. This demands continuing evolution." Frankel (1974) also foresaw the need for pervasive management of nature, noting that "[wild] species almost everywhere in the world will depend on organized protection for their survival" and such protection is predicated on management that facilitated "continuing evolution."

Frankel (1974) asked the important questions, including what were the conditions for continuing evolution. "The prime parameters," he said, "are the level and distribution of variation, the size of the **minimum viable population** (emphasis ours), and the optimal and minimal sizes of reserves. We need to know the minimum population size which is likely to yield a required level of variation." Harlan (1975) made a similar statement soon thereafter, and *in situ* conservation of traditional crop germplasm has been promoted more recently (e.g. Wilkes 1983; Nabhan 1985; Altieri and Merrick 1988). Soon thereafter, zoologists (Flesness 1977; Denniston 1978; Seal 1978) were to sound the alarm about the impact of genetic drift and bottlenecks on the amount of genetic variation in small, isolated, natural populations, although at that time there was little published data confirming these theoretical expectations.

As evidence for effects of genetic drift, Frankel cited our own work on island lizards (Gorman et al. 1975; Soulé 1980) that demonstrated the "small island effect." Lizard populations on islands smaller that five or ten hectares have lower than normal levels of heterozygosity. Based on these data, Frankel (1974) quotes Soulé (personal communication) as concluding that minimum population sizes assuring long-term survival (thousands of years) should be of the order of 10^4 or more, a number derived from the estimated population size of island lizard populations having levels of genetic variation typical of those of related mainland populations.

Frankel anticipated a controversy known later as the SLOSS debate among biogeographers and ecologists — the relative benefits of **Single Large Or Several Small** conservation units. He noted that "It has been argued that population subdivision maximizes the potential rate of evolution . . . . But it will be rash to conclude that many small reserves are more effective than fewer larger ones." Indeed, the subsequent discovery that persisting species in isolates are distributed as nested subsets (i.e. the species that persist in small patches are virtually always a sub-set of those persisting in larger patches, and those that disappear in small patches are virtually the same set from patch to patch), argues for larger reserves. This predictability in patterns of extinction has confirmed the advantages of large protected areas for some taxa (Patterson and Atmar 1986; Blake 1991; Bolger et al. 1991).

Frankel did not eschew philosophy either. Elaborating on the need for an ethical stance in genetics, Frankel came close to embracing a version of the intrinsic value argument for conservation. He asks if it is not our responsibility to keep evolutionary options open, so far as we can, for the sake of future humans? He went further in the direction of extending the ethical community to non-humans, hoping that this idea of evolutionary responsibility would deepen into an ethic "if and when men come to regard other species as an essential part of their own existence." This latter view expresses a concept of "self" that includes identifications with non-human species, and is not far from the deep ecology philosophy as enunciated by Arne Næss (1989, this volume).

The first international meeting concerning the application of conservation genetics to wild species occurred at a conference in La Jolla, California in 1978. The results of that conference (Soulé and Wilcox 1980) produced the first specific recommendations about genetical MVPs for animals, including the 50/500 numbers (effective population sizes) for short-term and long-term programs, respectively (Franklin 1980; Soulé 1980; Frankel and Soulé 1981; Hedrick and Miller this volume). An effective size of 50 was thought to provide some protection against short-term loss of fitness due to inbreeding; an effective size of 500 was thought to prevent loss of genetic variation, *per se*. These rules of thumb have fallen into disrepute for several reasons. First, they are based solely on genetical argumentation and data; since their publication in 1980, other fields, including demography and population ecology have started to advance their own recipes for the estimation of MVPs (Shaffer 1981; Soulé 1987). Nowadays, MVPs usually are based on assumptions or objectives about the probability of persistence (say 0.95) and time intervals (say 200 years). Some recent MVPs incorporate population, environmental, as well as genetical considerations (e.g. Gilpin 1989; Doak 1989).

Second, these numbers have been misused more often than they have been applied correctly. As pointed out by Harris and Allendorf (1989), many practitioners have ignored the difference between census size and effective size, thus grossly underestimating the actual MVP, especially for long-term conservation purposes.

Third, and related to the first point, it is currently fashionable to denigrate the significance of genetics in the estimation of MVPs, on the grounds that the effects of environmental variability on population persistence (Goodman 1987a; Belovsky 1987) will swamp any impact that inbreeding and loss of genetic fitness might have (Schwartz et al. 1986; Dawson et al. 1987; Lande 1988; see also Pimm et al. 1988, 1989). This conclusion, in our opinion, is premature (see below).

Just as it does in most other cultural phenomena, the pendulum in science swings back and forth. Thus, the current unpopularity of genetics is neither unprecedented nor unexpected. In the past, many conservationists have been extremely reluctant to accept the role of genetics in conservation. The reasons are not difficult to discern, and they had nothing to do with science, *per se*. Genetics is an arcane field, in part because of its difficult jargon, and in part because it is quantitative. It kindles fear in many biologists and laypersons alike, and fear creates hostility. Second, genetics is associated with theories of racial differences, human instincts, and similar deterministic ideas that were anathema to many left-leaning intellectuals during the middle decades of this century. Only recently has the role of genetics in human psychology, physiology, and disease been tolerated once more (e.g. Bouchard et al. 1990).

Another reason for the reluctance of some to accept the role of genetics in conservation is that acceptance would have meant the abandonment of cherished practices. In the zoo world, for example, many curators wished to maximize the number of species in their respective collections. A result was competition for species diversity; group sizes for many species were commonly in the range of two to six individuals; rarely were there interchanges between zoos. To acknowledge that inbreeding might be a factor in the frequent loss of these small groups would have necessitated the abandonment of many species in favor of larger population sizes for those retained. It took the encouragement of a few leaders (Benirschke et al. 1980; Conway 1986) and the inspired work of Katherine Ralls, Jon Ballou and their colleagues (Ralls and Ballou 1983; Ralls et al. 1988), not to mention the growing governmental red tape and economic costs of replacing lost groups, to convince many zoo curators that inbreeding was a leading cause of group morbidity and extinction.

The role of genetics in nature conservation was consolidated by the publication of two books (Frankel and Soulé 1981; Schonewald-Cox et al. 1983). In the latter volume, Frankel (1983) reiterated that a central concept of conservation genetics is fitness. These days when we ask such questions as (1) why do we census for heterozygosity?, (2) why do we calculate effective population size (N_e)?, (3) why do we do pedigree analysis?, (4) why do we do molecular genetic studies?, the answers are usually framed in terms of (1) either minimizing the loss of genetic variance which is thought to be associated with fitness, (2) minimizing the costs of inbreeding, or (3) avoiding trans-specific crosses between morphologically similar species. Conservation genetics exists for one reason only – to promote the fitness of targeted populations (e.g. Ralls and Ballou 1986). This is our commandment.

Conceptually, though, fitness is not as straightforward as it might seem. The concept of fitness is often dichotomized – always a danger sign in biology. The first category is immediate or short-term fitness, such as the average viability or fecundity of individuals living now. This kind of fitness is easy to define operationally and is relatively easy to measure. The second category is "potential" fitness. It represents the idea that genetic variation in a lineage is required to adapt, through natural selection, to changing conditions. It is based on the premises that (1) genetic variation can be limiting, and (2) any loss of genetic variation can compromise the future capacity of a population to adapt and survive. Long-term fitness, as represented by genetic variation, is like wealth, as represented by gold coins buried in the garden. It is a hedge against future disasters.

Potential **anything** is problematic, and potential or long-term fitness is no exception. First, it is a different kind of concept altogether from short-term fitness, because, unlike gold, there may not be a straightforward relationship between genetic variation and wealth. Whereas it may be assumed that two pieces of gold will buy twice as many pigs as a single piece, there is no guarantee that a population with 8 % heterozygosity has twice the probability of surviving 1,000 years as one with 4 % heterozygosity. Nevertheless, a proportionality between genetic variation and persistence time (long-term fitness) is assumed; this idea could be expressed as follows:

$$E_t \approx A\ [\ V_A\ ]$$

where E_t is the expected time of persistence of a population, A is a species-specific constant representing general "adaptiveness," and V_A is the current level of additive genetic variation. The evidence for such a relationship is remarkably thin, being limited to a few short-term studies (Frankel and Soulé 1981). Nevertheless, much of what we call conservation genetics is based on this premise.

IS GENETICS RELEVANT TO POPULATION PERSISTENCE?

As already mentioned, genetics has lost its short-lived hegemony in the MVP arena for natural populations. Will the pendulum ever swing back, or is conservation genetics banished to the computers of those managing gene banks, captive groups in zoos, and botanical gardens? This leads us to the subject of persistence, *per se*, and our ability to predict it.

The issue of population persistence or viability is intellectually fascinating, in part because it requires a multidisciplinary, systems approach (Gilpin and Soulé 1986). Consider a population restricted to a patch of habitat; assume there is negligible immigration. Given that one knew enough about this population's demography, its environment, and the relationships between the two, one could, at least in theory, estimate the mean and variance of its persistence time (Goodman 1987a; Belovsky 1987). In spite of the complexity and elegance of such an analysis, however, the result could be very misleading.

At least two factors would have been left out. One of these is the possibility of a catastrophe (Ewens et al. 1987); we won't elaborate on this subject here. The other factor is the interplay between population dynamics and the loss of fitness due to inbreeding or genetic drift (Soulé 1987). Returning to the population mentioned above, could it persist at a small size (say, $N \leqslant 50$) long enough to lose about 10 % of its genetic variation? If it could, would this amount of genetic attrition produce detectable and ecologically significant impacts on the fecundity and viability of its members? For example, Frankel and Soulé (1981: p. 68) predicted on the basis of laboratory results with domesticated animals and plants that a 10 % increase in the inbreeding coefficient is often likely to produce a 10 to 25 % lowering of total reproductive performance. It has been suggested that reproductive depression of this magnitude could significantly compromise r, the intrinsic rate of increase, thus compromising a small population's ability to persist in a fluctuating environment (Soulé 1987).

Simulating the effects of inbreeding

One of us (LSM) is exploring this question by simulating the impact of inbreeding on population growth. Only a brief overview of preliminary results are presented here. Consider two small, recently isolated populations with three age classes and no density dependence. One has an r near zero, but positive (approximately 0.06), a reasonable rate for many natural mammal populations; the other has an r of approximately 0.75, near the high end observed for mammals in the wild (Pianka 1970; Robinson and Redford 1986). The demographic costs of inbreeding on survivorship and fecundity may be obtained from the literature (e.g. Ralls et al. 1988). These costs may then be incorporated into a Leslie matrix growth model by weighting survivorship and fecundity by the fitness reduction due to inbreeding. For example, p_x (survivorship during age x) values are replaced by $p_x(F,t)$, where F is the average inbreeding coefficient for individuals born during time period t, and m_x (fecundity at year x) values are replaced by $m_x(F,t)$, age-specific reductions in these parameters due to inbreeding. In essence then, age-specific demographic values are reduced according to empirically measured inbreeding costs at each age. Based on the modified values of survivorship and fecundity, the constant Leslie matrix L is replaced by a variable matrix $L(F,t)$, which can be projected to estimate population sizes in the future.

The results of two such population projections are shown in Figures 1 and 2. Both are initiated with a population of 30; four projection intervals (years) are shown. The population with $r = 0.06$ (Figure 1) grows to over 40 in four years with no inbreeding effects. With inbreeding, however, the population declines to 15. For the population with a very high intrinsic rate of increase, the size reached in four years is large, even with inbreeding costs, but it is less than half that of the same population without inbreeding costs (Figure 2).

As mentioned, it is currently fashionable to dismiss the role of inbreeding depression when projecting the viability of small populations. Some experts (Schwartz et al. 1986; Dawson et al. 1987; Lande 1988; see also Pimm et al. 1988, 1989) argue that when a population is so small that its fitness may be compromised by inbreeding, that it is in much greater danger of crashing from demographic and environmental stochasticity. And such a population, they argue, is unlikely to persist long enough to experience enough inbreeding to do it significant damage. We believe, to the contrary, that theoretical and empirical studies such as the one just summarized will demonstrate that inbreeding can affect population size, and therefore persistence, in the face of random demographic events and environmental fluctuations.

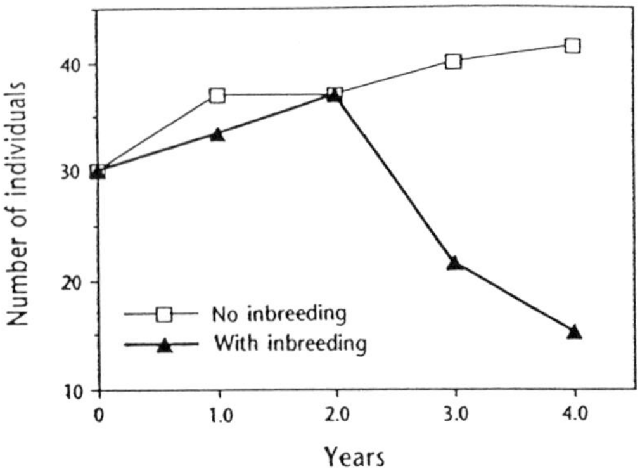

Figure 1. The effect of imposing an inbreeding depression cost on a population with a low but positive intrinsic growth rate ($r_m = 0.06$) typical of mammalian species in nature. The model is a deterministic Leslie Matrix based on a population with two age classes. See text.

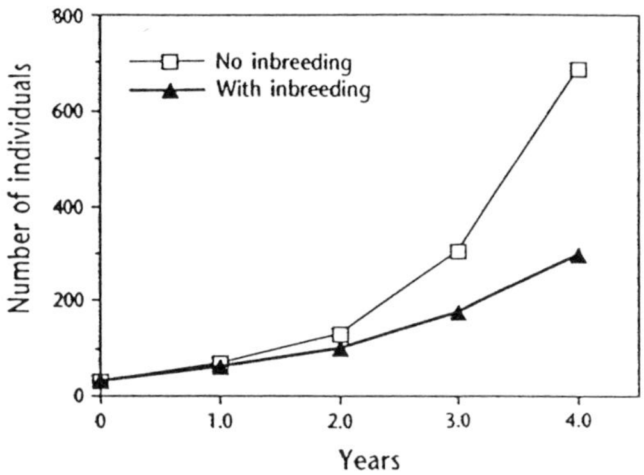

Figure 2. The effect of imposing an inbreeding depression cost on a population with a very high intrinsic growth rate ($r_m = 0.75$). The model is a deterministic Leslie Matrix based on a population with two age classes. See text.

Current simulation studies in collaboration with Peter Smouse at Rutgers University are incorporating age-specific costs of inbreeding as variables in a stochastic population growth model (L. S. Mills and P. Smouse, in preparation). In addition, LSM is testing the implications of the model using natural populations of the California red-backed vole (*Clethrionomys californicus*) in forest remnants.

MVP HYPERINFLATION

The second problem we would like to mention derives from the success of conservation biology as reflected in the startling changes in the last 25 years. Recall that in the sixties, MVPs (demographic) were in the range of 25 to 50 (MacArthur and Wilson 1967; Richter-Dyn and Goel 1972). A population larger than this is not much threatened by demographic stochasticity (Leigh 1981; Goodman 1987a). The next stage in MVP studies, the genetic era, could be characterized by the 50/500 criteria. The 500 number refers to a genetically effective size of 500, which often translates to a census size of 2,000 to 5,000 or more (Nei and Graur 1984; Harris and Allendorf 1989; Michael Smith, personal communication; R. Frankham, personal communication). Then came the era of MVP hyperinflation, with values in the range of 10^4 to 10^6 based on analyses of persistence based, in turn, on the effects of random environmental fluctuations on the dynamics for single populations (Goodman 1987a; Belovsky 1987). (Note that the numerical criteria used for germplasm collections is not covered in this discussion.) These large numbers, however, are problematic for both pragmatic and scientific reasons. The pragmatic problem is that administrators and managers are confronted with unreachable population targets; to follow expert advice in many cases would not be economically or politically expedient. Consider, for example, the home range of a mountain lion (*Felis concolor*) in North America may be 100 km^2 or more (Hempker et al. 1984; Hopkins et al. 1986). If the MVP is 10^5, a viable population in the USA would require more than all the land in the contiguous 48 states.

In the United States, let alone in nations where wildlife is competing for space with many landless or poor, only a very single-minded conservationist would stand up in public to argue that an area of these dimensions be given back to nature. But neither should we disguise the truth as we see it. How do we escape from this dilemma? The only answer, we believe, is to accompany the arguments for such huge numbers with practical management alternatives. Among these would be (1) the establishment or enhancement of multiple populations, (2) a combination of *in situ* and *ex situ* approaches, including the storage of germplasm (e.g. seed banks, tissue culture, cryopreservation of sperm, embryos, and larvae), and even, at some risk to the species, (3) accepting proposals that buy time instead of achieving an ideal solution in one step.

Finally, an MVP in the range of 10^5 to 10^6 may not be defensible on purely biological grounds (Soulé 1987; Goodman 1987b) for the simple reason that a single, very large population is neither the prudent strategy nor the natural solution. The total number needed may be much less if the individuals are living in two or more places, the environments of which are not perfectly correlated. In nature, many species are distributed as metapopulations – a patchwork of interacting populations over a wide and heterogeneous geographic region; at any given time some of the patches are extinct and some are newly colonized. Partly as a consequence of the environmental differences that these manifold, dispersed population units encounter, the persistence of a metapopulation of a given N or N_e is expected to be superior to that of a single population of the same numerical order (Soulé 1987; Goodman 1987b; Gilpin and Hanski 1991). It seems probable, therefore, that the incorporation of both spatial and temporal heterogeneity along with metapopulation dynamics into MVP analysis will reduce the number of individuals required to ensure long-term viability. Metapopulation dynamics could either increase or decrease the rate of loss of genetic variation, however, depending on the extinction and recolonization rates (Gilpin 1987), as well as the population growth rates.

THE "FAILURE" OF SPECIES CONSERVATION

Another potential cloud on the horizon is the growing reaction against species-based conservation. More and more these days, we hear voices raised against the emphasis on species rather than on habitat, ecosystem, or landscape approaches (e.g. Harris 1984; Scott et al. 1987; Simberloff 1987; Kohm 1991). Because conservation genetics is inherently a species or population-based discipline, conservation genetics is guilty by association with so-called species chauvinists.

The anti-species movement could produce a backlash against endangered species and the scientists and managers who work to protect them. The tendency among some conservation biologists to dichotomize nature into species and ecosystems almost guarantees that many innocent species and dedicated scientists and professionals will be characterized as "bad guys." The new heroes are the plant geographers, the ecosystem ecologists, restoration ecologists, and landscape ecologists who favor a more holistic, comprehensive and proactive conservation strategy.

The growing backlog of endangered species

It is easy to see why many people are becoming frustrated, indeed infuriated with species-based conservation. Consider the statistics on endangered species in the US. There are about 570 officially listed threatened and endangered species, but only about half of these have recovery plans. Recovery plans establish the criteria for recovery and recommend the management interventions that should be implemented. But these plans have no teeth. No funds accompany the approval of recovery plans, and rarely do the plans identify critical habitat, let alone protect it. In fact, most plans are nothing more than a promise cloaked in a prayer. Granted, the provisions of the Endangered Species Act prohibit the so-called "taking" of individuals or the destruction of its habitat, but a determined developer can often find a way to circumvent the recommendations in a recovery plan.

These problems of implementation and enforcement, however, are not the only reasons for the growing hostility to species-based approaches. Even if recovery plans were efficient and well funded, this strategy would still be unpopular. One excuse for attacking the species approach is that many of the species that receive special attention are large and attractive, ignoring many species-rich groups such as plants and insects (Wilson 1988). Another reason for the growing concern is that species are listed more slowly than they are nominated, and the backlog increases yearly. In the US, there are now more than 4,000 candidate species. Recovery and delisting of species are rare, and recovery plans are produced and implemented at a much slower rate than species are nominated and listed. Many people are now saying that the system has broken down (Scott et al. 1991).

Endangered species are symptoms, not causes

There are other problems. A question asked by growing numbers of conservationists throughout the world is "what is the point of adding a name to a list of moribund species if little or nothing is being done to counteract the causes that produce so many species crises?" These causes are all too familiar: (1) the loss and disturbance of habitat, (2) the fragmentation of habitat, (3) air and water pollution, (4) the spread of exotic species and diseases, (5) climate change, and (6) overexploitation of populations.

Simply put, **the species approach to conservation is a crisis management approach** (Salwasser 1991). Even those activists who are successful in obtaining funding for their favorite endangered species may perceive the futility of trying to rescue a particular taxon where nothing is being done to prevent many more species from becoming endangered. Of course, endangered species are very popular, and organizations can raise a lot of money for them, but the supporters of less popular species or of whole systems are frustrated when millions of dollars are spent on a few flagship species, leaving little or nothing for less charismatic creatures, nor preventing further environmental deterioration.

IN DEFENSE OF PLURALISM

We are concerned, though, that the conservation crisis could evolve into a very destructive conflict. On the one side are the species enthusiasts who generally defend large, charismatic vertebrates and keystone species; on the other side are the ecosystem enthusiasts who favor a habitat approach based on (1) the identification, classification, and protection of representative samples of all biotic communities (e.g. Jenkins 1988; Huntley 1988; Crumpacker et al. 1989) and (2) the management of these protected areas at the appropriate physical and temporal scales (Clark and Harvey 1988). In theory, there should be no conflict. Just as assessment of impacts at the ecosystem level must depend on the monitoring of individual species (Van Horne 1983), so the conservation of species requires attention to ecosystem processes and disturbance dynamics.

In practice, though, supporters of a species approach are not convinced that a habitat approach can be implemented and maintained for many decades or centuries, particularly in the many parts of the tropics; the list of protected areas that are being overrun, overexploited or destroyed by mining, agriculture, forestry and water projects is growing at an alarming rate, according to the Threatened Parks Register of the International Union for Conservation of Nature and Natural Resources' Commission on National Parks and Protected Areas. Thus, while supporters of systems approaches see money being "wasted" on a few charismatic vertebrates, the supporters of endangered species and captive propagation point to the increasing rate of park failure and the need for a backup system that is more secure, even if less natural.

A naive expression of the ecosystem approach is one that holds the extremist assumption that design and management of protected areas can ignore species altogether, focusing instead on ecosystem processes and patch dynamics. Clearly, though, ecosystem management must attend to the viability of certain, critical species, even if they are not yet endangered. Any protected area will undergo biotic collapse unless designers and managers understand the natural history and viability of certain species. A corollary of this premise is that the failure to manage or maintain these species will lead to the collapse of the system, or at least to a significant reduction in species richness (Frankel and Soulé 1981; Botkin 1990).

Table 1 presents three sets of questions that may clarify this point. The first question is the amount of area needed to prevent the loss of biological diversity in the system. Strictly speaking, ecosystem approaches do not explicitly identify this question, nor can they approximate a general answer or identify the criteria that should be considered. A biologically defensible solution must be based in part on the minimum genetical and spatial requirements of certain, strongly interacting, species (L. S. Mills and M. E. Soulé, in preparation). For example, if trees of the genus *Ficus*, invertebrates such as starfish or termites, fishes such as groupers, or vertebrates such as beavers, tapirs, alligators or elephants are important in the long-term maintenance of habitat diversity in the system, special attention must be devoted to the needs and viability of their populations within the protected area or the system of protected areas. The point is that in most places, **system viability cannot be disassociated from species viability**.

A related point is that the fragmentation of wildlands will usually threaten elements of biodiversity that depend on long-range dispersal movements, migratory movements, or foraging movements. Some of these species may be important mutualists (e.g. pollinators, mycorrhizal fungi). For example, if bats are important pollinators of many tree species, it will be necessary to insure that their roosting and breeding sites are protected within the system of protected areas, and that human activities around and in the reserves do not compromise their long-term viability. If deep forest birds or sphingid moths must migrate between wet-

Table 1. Some questions for the designers and managers of protected areas that demonstrate the roles of species and process management in an integrated, "ecological" approach to conservation.

1. How much area is needed to prevent the loss of biological diversity in the system?
   - What constitutes a substantial loss of biodiversity for the system?
   - What criteria can be used to predict whether such losses will occur?
     - Does the system contain viable populations of all the strongly interacting ("keystone") species? How large an area is needed for these viable populations?
     - Is the system sufficient in size and topographic heterogeneity to generate an appropriate level of disturbance to maintain the necessary diversity of habitats under foreseeable changes in climate and other conditions?

2. What sensitive species or processes are threatened by loss of connectivity?
   - Would the loss of any species (e.g. mutualist or migratory species) lead to extinction cascades and compromise biodiversity?
     - What interventions or new design features would mitigate such threats?

3. What management interventions will be needed to prevent the spread of deleterious exotic species?
   - What plants, animals, pathogens are contributing to habitat changes, to decreases in the diversity of native species and assemblages, and to deleterious edge effects?
     - How will these species and changes compromise the viability of critical species or processes?
     - How can these deleterious species and changes be managed/controlled?

season and dry-season sites (Karr 1982; Janzen 1986), conservation corridors may have to be established.

The third question in Table 1 points to another category of species – those that are harmful to system stability and species richness. Many of these deleterious species will be exotics, often but not always originating in other continents (e.g. Soulé 1990). Exotic or expanding species of plants, predators, herbivores, or nest parasites (such as cowbirds in the US) often create the conditions for wildfire, for the invasion of weeds, for the mortality of critical pollinators or seed dispersers, for the spread of disease, and the inviability of sensitive species. Management protocols tailored to specific species are a necessity for most such situations.

In conclusion, the design and management of conservation areas will rarely succeed without attending to the needs and control of critical or problem species. The species vs. ecosystem dichotomy, like almost every other dichotomy in biology, is fatally flawed conceptually and operationally. A pure ecosystem approach is just as contradictory as a pure species approach. As we write, the popularity of genetical and species approaches to conservation is waning. So-called holistic approaches are in the ascendency. To the extent that these trends bring about a biologically realistic synthesis, they are welcomed. A purely genetic approach to population viability is absurd (Frankel and Soulé 1981). A conservation strategy blind to anything except species on the brink of annihilation is absurd. The danger is that the pendulum could swing too far in the other direction. The deep-seated antagonism against genetics, and the hostility that sometimes exists between ecosystem ecologists and species-oriented biologists could undermine the emergence of new, synthetic approaches and technologies.

ACKNOWLEDGMENTS

Vernon Heywood and Tony Brown are thanked for key references.

LITERATURE CITED

Altieri, M. A. and L. C. Merrick. 1988. Agroecology and in situ conservation of native crop diversity in the Third World, pp. 361–369. In: E. O. Wilson (ed) *Biodiversity*. National Academy Press, Washington, DC.

Belovsky, G. E. 1987. Extinction models and mammalian persistence, pp. 35–58 In: M. E. Soulé (ed) *Viable Populations for Conservation*. Cambridge University Press, Cambridge.

Benirschke, K., B. Lasley and O. Ryder. 1980. The technology of captive propagation, pp. 225–242. In: M. E. Soulé and B. A. Wilcox (eds) *Conservation Biology: An Evolutionary – Ecological Perspective*. Sinauer, Sunderland, MA.

Bennett, E. 1965. Plant introductions and genetic conservation: genecological aspects on an urgent world problem. Scottish Plant Breeding Station Record: 27–113.

Blake, J. G. 1991. Nested subsets and the distribution of birds on isolated woodlots. Conserv. Biol. 5: 58–66.

Bolger, D. T., A. C. Alberts and M. E. Soulé. 1991. Occurrence patterns of bird species in habitat fragments: sampling, extinction, and nested species subsets. Amer. Nat. 137: 155–166.

Botkin, D. B. 1990. *Discordant Harmonies: A New Ecology for the Twenty-First Century*. Oxford University Press, Oxford.

Bouchard jr., T. J., D. T. Lykken, M. McGue, N. L. Segal and A. Tellegen. 1990. Sources of human psychological differences. The Minnesota study of twins reared apart. Science 250: 223–228.

Clark, T. W. and A. H. Harvey. 1988. Management of the greater Yellowstone ecosystem: an Annotated Bibliography. Northern Rockies Conservation Cooperative: 52 pp. P.O. Box 2705, Jackson, WY 83001.

Conway, W. 1986. The practical difficulties and financial implications of endangered species breeding programs. Int. Zoo Yearb. 24/25: 210–219.

Crumpacker, D. W., S. W. Hodge, D. Friedley and W. P. Grepp. 1988. A preliminary assessment of the status of major terrestrial and wetlands ecosystems on federal and Indian lands in the United States. Conserv. Biol. 2: 103–115.

Dawson, W. R., J. D. Ligon, J. R. Murphy, J. P. Myers, D. Simberloff and J. Verner. 1987. Report of the scientific advisory panel on the spotted owl. Condor 89: 205–229.

Denniston, C. 1978. Small population size and genetic diversity. Implications for endangered species, pp. 281–290. In: S. A. Temple (ed) *Endangered Birds: Management Techniques for Preserving Threatened Species.* University of Wisconsin, Madison, WI.

Doak, D. 1989. Spotted owls and old growth logging in the Pacific Northwest. Conserv. Biol. 3: 389–396.

Ewens, W. J., P. J. Brockwell, J. M. Gani and S. I. Resnick. 1987. Minimum viable populations size in the presence of catastrophe, pp. 59–69. In: M. E. Soulé (ed) *Viable Populations for Conservation.* Cambridge University Press, Cambridge.

Flesness, N. R. 1977. Gene pool conservation and computer analysis. Int. Zoo Yearb. 17: 62–68.

Frankel, O. H. 1970. Variation – the essence of life. Proc. Linn. Soc. New South Wales 95: 158–169.

Frankel, O. H. 1974. Genetic conservation: Our evolutionary responsibility. Genetics 78: 53–65.

Frankel, O. H. 1983. The place of management in conservation pp. 1–14. In: C. M. Schonewald-Cox, S. M. Chambers, B. MacBryde and W. L. Thomas (eds) *Genetics and Conservation: A Reference for Managing Wild Animal and Plant Populations.* Benjamin/Cummings, Menlo Park, CA.

Frankel, O. H. and E. Bennett (eds) 1970. *Genetic Resources in Plants.* IBP Handbook No. 11. Blackwell Scientific Publications, Oxford.

Frankel, O. H. and J. G. Hawkes (eds) 1975. *Crop Genetic Resources for Today and Tomorrow.* International Biological Programme 2. Cambridge University Press, Cambridge.

Frankel, O. H. and M. E. Soulé. 1981. *Conservation and Evolution.* Cambridge University Press, Cambridge.

Franklin, I. R. 1980. Evolutionary change in small populations, pp. 135 -149. In: M. E. Soulé and B. A. Wilcox (eds) *Conservation Biology: An Evolutionary – Ecological Perspective.* Sinauer, Sunderland, MA.

Gilpin, M. E. 1987. Spatial structure and population vulnerability, pp. 125–140. In: M. E. Soulé (ed) *Viable Populations for*

Conservation. Cambridge University Press, Cambridge.

Gilpin, M. E. 1989. Population viability analysis. Endangered Species Update 6(10): 15–18.

Gilpin, M. E. and I. Hanski. 1991. *Metapopulation Dynamics: Empirical and Theoretical Investigations*. Academic Press, San Diego.

Gilpin, M. E. and M. E. Soulé. 1986. Minimum viable populations: Process of species extinctions, pp. 19–34. In: M. E. Soulé (ed) *Conservation Biology: The Science of Scarcity and Diversity*. Sinauer, Sunderland, MA.

Goodman, D. 1987a. The demography of chance extinction, pp. 11–34. In: M. E. Soulé (ed) *Viable Populations for Conservation*. Cambridge University Press, Cambridge.

Goodman, D. 1987b. How do any species persist? Lessons from conservation biology. Conserv. Biol. 1: 59–62.

Gorman, G. L., M. E. Soulé and S. Y. Yang. 1975. Evolutionary genetics of insular Adriatic lizards. Evolution 29: 52–71.

Harlan, J. R. 1975. Our vanishing genetic resources. Science 188: 618–622.

Harris, L. D. 1984. *The Fragmented Forest: Island Biogeography Theory and the Preservation of Biotic Diversity*. University of Chicago Press, Chicago.

Harris, R. B. and F. W. Allendorf. 1989. Genetically effective population size of large mammals: an assessment of estimators. Conserv. Biol. 3: 181–191.

Hedrick, P. W. and P. Miller. 1992. Conservation genetics: theory and management of captive populations. This volume.

Hempker, T. P., F. G. Lindzey and B. B. Ackerman. 1984. Population characteristics and movement patterns of cougars in southern Utah. J. Wildl. Mgmt. 48: 1275–1284.

Hopkins, R. A., J. J. Kutilek and G. L Shreve. 1986. Density and home range characteristics of mountain lions in the Diablo Range in California, pp. 223–235. In: S. D. Miller and D. D. Everett (eds) *Cats of the World: Biology, Conservation, and Management*. National Wildlife Federation, Washington, DC.

Huntley, B. J. 1988. Conserving and monitoring biotic diversity: some African examples, pp. 248–260. In: E. O. Wilson (ed) *Biodiversity*. National Academy Press, Washington, DC.

Janzen, D. H. 1986. The eternal external threat, pp. 286–303. In: M. E. Soulé (ed) *Conservation Biology: The Science of Scarcity and Diversity*. Sinauer, Sunderland, MA.

Jenkins, R. E. 1988. Information management for the conservation of biodiversity, pp. 231–239. In: E. O. Wilson (ed) *Biodiversity*. National Academy Press, Washington, DC.

Karr, J. R. 1982. Avian extinction on Barro Colorado Island, Panama: A reassessment. Amer. Nat. 119: 220–239.

Kohm, K. A. (ed) 1991. *Balancing on the Brink of Extinction: The Endangered Species Act and Lessons for the Future*. Island Press, Washington, DC.

Lande, R. 1988. Genetics and demography in biological conservation. Science 241: 1455–1460.

Leigh, E. G. jr. 1981. The average lifetime of a population in a varying environment. J. Theor. Biol. 90: 231–239.

MacArthur, R. H. and E. O. Wilson. 1967. *The Theory of Island Biogeography*. Princeton University Press, Princeton, NJ.

Marshall, D. R. and A. H. D. Brown. 1975. Optimum sampling strategies in genetic conservation, pp. 53–80. In: O. H. Frankel and J. G. Hawkes (eds) *Crop Genetic Resources for Today and Tomorrow*. International Biological Programme 2. Cambridge University Press, Cambridge.

Nabhan, G. B. 1985. Native crop diversity in Aridoamerica: Conservation of regional gene pools. Econ. Bot. 39: 387–399.

Næss, A. 1989. *Ecology, Community and Lifestyle*. Cambridge University Press, Cambridge.

Næss, A. 1992. Sustainability! The integral approach. This volume.

Nei, M. and D. Graur. 1984. Extent of protein polymorphism and the neutral mutation theory. Evol. Biol. 17: 73–118.

Patterson, B. D. and W. Atmar. 1986. Nested subsets and the structure of insular mammalian faunas and archipelagos, pp. 65–82. In: L. R. Heaney and B. D. Patterson (eds) *Island Biogeography of Mammals*. Academic Press, New York.

Pianka, E. R. 1970. On r and K selection. Amer. Nat. 104: 592–597.

Pimm, S. L., H. L. Hones and J. M. Diamond. 1988. On the risk of extinction. Amer. Nat. 132: 757–785.

Pimm, S. L., J. L. Gittleman, G. F. McCracken and M. Gilpin. 1989. Plausible alternatives to bottlenecks to explain reduced genetic diversity. Trends Ecol. Evol. 4: 176–177.

Ralls, K. and J. Ballou. 1983. Extinction: lessons from zoos, pp. 164–184. In: C. M. Schonewald-Cox, S. M. Chambers, B. MacBryde and W. L. Thomas (eds) *Genetics and Conservation: A Reference for Managing Wild Animal and Plant Populations*. Benjamin/Cummings, Menlo Park, CA.

Ralls, K. and J. Ballou. 1986. Genetic management of captive populations. Zoo Biol. 5: 127–138.

Ralls, K., J. D. Ballou and A. Templeton. 1988. Estimates of lethal equivalents and the cost of inbreeding in mammals. Conserv. Biol. 2: 185–193.

Richter-Dyn, N. and N. S. Goel. 1972. On the extinction of a colonizing species. Theor. Pop. Biol. 3: 406–433.

Robinson, J. G. and K. H. Redford. 1986. Intrinsic rate of natural increase in Neotropical forest mammals: relationship to phylogeny and diet. Oecologia 68: 516–520.

Salwasser, H. 1991. In search of an ecosystem approach to endangered species conservation, pp. 247–265. In: K. A. Kohm (ed) *Balancing on the Brink of Extinction: The Endangered Species Act and Lessons for the Future*. Island Press, Washington, DC.

Schonewald-Cox, C. M., S. M. Chambers, B. MacBryde and W. L. Thomas (eds) 1983. *Genetics and Conservation: A Reference for Managing Wild Animal and Plant Populations*. Benjamin/Cummings, Menlo Park, CA.

Schwartz, O. A., V. C. Bleich and S. A. Holl. 1986. Genetics and the conservation of mountain sheep *Ovis canadensis nelsoni*. Biol. Conserv. 37: 179–190.

Scott, J. M., B. Csuti, J. D. Jacobi and J. E. Estes. 1987. Species richness: a geographic approach to protecting future biological diversity. BioScience 37: 782–788.

Scott, J. M., B. Csuti, K. Smith, J. E. Estes and S. Caicco. 1991. Gap analysis of species richness and vegetation cover: an integrated biodiversity conservation strategy, pp. 282–297. In: K. A. Kohm (ed) *Balancing on the Brink of Extinction: The Endangered Species Act and Lessons for the Future*. Island Press, Washington, DC.

Seal, U. S. 1978. The Noah's Ark problem: multigeneration management of wild species in captivity, pp. 303–319. In: S. A. Temple (ed) *Endangered Birds: Management Techniques for Preserving Threatened Species*. University of Wisconsin, Madison, WI.

Shaffer, M. L. 1981. Minimum population sizes for species conservation. BioScience 31: 131–134.

Simberloff, D. 1987. The spotted owl fracas: mixing academic, applied, and political ecology. Ecology 68: 766–772.

Soulé, M. E. 1980. Thresholds for survival: criteria for maintenance of fitness and evolutionary potential, pp. 151 -170. In: M. E. Soulé and B. A. Wilcox (eds) *Conservation Biology : An Evolutionary – Ecological Perspective*. Sinauer, Sunderland, MA.

Soulé, M. E. (ed) 1987. *Viable Populations for Conservation*. Cambridge University Press, Cambridge.

Soulé, M. E. 1990. The onslaught of alien species, and other challenges in the coming decades. Conserv. Biol. 4: 233–240.

Soulé, M. E. and B. M. Wilcox (eds) 1980. *Conservation Biology: An Evolutionary – Ecological Perspective*. Sinauer, Sunderland, MA.

Van Horne, B. 1983. Density as a misleading indicator of habitat quality. J. Wildl. Mgmt. 47: 893–901.

Wilkes, H. G. 1983. Current status of crop plant germplasm. CRC Crit. Rev. Plant Sci. 1: 133–181.

Williams, J. T. 1988. Identifying and protecting the origins of our food plants, pp. 240–247. In: E. O. Wilson (ed) *Biodiversity*. National Academy Press, Washington, DC.

Wilson, E. O. 1988. The current state of biological diversity, pp. 3–18. In: E. O. Wilson (ed) *Biodiversity*. National Academy Press, Washington, DC.

4 Conservation genetics: theory and management of captive populations

Philip W. Hedrick and **Philip Miller**
Pennsylvania State University,
Department of Biology,
University Park, Pennsylvania 16802,
USA

ABSTRACT

Conservation genetics utilizes the tools and concepts of genetics and applies them to problems in conservation biology. For captive animal populations, pedigree analysis has become the basic approach to evaluate breeding priority of particular individuals. Several pedigree analysis techniques are commonly used, but peeling and gene dropping give the most information. We compare these techniques and illustrate their value with applications to the Guam rail, Przewalski's horse and other endangered captive animals. The rationale for much conservation genetic interpretation is based in evolutionary genetics. We discuss the avoidance of inbreeding depression and the maintenance of genetic variation, both primary conservation genetic goals, from this perspective. In addition, we suggest aspects of these factors that deserve greater attention in their overall application to conservation planning.

INTRODUCTION

Because of the increase in the human population and expansion of human-related activity in recent decades, many species have become extinct or are on the verge of extinction (e.g. Ehrlich and Ehrlich 1981). Increased human activity has led to dramatic habitat fragmentation and loss, particularly in the tropics; hence, extinction rates have risen alarmingly over the last 15 years (Wilson 1988). Although the preservation of most endangered species depends primarily on the protection of the species and the maintenance of adequate natural habitat, in the past decade substantial attention has focused on the conservation genetics of endangered species and captive animal populations (e.g. Ralls and Ballou 1986; Soulé 1986, 1987; Ballou et al. in press). Conservation genetics may become more important because some species (and probably many more in the future) exist only in captive situations.

Because theoretical ideas from population and quantitative genetics have been widely applied to conservation genetics, our

Conservation of Biodiversity for Sustainable Development.
Edited by O. T. Sandlund, K. Hindar and A. H. D. Brown. 1992. Scandinavian University Press, Oslo.

purpose here is to evaluate and put in a reasonable perspective some of these suggestions. First, we will discuss developments in pedigree analysis with emphasis on gene drop analysis and peeling. Second, we will introduce some ideas about inbreeding depression and its relation to conservation genetics. Finally, we will discuss suggestions concerning the population size necessary to conserve a given amount of variation and other factors that may influence genetic variation.

PEDIGREE ANALYSIS

The catastrophic decline in global biodiversity has led some conservation biologists to conclude that the last refuge for a number of organisms, primarily birds and mammals, may be zoological parks. As a result, research into conserving genetic diversity within captive populations has recently received significant scientific attention (e.g. Ralls and Ballou 1986; Ballou et al. in press) and has resulted in the Species Survival Plan program (AAZPA 1983), a cooperative effort among zoos to strengthen and coordinate captive breeding programs of a selected group of critically endangered species. One of the important tools available to those involved in the genetic management of captive populations is pedigree analysis. With detailed pedigrees, the genetic structure of the population can be evaluated including levels of inbreeding in the populations and the distribution of founder genes within the current members of the population. The identification of genetically important individuals (e.g. Geyer et al. 1989, and discussion below) within a captive population through pedigree analysis has become a vital component of any breeding program because lack of space for animals is a serious problem in zoos. In addition, minimization of inbreeding levels and, consequently, minimization of inbreeding depression is a primary short-term goal in captive breeding.

The inbreeding coefficient of individual X when lines of descent for that individual are known can be calculated as

$$f_X = \sum_{i=1}^{k} (1/2)^{n_i + m_i + 1} (1 + F_{CA_i}) \qquad (1)$$

where n_i and m_i are the number of generations from male and female parent, respectively, to a common ancestor CA_i, F_{CA_i} is the inbreeding coefficient present in common ancestor i, and k is the number of common ancestors. One of the most common algorithms used to calculate f is the path analysis technique (e.g. Stevens 1975; Boyce 1983) in which all distinct paths which ascend from one parent of the individual in question to the common ancestor and then descend to the same individual through the other parent are identified. The major stipulations are that no individual can be counted more than once (thereby constituting a "stem" in the path), only ascending links can be constructed while travelling from the first parent to the common ancestor, and once the descent to the other parent has begun, only descending links can be constructed. Another technique to calculate inbreeding is the numerator relationship method defined by Quaas (1976) and applied to endangered species pedigrees through the use of the additive relationship matrix (Ballou 1983). The additive relationship matrix is that matrix with the ith diagonal element (equivalently, the ith individual) equal to $1 + f_i$ (for details about this procedure, see Quaas 1976; Ballou 1983).

Both of these computational methods were used by MacCluer et al. (1983) in their study of the extent of inbreeding in North American Standardbred horses. In the pedigree for these horses a total of 5,207 animals were represented, with paths of co-ancestry tracing back as far as 25 generations. It was determined that remote inbreeding occurring early in the pedigree contributed significantly to f in the current population, suggesting that the common practice of truncating the search for common ancestors after just a few generations may result in a serious underestimate of the true inbreeding coefficients. The numerator relationship matrix method of Quaas (1976) is significantly faster and easier to apply to complex extended pedigrees than is the path analysis technique of Stevens (1975).

One of the primary goals of any captive breeding program is the maximization of founder allele survival. As reintroduction of the species to its original habitat is a high priority, it is crucial to maintain the genetic representation of all wild-caught founders. With information about genetically important individuals (or important groups of individuals) among the current population, subsequent breeding recommendations can be made with the short-term goal of equalizing founder contributions or achieving particular target founder contributions within the current population (for a discussion of these and other criteria, see Haig et al. 1990).

One of the more direct ways to determine the genetic contribution of each founder to the current population is by way of the additive relationship matrix (Ballou 1983). Using this matrix, the coefficient of relationship between an individual and a particular founder can be calculated and the summation of these values over all living descendents in the population gives the expected contribution for that particular founder. An example of the founder contribution of the 16 wild-caught Guam rails (*Rallus owstoni*) to the current population is given in Figure 1 (Haig et al. 1990). Five of the potential founders (8, 10, 32, 33 and 36) have not contributed at all. If all the birds had equal representation (parity), they would each have values of 0.0625. In fact, a number have much higher representation (1, 11 and 15) and several besides the ones that have not contributed at all have much lower representation (7, 9, 31 and 98). Unequal representation will lead to lower levels of genetic variation in the population relative to a descendent population derived from the same number of founders, all with equal representation. Lacy (1989) has defined the number of founder equivalents, f_e, to be that number of founders required to produce the same level of genetic variation as in the pedigree population if all founders had contributed equally to each generation,

$$f_e = \frac{1}{\Sigma p_i^2} \qquad (2)$$

where p_i is the proportion of the alleles of the current population that can be traced to founder i. Calculations on the pedigrees of the okapi (*Okapia johnstoni*) and Goeldi's monkey (*Callimico goeldii*) show the number of founder equivalents to be 60–65 % of the true number of founders (Lacy 1989).

An extension of the simple founder contribution calculations is the determination of

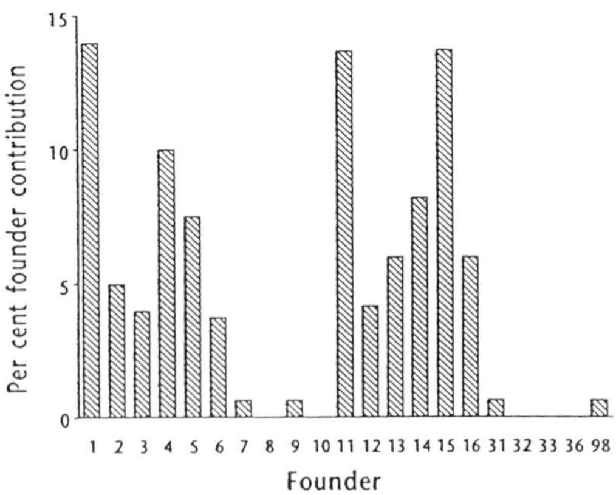

Figure 1. The contributions of the 16 founders to the current captive Guam rail population (Haig et al. 1990).

probabilities of founder allele extinction or survival in a given pedigree. In other words, we can calculate the probability that particular founder alleles are still present in the descendent population. Such an analysis can not only reveal those founders that are underrepresented in the pedigree, but can identify founders whose alleles are probably in very low frequency in the descendent population, and are therefore at high risk of subsequent loss. These calculations can either be done by using computer simulation (gene dropping) or analytically (peeling).

Gene dropping (MacCluer et al. 1986) is a Monte Carlo simulation method that applies the principle of Mendelian segregation to a given pedigree. At the beginning of the procedure, each founder is assigned two unique alleles at a hypothetical locus. A pseudo-random number generator is used to determine which of each founder's alleles is passed down to each offspring, a process that continues for each generation through the pedigree. Table 1 shows the kind of information that can be obtained using this method for the Przewalski's horse (*Equus przewalskii*) pedigree (MacCluer et al. 1986). In the first column is the contribution of each of the 13 founders to the current population, values quite different from parity which in this case would be 1/13 = 0.077. Two other measures, the estimated proportion of the genome lost or at high risk are also given in Table 1. Note that although these measures are generally associated, there are some exceptions. For example, founders 39 and 40 have the third and fourth highest founder contributions but also the highest proportions of alleles at high risk. The straightforward, simple nature of the gene drop analysis makes it an attractive tool for managing zoo populations.

The technique of peeling a pedigree (Cannings et al. 1978) gives exact probabilities for the possible genotypes in the set of current individuals in the pedigree given a particular distribution of founder geno-

Table 1. Founder contribution and loss of genetic variation in Przewalski's horses determined by gene dropping (MacCluer et al. 1986).

		Estimated proportion of	
Founder	Founder contribution	Founder genome lost	Surviving alleles at high risk of loss[a]
1	0.038	0.663	0.060
5	0.019	0.817	0.077
11	0.126	0.504	0.035
12	0.063	0.710	0.045
17	0.061	0.726	0.019
18	0.061	0.731	0.026
39	0.110	0.338	0.200
40	0.098	0.442	0.165
52	0.019	0.818	0.074
DOM	0.063	0.713	0.040
211	0.067	0.751	0.010
212	0.201	0.377	0.008
231	0.074	0.123	0.053

[a]Proportion of replicates in which each allele was present at a frequency of less than 0.01

types. For example, consider a gene with alleles A_1 and A_2 and genotypes A_1A_1, A_1A_2, and A_2A_2. The random vectors C and F give the genotypes of the current population and the founders, respectively. Particular states for C and F are c and f, such as $c = \{A_1A_1, A_1A_1, A_1A_2, A_2A_2, ...\}$. The relevant conditional probability is then $P(C = c \mid F = f)$ (Thomas 1990). In practice, starting with a distribution of all current individual genotypes, one can work backwards through the pedigree to evaluate the above expression for all possible founder numbers of A_1 and A_2 genotype combinations. When this is done, the joint extinction probabilities for all combinations of founder alleles is obtained, and the probability distribution for the number of distinct founder alleles can then be computed.

Such techniques have been used extensively on human pedigrees (e.g. Thompson 1978) and are now being applied to captive populations of endangered species (Thompson 1986; Geyer and Thompson 1988; Geyer et al. 1989). Of particular interest is the work on the pedigree of Przewalski's horse, now extinct in the wild and the subject of a global captive breeding program (Geyer and Thompson 1988; Geyer et al. 1989). Peeling techniques have shown that there is a complex dependency on founder allele survival as determined from evaluation of the joint probability distributions. Furthermore, peeling can readily identify genetically important individuals in the Przewalski's horse pedigree – those animals with high probabilities of possessing single-copy alleles (Geyer et al. 1989). Information like this is extremely important to those involved in the captive breeding program for this animal where equalization of founder allelic representation is a high priority.

Thompson (1986) graphically illustrates this point with data from the human population on the island of Tristan da Cunha (Figure 2). Ten of the eleven founders of the population could be divided into two genealogical groups, each contributing approximately equally to the descendent population; as determined by the method of Ballou (1983), group A contributes 33 % and group B contributes 36 %. However, peeling revealed founder group B contributed a significantly greater number of distinct alleles to the descendents (approximately seven) than did group A (approximately five). Notice that group B must have at least four distinct alleles surviving while group A may have as few as two. The basis for this difference in allele survival is that group A is dominated by multiple copies of two or three genes from one founder couple.

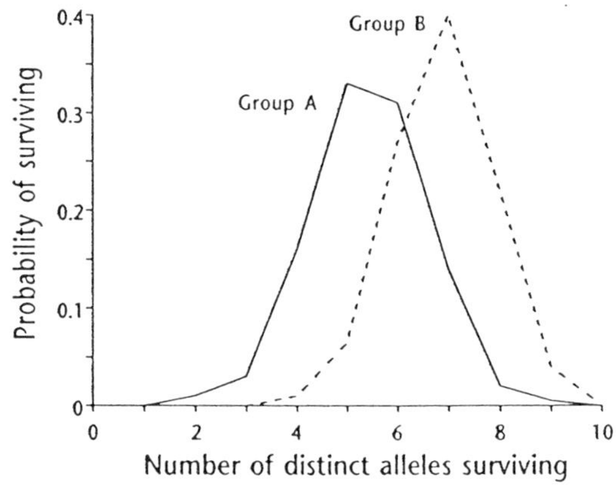

Figure 2. Allele survival probabilities for two founder groups (each of five individuals) out of the 11 founders of the human Tristan da Cunha population (Thompson 1986).

How do the methods of gene dropping and peeling directly compare with one another regarding the calculation of founder gene representation? In general, one might expect that the calculation of exact probabilities of founder gene survival using the peeling procedure is preferable over approximations obtained through gene dropping. However, peeling is extremely intensive with respect to computer memory requirements; in fact, this requirement increases exponentially with the size of the pedigree, compared to a linear increase with pedigree size in the gene drop procedure. Consequently, gene dropping may be the preferred method of analysis on large, complex pedigrees provided the resulting probability distribution is comparable to that produced by peeling.

A direct comparison of the two procedures is given in Figure 3 in the analysis of the Oman oryx pedigree (*Oryx leucoryx*) (Thomas 1990). Obviously there is a very close correlation between those results obtained by peeling and by gene dropping with the largest difference for a given number of alleles being 0.005. It should be noted, however, that in order to achieve the same level of precision as through peeling, 40,000 gene drop iterations were necessary, making gene dropping much more time intensive than peeling. The lack of precision associated with the gene dropping technique with small numbers of replicates is its major weakness, particularly when events of very small probability are the subject of study. Thomas (1990) concluded that both techniques have their respective place in pedigree analysis; a practical mode of operation might be to attempt peeling on the pedigree of interest and, if this turns out to be prohibitively complex, gene dropping with a sufficient (but manageable) number of iterations can be substituted.

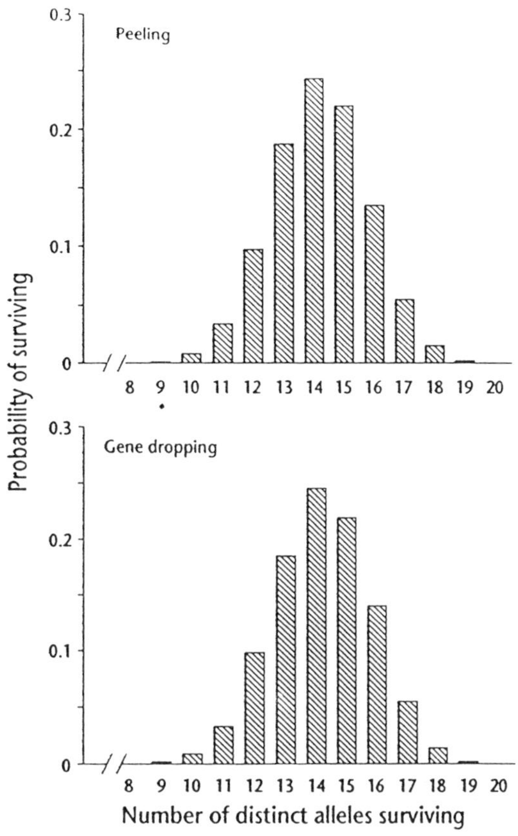

Figure 3. The probability of the number of alleles surviving from peeling (top) and gene dropping (bottom) for the Oman oryx captive population (Thomas 1990).

INBREEDING DEPRESSION

Early in the development of Mendelian genetics, researchers realized that the increased homozygosity resulting from inbreeding caused a loss in fitness, a phenomenon termed inbreeding depression (see Lerner 1954, for an extensive discussion). There has been debate for many decades over the genetic basis of inbreeding depression (and its complement, heterosis), some advocating a dominance explanation and others an overdominance (heterozygous advantage) hypothesis (e.g. Wright 1977). Current evidence generally indicates that homozygosity for unfavorable recessive alleles is the major cause of inbreeding depression (e.g. Charlesworth and Charlesworth 1987). More specifically, detailed studies in *Drosophila melanogaster* indicate that homozygosity of recessive lethals in inbred offspring accounts for approximately half of the observed inbreeding depression in viability (for reviews, see Simmons and Crow 1977; Crow and Simmons 1983). The majority of the remainder appears to be caused by homozygosity of detrimental alleles ranging from semi-lethals to mildly deleterious variants.

For a single locus, inbreeding depression can be defined simply as the difference in the mean fitness between an outbred and an inbred population, $\bar{w}$ and $\bar{w}_f$, respectively, or

$$\bar{w} - \bar{w}_f = sfpq\ (1 - 2h) \tag{3}$$

where f is the inbreeding coefficient, p and q are the frequencies of alleles A_1 and A_2 and 1, $1 - hs$, and $1 - s$ are the relative fitnesses of genotypes A_1A_1, A_1A_2, and A_2A_2, respectively, and h is a measure of dominance (e.g. Hedrick 1983). In other words, the inbreeding depression is expected to increase linearly with both inbreeding and the selection coefficient (s) and is highest for loci with intermediate allelic frequencies. Furthermore, for there to be inbreeding depression, the fitness of the heterozygote must be closer to that of the favorable homozygote than the deleterious one ($h < 1/2$).

A model to describe the general impact of inbreeding (for all loci) on the fitness in a population was given by Morton et al. (1956) as

$$\bar{w}_f = e^{-(A + Bf)} \tag{4}$$

where A indicates the impact of the environment and genetics (in a random-mating population) on fitness and B is the rate at which fitness decreases with a change in inbreeding. Generally, equation (4) is applied to viability measures, such as juvenile survival or egg-to-adult survival. In this case, B is a fairly good measure of the number of lethal equivalents per gamete (Crow and Kimura 1970; Cavalli-Sforza and Bodmer, 1971).

This approach is useful when matings producing offspring with different inbreeding levels (f_i) are available. If we take the natural logarithms of expression (4) and assuming that the inbreeding level f_i results in viability v_i, then

$$\ln\ (v_i) = -A - Bf_i \tag{5}$$

and B becomes the regression coefficient in a linear regression equation and can be used to estimate the number of lethal equivalents (e.g. Templeton and Read 1984). As an example of this approach, Table 2 gives the estimated number of lethal equivalents for the ten species with the most extensive data sets in Ralls et al. (1988). The estimate of the number of lethal equivalents ranges from 0.01 in the Sumatran tiger to a high of 9.17 in the brown lemur. Also given in Table 2 is the cost of inbreeding i (reduction in survival) for $f = 0.25$ (matings between first-degree relatives such as full sibs) which is

$$i = 1 - e^{-0.25B} \tag{6}$$

The value ranges from 0.003 (Sumatran tiger) to 0.90 (brown lemur) with an average of 0.31. To give the cost of inbreeding some perspective, note that in this data set it is on average 1.24 times the inbreeding coefficient for this single aspect of fitness (juvenile survival). In fact, Soulé (1980) suggested as a general rule that a 10 % reduction in a given fitness trait occurs with a 10 % increase in the inbreeding coefficient and that when total fitness is considered a 10 % increase in f results in a 25 % reduction in fitness. Obviously, from Table 2, some animals have much more inbreeding depression than this while other species have much less.

Table 2. The number of lethal equivalents and cost of inbreeding when f = 0.25 for 10 species (Ralls et al. 1988).

Species	N	B (lethal equivalents)	Cost of inbreeding (f=0.25)
Short bare-tailed opossum (*Monodelphis domestica*)	251	0.43[a]	0.10
Elephant shrew (*Elephantulus rufescens*)	218	2.12	0.41
Golden lion tamarin (*Leontopithecus r. rosalia*)	974	2.15[a]	0.42
Brown lemur (*Lemur fulvus*)	136	9.17[a]	0.90
Greater galago (*Galago c. crassicaudatus*)	251	1.69[a]	0.34
Gaur (*Bos gaurus*)	182	0.51	0.12
Dorcas gazelle (*Gazella dorcas*)	143	1.85[a]	0.37
Bush dog (*Speothos venaticus*)	176	0.24	0.06
Sumatran tiger (*Panthera tigris sumatrae*)	427	0.01	0.003
Pygmy hippopotamus (*Choeropsis liberiensis*)	419	1.59[a]	0.33

[a]Slope significant at 0.05 level.

Although inbreeding depression has been a topic of interest in animal breeding (Wright 1977), plant mating systems (Charlesworth and Charlesworth 1987), and determination of the genetic load of mutations (Crow and Simmons 1983), there are several aspects of inbreeding depression critical to conservation genetics that have not been adequately addressed. First, most studies of inbreeding depression have focused on aspects of viability. However, viability is only one component of fitness, with total fitness resulting from viability, fecundity and mating components (e.g. Christiansen and Frydenberg 1973; Hedrick and Murray 1983). In fact, other aspects of fitness such as fecundity (Marinkovic 1967), male mating success (Sharp 1984) and seed yield (Levin and Bulinska-Radomska 1988), may be greatly influenced by inbreeding. For example, using a competitive index to estimate fitness in *Drosophila*, Latter and Robertson (1962) estimated that for small inbreeding coefficients, the reduction in overall fitness was 2.7 times the inbreeding coefficient.

An ingenious approach has been used in *Drosophila* to determine the total (or net) fitness of chromosomes. For example, Sved (1971) estimated that the average relative viability of 24 individual isogenic (completely homozygous) chromosomes was 0.73. He then estimated total fitness of the chromosomes from their equilibrium frequency when competing with a mutant chromosome. The average total fitness value was only 0.28. This large difference, 62.5 % of the total inbreeding effect, he attributed to other components of fitness, specifically female fecundity and male mating success.

Second, in general, the effect of inbreeding on components of fitness has been estimated for populations in a benign or optimal environment. For example, the captive animals analyzed for inbreeding depression by Ralls et al. (1988) were in situations in which there were no predators or competitors, few pathogens, and no severe physical environments, such as heat or cold. When examining isogenic chromosomes in *Drosophila*, usually optimum temperature and low densities are utilized. In other words, the effects of inbreeding on fitness may be much more severe than generally suggested if inbred individuals are more sensitive to stressful or extreme environments (e.g. Lerner 1954; Schmitt and Ehrhardt 1990; Dudash 1990).

Third, a simple analysis based on expression (3) implies that the amount of inbreeding depression is a linear function of the inbreeding coefficient and that the extent of inbreeding depression is a static one. Given a certain level of inbreeding for a period of time, the frequency of lethal or detrimental alleles will decline from the level before inbreeding because inbreeding exposes more recessive lethal or detrimental alleles to selection in homozygotes, thereby causing a reduction in their frequency. Assuming that inbreeding continues at a given level for a long time, the new equilibrium allele frequency of A_2 is

$$q_e = \frac{-f + [f^2 + 4\,(1-f)\frac{u}{s}]^{\frac{1}{2}}}{2\,(1-f)} \qquad (7)$$

where u is the mutation rate to A_2 (Haldane 1940). In other words, expression (3) is not appropriate because a noninbred and an inbred population generally have different allelic frequencies at selected loci.

Finally, the extent of inbreeding depression is expected to be greater in a small population than a large population for a given inbreeding coefficient because of the reduced effectiveness of selection relative to genetic drift. For example, the equilibrium mutation load is larger in smaller finite populations (Kimura et al. 1963). A genetic drift–selection–mutation equilibrium may not occur in very small populations resulting in a rapid decline of fitness or "mutational meltdown" (Lynch and Gabriel 1990).

Proposals for viable population size and breeding plans for captive populations suggest that inbreeding depression can be minimized by various breeding strategies that may reduce the inbreeding depression. For example, Franklin (1980) states that "animal breeders accept inbreeding coefficients as high as a one percent increase per generation (i.e. an effective population size, N_e, of 50) in domestic animals without great concern" because such slow inbreeding allows selection to remove deleterious alleles without endangering the population (Soulé 1980, also made the same proposal). In fact, Tantawy and Reeve (1956) and Latter and Robertson (1962) demonstrated that a slow increase in inbreeding results in less inbreeding depression than a fast increase.

We should note that this number of 50, based on inbreeding depression, and that of 500, based on maintenance of genetic variation (see below) were only intended as general guideposts (for further discussion of N_e and these numbers, see Lande and Barrowclough 1987). For example, a number of endangered species with successful breeding programs have had quite low founder numbers, e.g. Przewalski's horse, 13 founders; Speke's gazelle (*Gazella spekei*), 4 founders; black-footed ferret (*Mustela nigripes*), 6 founders; red-ruffed lemur (*Varecia variegata variegata*), 7 founders; Guam rail, 15 founders, etc. In other words, just because the founder number is low does not mean that all attempts should not be made to save the species because eventually the number may be much higher, enough to overcome the initial population size restriction.

Templeton and Read (1983, 1984) have suggested a radical breeding program for use in certain situations in which the inbreeding coefficient is rapidly increased in an effort to quickly purge a population of lethals and avoid the effects of inbreeding depression over many generations (see also Hollingsworth and Maynard Smith 1955). The success of a breeding scheme to purge lethals would depend upon the reproductive potential in the species with the probability of success being much less in species with low intrinsic rates of increase. Furthermore, intentional inbreeding and selection may greatly reduce the genetic variation at other genes (non-lethal genes). In other words, the goal of reducing inbreeding depression by purging lethals may cause a reduction in the genetic variation necessary for future adaptation and successful reintroduction as well as possibly increase the short-term probability of extinction.

It has been suggested that concern over inbreeding depression is overstated based on the observation that highly inbred lines of mice, rats, and *Drosophila* exist and that Pere David's deer (*Elaphurus davidianus*) and the European bison (*Bison bonatus*) both appear to have healthy populations even though they both were initiated from very small numbers. However, often the surviving inbred lines are very few (in a survey of seven studies in five species, Soulé (1980) found that between 0 and 14.3 % of the initial lines survived intense inbreeding), allowing strong selection among lines to occur. In the two endangered species mentioned above, it is likely that many of the deleterious alleles were eliminated by a slow increase in inbreeding combined with selection (Frankel and Soulé 1981).

It therefore appears necessary to consider the joint dynamic effects of inbreeding and selection on fitness. Organisms in populations that have a long history of inbreeding appear to have lower inbreeding depression (e.g. Charlesworth and Charlesworth 1987; Hedrick 1987; Rao and Inbaraj 1977) although even highly self-fertilizing populations should still have some inbreeding depression (Charlesworth et al. 1990). Most endangered species are outbred and are only now with low numbers experiencing inbreeding and, therefore, generally have high inbreeding depression. Obviously, there will be some species with low values of inbreeding depression, as illustrated in Table 2.

MAINTENANCE OF GENETIC VARIATION

One of the major long-term goals of conservation genetics is the retention of enough genetic variation so that future adaptation, successful expansion, or reestablishment in natural populations is possible. Franklin (1980) suggested that in general a population size of 500 would be necessary to retain the genetic variation at a level found in a natural population. This size estimate was obtained by determining the mutation-genetic drift equilibrium for a quantitative trait, i.e. the value at which the additive genetic variance is increased by mutation to the same extent it is reduced by genetic drift. Genetic drift acts through the effective population size (N_e) such that

$$V_{a(t+1)} = V_{a(t)} \left(1 - \frac{1}{2N_e}\right) \qquad (8)$$

where $V_{a(t)}$ is the additive genetic variance in the tth generation. The same relationship between generations, reduction by a factor of $(1 - 1/2N_e)$ is also true for heterozygosity (e.g. Crow and Kimura 1970; Hedrick 1983).

Soulé et al. (1986) expanded this initial framework to consider differences that occur between species and to focus on more exact goals. They suggested that a principle aim of captive breeding should be the maintenance of 90 % of the genetic variation in the source (wild) population over a period of 200 years (or some such similar target). In their treatment, the actual population size necessary to achieve this level is primarily a function of the size of the initial founder population, the rate of population growth to the final population size, and the generation length.

The generation of additional additive genetic variance by mutation, V_m / V_e, where V_m is the additive genetic variation due to mutation and V_e is the environmental variance, has been estimated to be approximately 0.001 per generation (Lande 1975; Lynch 1988). For mutation to increase the genetic variance by the amount that genetic drift reduces it, then

$$\frac{V_\mathrm{m}}{V_\mathrm{e}} - V_\mathrm{a}\left(\frac{1}{2N_\mathrm{e}}\right) = 0 \qquad (9)$$

(e.g. Clayton and Robertson 1955). In other words, if $V_m / V_e = 0.001$ (and $V_a = 1$), then $N_e = 500$. Although this number and the value of 50 related to inbreeding depression have been widely cited and somewhat abused, their original intent was to give general guideposts for the genetic management of endangered species (see discussion earlier).

In order to focus on these problems, let us introduce a basic model that allows predictions concerning these population parameters (this follows the continuous time treatment of Lande in press). Assume the population grows geometrically so that given the effective number of founders is $N_{e(0)}$, the effective number of individuals in generation t is

$$N_{\mathrm{e}\,(\mathrm{t})} = N_{\mathrm{e}\,(0)}\, R_0^{\mathrm{t}} \qquad (10a)$$

where R_0 is the finite rate of growth per generation ($R_0 = e^r$ where r is the intrinsic rate of increase). If we are concerned with a seasonal breeder, then the effective population size in year y is

$$N_{\mathrm{e}\,(\mathrm{y})} = N_{\mathrm{e}\,(0)}\, \lambda^{\mathrm{y}} \qquad (10b)$$

where λ is the finite rate of growth per year ($\lambda = e^{r/T}$ where T is the generation length). If we assume that V_a is reduced by genetic drift and increased by mutation, then the recursion relation for V_a between generations is

$$V_{\mathrm{a}\,(\mathrm{t}+1)} = V_{\mathrm{a}\,(\mathrm{t})}\left(1 - \frac{1}{2N_{\mathrm{e}\,(\mathrm{t})}}\right) + \frac{V_\mathrm{m}}{V_\mathrm{e}} \qquad (11a)$$

and between years is approximately

$$V_{\mathrm{a}\,(\mathrm{y}+1)} \approx V_{\mathrm{a}\,(\mathrm{y})}\left(1 - \frac{1}{2TN_{\mathrm{e}\,(\mathrm{t})}}\right) + \frac{V_\mathrm{m}}{V_\mathrm{e}T} \qquad (11b)$$

when the influence of genetic drift is small so that the effects are linear over one generation (following Hill 1979).

If we also assume $V_m / V_e = 0$ and that $r = 0.5$, the final effective population size, K_e, necessary to retain 90 % of the variation after 200 years is given in the top half of Figure 4 (Lande in press). For example, when the effective number of founders is 20 and the generation time is five years, the final population size must be 280 or greater to retain 90 % of the variation (indicated by the closed circle). As also shown by Soulé et al. (1986) for heterozygosity (which has the same dynamics as V_a in this case) if the generation time is short even with a final effective size of 500, 90 % of the variation cannot be retained in this case.

If mutation is allowed to contribute to the genetic variance, then the generation of new variation via mutation may overcome the loss of variation from genetic drift when the population size is relatively large. The

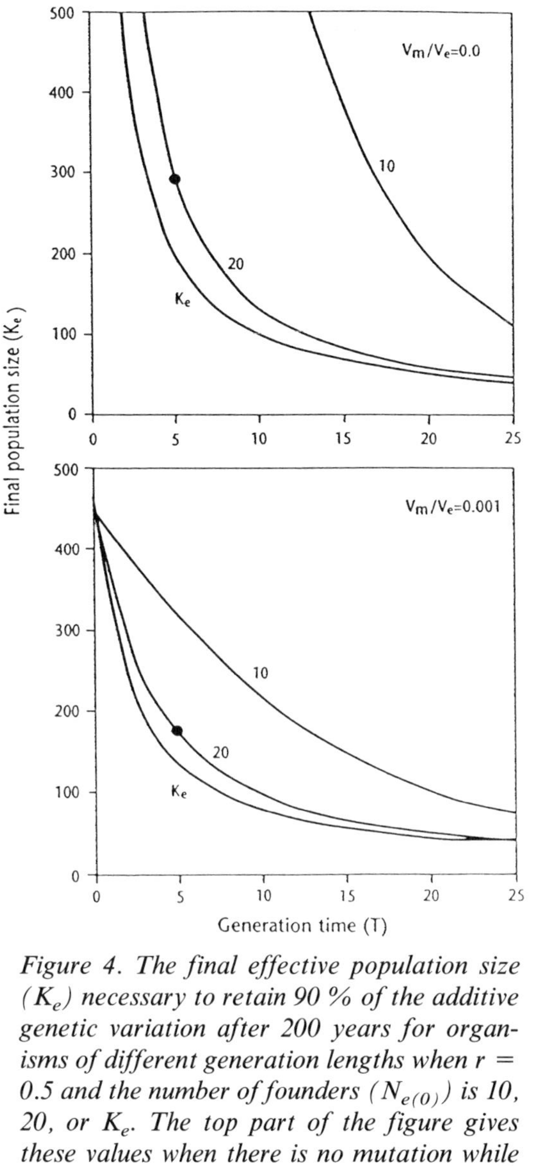

Figure 4. The final effective population size (K_e) necessary to retain 90 % of the additive genetic variation after 200 years for organisms of different generation lengths when $r = 0.5$ and the number of founders ($N_{e(0)}$) is 10, 20, or K_e. The top part of the figure gives these values when there is no mutation while the bottom gives the value when $V_m/V_e = 0.001$ (Lande in press).

bottom half of Figure 4 illustrates that given $V_m / V_e = 0.001$ when, for example, the effective number of founders is 20 and the generation time is five years, the final population size must be only 180 (closed circle). Mutation has the most substantial effect when the effective population is near the mutation-genetic drift balance. For example if $V_m / V_e = 0.001$ and if $N_{e(0)} = K_e = 450$, 90 % of V_a will be retained.

These findings when mutation is included are somewhat reassuring with higher N_e values. Although it is thought that generation of additive genetic variance by mutational input is generally independent of population size (e.g. Lynch 1988), for heterozygosity when N_e is lower, mutation will have a much less significant influence generating genetic variation. Furthermore, generation of variation by mutation is probably variable over time and may not occur at the regular rate suggested here. In addition, the new mutants have not gone through the sorting process of natural selection and they may, for example, reduce fitness (e.g. Simmons and Crow 1977 find that most new mutations are deleterious and not completely recessive) as well as increase variance. Such concerns have been widely contested in the recent quantitative genetics literature (e.g. Turelli 1986), but because we are concerned with avoiding extinction of endangered species, relying on variation generated by mutation seems somewhat risky.

In addition, Lande (in press) has examined the impact of population subdivision on genetic variation. If a species is split into separate subpopulations with no gene flow among them, then alternate alleles may be fixed in different subpopulations, permanently retaining them in the total population. However, extreme inbreeding depression may occur in these subpopulations and the character of genetic variation may be irreversibly changed, e.g. from a natural degree of heterozygosity to a very low level within a subpopulation (see Hedrick et al. 1986).

If subpopulations go extinct and are recolonized, then the rate and pattern of loss of genetic variation may be quite different than that predicted by knowing only subpopulation size (e.g. Gilpin 1991). Maruyama and Kimura (1980) and Ewens (1989) analytically examined the loss of genetic variation in such metapopulations under certain conditions. Gilpin (1991), on the other hand, using computer simulation has attempted to evaluate the influence of metapopulation dynamics on genetic variation in a model that "emphasizes both ecological and genetic realism."

Investigating an array of extinction and recolonization rates of patches (subpopulations), Gilpin (1991) found a broad range of conditions in which patch coalescence occurs, i.e. all occupied patches of the metapopulation descend from individuals in a single patch in the past. As a result, even though the numbers in the metapopulation have always been and are still large, there is no genetic variation left. To illustrate this phenomenon, a diagrammatic representation of the results of a three-patch metapopulation simulation is given in Figure 5 (after Gilpin 1991). Initially all subpopulations have high heterozygosity (they each have $N_e = 500$). The important sequence of events start in generation 48 when patch 2 goes extinct and is recolonized from patch 3 with a reduction in heterozygosity. Next patch 1 is colonized from patch 2 with a founder population having no variation. When patch 2 goes extinct in generation 71, the metapopulation has no variation although there are still 500 individuals in patch 1. All of these individuals can be traced back to some individuals in patch 3 before generation 51.

Gilpin (1991) has found the general range of colonization and extinction probabilities in which patch coalescence occurs (Figure 6). Under these conditions, the species (metapopulation) would not become extinct (region shown as hatched) but all variation would be lost. Perhaps such a scenario could explain the low genetic variation in cheetahs or at least it may be a reasonable alternative to the bottleneck hypothesis suggested by O'Brien et al. (1985). In any case, the loss of genetic variation with these metapopulation dynamics is much greater than predicted from knowing the current population size. To make these findings relevant for quantitative traits, mutation would need to be included.

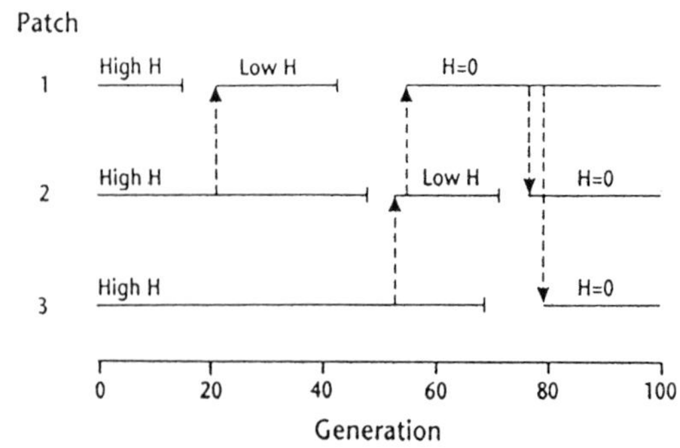

Figure 5. The level of heterozygosity (H) over time in a simulation of a population existing in three patches (after Gilpin 1991). Short vertical lines indicate extinctions in a patch and the arrows indicate recolonization.

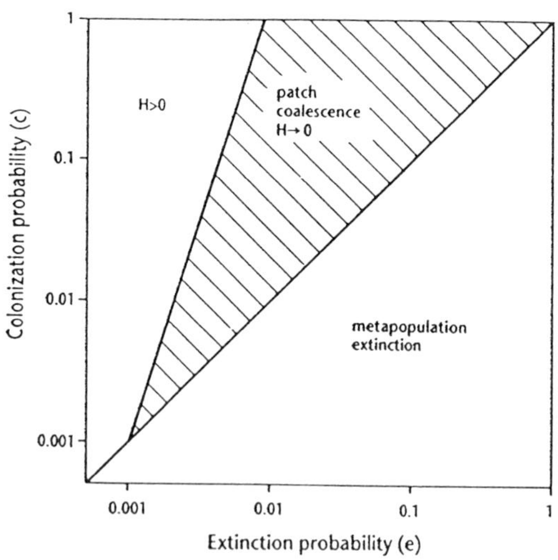

Figure 6. The general combinations of colonization and extinction probabilities that result in H > 0, patch coalescence (H approaching 0) or extinction of the metapopulation (after Gilpin 1991).

CONCLUSIONS

Preventing further degradation of our environment, such as the prevention of the extinction of endangered plants and animals, should be of major concern for everyone. Although political, economic, and social factors may often be given major roles in decisions affecting endangered species, hopefully reasoned scientific information, such as the conservation genetic reasoning we have attempted to present here, will aid in preventing further extinctions.

In our discussion of the theoretical basis of conservation genetics, we have attempted to be circumspect. To err in the application of genetics to an endangered species may result in extinction. For example, we agree with Lande (1988) that demographic factors, not genetic ones, are most critical in the preservation of the northern spotted owl. Only by thoroughly understanding pedigree analysis, inbreeding depression, maintenance of genetic variation, etc., can responsible application to endangered species be carried out. The simple genetic guideposts suggested a decade ago cannot be taken as absolute dogma and must be evaluated for each species in order for conservation genetics to gain the respectability it merits.

ACKNOWLEDGMENTS

We appreciate the comments of Bob Lacy, Russ Lande, Mike Lynch, Mike Soulé, and Bob Vrijenhoek on the manuscript. This research was supported by NSF BSR-8923007.

LITERATURE CITED

American Association of Zoological Parks and Aquariums. 1983. *Species Survival Plan*. AAZPA, Wheeling, WV.

Ballou, J. D. 1983. Calculating inbreeding coefficients from pedigrees, pp. 509–520. In: C. M. Schonewald-Cox, S. M. Chambers, B. MacBryde and W. L. Thomas (eds) *Genetics and Conservation: A Reference for Managing Wild Animal and Plant Populations*. Benjamin/Cummings, Menlo Park, CA.

Ballou, J. D., T. Foose and M. Gilpin (eds) (in press). *Population Management for Survival and Recovery*. University of Chicago Press, Chicago.

Boyce, A. J. 1983. Computation of inbreeding and kinship coefficients on extended pedigrees. J. Hered. 74: 400–404.

Cannings, C., E. A. Thompson and M. H. Skolnick. 1978. Probability functions on complex pedigrees. Adv. Appl. Prob. 10: 26–61.

Cavalli-Sforza, L. L. and W. F. Bodmer. 1971. *The Genetics of Human Populations*. Freeman, San Francisco.

Charlesworth, D. and B. Charlesworth. 1987. Inbreeding depression and its evolutionary consequences. Ann. Rev. Ecol. Syst. 18: 237–268.

Charlesworth, D., M. T. Morgan and B. Charlesworth. 1990. Genetic loads and estimates of mutation rates in highly inbred plant populations. Nature 347: 380–382.

Christiansen, F. B. and O. Frydenberg. 1973. Selection component analysis of natural polymorphisms using population samples including mother–offspring combinations. Theor. Pop. Biol. 4: 425–445.

Clayton, G. A. and A. Robertson. 1955. Mutation and quantitative variation. Amer. Nat. 89: 151–158.

Crow, J. F. and M. J. Simmons. 1983. The mutation load in *Drosophila*, pp. 1–35. In: M. Ashburner, H. L. Carson and J. N. Thompson (eds) *The Genetics and Biology of* Drosophila, Vol. 3c. Academic Press, London.

Crow, J. F. and M. Kimura. 1970. *An Introduction to Population Genetics Theory*. Harper and Row, New York.

Dudash, M. R. 1990. Relative fitness of selfed and outcrossed progeny in a self-compatible, protandrous species, *Sabatia angularis* (Gentianaceae): a comparison in three environments. Evolution 44: 1129–1139.

Ehrlich, P. R. and A. H. Ehrlich. 1981. *Extinction: The Causes and Consequences of the Disappearance of Species*. Random House, New York.

Ewens, W. J. 1989. The effective population sizes in the presence of catastrophes, pp. 9–25. In: M. W. Feldman (ed) *Mathematical Evolutionary Theory*. Princeton University Press, Princeton, NJ.

Frankel, O. H. and M. E. Soulé. 1981. *Conservation and Evolution*. Cambridge University Press, Cambridge.

Franklin, I. R. 1980. Evolutionary changes in small populations, pp. 135–149. In: M. Soulé and B. A. Wilcox (eds) *Conservation Biology: An Evolutionary – Ecological Perspective*. Sinauer, Sunderland, MA.

Geyer, C. J. and E. A. Thompson. 1988. Gene survival in the Asian wild horse (*Equus przewalskii*): I. Dependence of gene survival in the Calgary Breeding Group. Zoo Biol. 7: 313–327.

Geyer, C. J., E. A. Thompson and O. A. Ryder. 1989. Gene survival in the Asian wild horse (*Equus przewalskii*): II. Gene survival in the whole population, in subgroups, and through history. Zoo Biol. 8: 313–329.

Gilpin, M. 1991. The genetic effective size of a metapopulation. Biol. J. Linn. Soc. 42: 165–175.

Haldane, J. B. S. 1940. The conflict between selection and mutation of harmful recessive genes. Amer. J. Eugen. 10: 415–521.

Haig, S. M., J. D. Ballou and S. R. Derrickson. 1990. Management options for preserving genetic diversity: Reintroduction of Guam rails to the wild. Conserv. Biol. 4: 290–300.

Hedrick, P. W. 1983. *Genetics of Populations*. Jones and Barlett, Boston.

Hedrick, P. W. 1987. Genetic load and the mating system in homosporous ferns. Evolution 41: 1282–1289.

Hedrick, P. W., P. F. Brussard, F. W. Allendorf, J. A. Beardmore and S. Orzack. 1986. Protein variation, fitness and captive propagation. Zoo Biol. 5: 91–99.

Hedrick, P. W. and E. Murray. 1983. Selection and measures of fitness, pp. 61–104. In: M. Ashburner, H. L. Carson and J. N. Thompson (eds) *The Genetics and Biology of* Drosophila, Vol. 3d. Academic Press, New York.

Hill, W. G. 1979. A note on effective population size with overlapping generations. Genetics 92: 317–322.

Hollingsworth, M. J. and J. Maynard Smith. 1955. The effects of inbreeding on rate of development and on fertility in *Drosophila subobscura*. J. Genet. 53: 295–314.

Kimura, M., T. Maruyama and J. F. Crow. 1963. The mutation load in small populations. Genetics 48: 1303–1312.

Lacy, R. C. 1989. Analysis of founder representation in pedigrees: Founder equivalents and founder genome equivalents. Zoo Biol. 8: 111–123.

Lande, R. 1975. The maintenance of genetic variability by mutation in a polygenic character with linked loci. Genet. Res. 26: 221–235.

Lande, R. 1988. Genetics and demography in biological conservation. Science 241: 1455–1460.

Lande, R. (in press). Breeding plans for small populations, based on the dynamics of quantitative genetic variance. In: J. D. Ballou, T. M. Foose and M. Gilpin (eds) *Population Management for Survival and Recovery*. University of Chicago Press, Chicago.

Lande, R. and G. R. Barrowclough. 1987. Effective population size, genetic variation, and their use in population management, pp. 87–124. In: M. E. Soulé (ed) *Viable Populations for Conservation*. Cambridge University Press, Cambridge.

Latter, B. D. H. and A. Robertson. 1962. The effects of inbreeding and artificial selection on reproductive fitness. Genet. Res. 3: 110–138.

Lerner, I. M. 1954. *Genetic Homeostasis*. Oliver and Boyd, Edinburgh.

Levin, D. A. and A. Bulinska-Radomska. 1988. Effects of hybridization and inbreeding on fitness in phlox. Amer. J. Bot. 75: 1632–1639.

Lynch, M. 1988. The rate of polygenic mutation. Genet. Res. 51: 137–148.

Lynch, M. and W. Gabriel. 1990. Mutation level and the survival of small populations. Evolution 44: 1725–1737.

MacCluer, J. W., A. J. Boyce, B. Dyke, L. R. Weitkamp, D. W. Pfennig and C. J. Parsons. 1983. Inbreeding and pedigree structure in Standardbred horses. J. Hered. 74: 394–399.

MacCluer, J. W., J. L. Vandeberg, B. Read and O. A. Ryder. 1986. Pedigree analysis by computer simulation. Zoo Biol. 5: 148–160.

Marinkovic, D. 1967. Genetic loads affecting fecundity in natural populations of *D. pseudoobscura*. Genetics 56: 61–71.

Maruyama, T. and M. Kimura. 1980. Genetic variability and effective population size when local extinction and recolonization are frequent. Proc. Natl. Acad. Sci. USA 77: 6710–6714.

Morton, N., J. Crow and H. J. Muller. 1956. An estimate of the mutational damage in man from data on consanguinous marriages. Proc. Natl. Acad. Sci. USA 42: 855–863.

O'Brien, S. J., M. E. Roelke, L. Marker, A. Newman, C. A. Winkler, D. Meltzer, L. Colly, J. F. Evermann, M. Bush and D. E. Wildt. 1985. Genetic basis for species vulnerability in the cheetah. Science 227: 1428–1434.

Quaas, R. L. 1976. Computing the diagonal elements and inverse of a large numerator relationship matrix. Biometrics 32: 949–953.

Ralls, K. and J. D. Ballou (eds) 1986. Proceedings of the workshop on genetic management of captive populations. Zoo Biol. 5: 84–238.

Ralls, K., J. D. Ballou and A. Templeton. 1988. Estimates of lethal equivalents and the cost of inbreeding in mammals. Conserv. Biol. 2: 185–193.

Rao, P. S. S. and S. G. Inbaraj. 1977. Inbreeding in Tamil Nadu, South India. Soc. Biol. 24: 281–288.

Schmitt, T. and D. W. Ehrhardt. 1990. Enhancement of inbreeding depression by dominance and suppression in *Impatiens capensis*. Evolution 44: 269–278.

Sharp, P. M. 1984. The effect of inbreeding on competitive male-mating ability in *Drosophila melanogaster*. Genetics 106: 601–612.

Simmons, M. J. and J. F. Crow. 1977. Mutations affecting fitness in *Drosophila* populations. Ann. Rev. Genet. 11: 49–78.

Soulé, M. E. 1980. Thresholds for survival: maintaining fitness and evolutionary potential, pp. 151–168. In: M. E. Soulé and B. A. Wilcox (eds) *Conservation Biology: An Evolutionary – Ecological Perspective*. Sinauer, Sunderland, MA.

Soulé, M. E. (ed) 1986. *Conservation Biology: The Science of Scarcity and Diversity*. Sinauer, Sunderland, MA.

Soulé, M. E. (ed) 1987. *Viable Populations for Conservation*. Cambridge University Press, Cambridge.

Soulé, M. E., M. Gilpin, N. Conway and T. Foose. 1986. The Millenium ark: How long a voyage, how many staterooms, how many passengers? Zoo Biol. 5: 101–114.

Stevens, A. 1975. An elementary computer algorithm for the calculation of the coefficient of inbreeding. Info. Proc. Let. 3: 163.

Sved, J. A. 1971. An estimate of heterosis in *Drosophila melanogaster*. Genet. Res. 18: 91–105.

Tantawy, A. O. and E. C. R. Reeve. 1956. Studies in quantitative inheritance. IX. The effects of inbreeding at different rates in *Drosophila melanogaster*. Z. Indukt. Abstamm.- u VererbLehre 87: 648–667.

Templeton, A. R. and B. Read. 1983. The elimination of inbreeding depression in a captive herd of Speke's gazelle, pp. 241–261. In: C. M. Schonewald-Cox, S. M. Chambers, B. MacBryde, and W. L. Thomas (eds) *Genetics and Conservation: A Reference for Managing Wild Animal and Plant Populations*. Benjamin/Cummings, Menlo Park, CA.

Templeton, A. R. and B. Read. 1984. Factors eliminating inbreeding depression in a captive herd of Speke's gazelle. Zoo Biol. 3: 177–199.

Thomas, A. 1990. Comparison of an exact and a simulation method for calculating gene extinction probabilities in pedigrees. Zoo Biol. 9: 259–274.

Thompson, E. A. 1978. Ancestral inference. II. The founders of Tristan da Cunha. Ann. Hum. Genet. 42: 239–253.

Thompson, E. A. 1986. Ancestry of alleles and extinction of genes in populations with defined pedigrees. Zoo Biol. 5: 161–170.

Turelli, M. 1986. Gaussian versus non-gaussian genetic analyses of polygenic mutation-selection balance, pp. 607–628. In: S. Karlin and E. Nevo (eds) *Evolutionary Processes and Theory*. Academic Press, New York.

Wilson, E. O. (ed) 1988. *Biodiversity*. National Academy Press, Washington, DC.

Wright, S. 1977. *Evolution and Genetics of Populations*, Vol. 3. *Experimental and Evolutionary Deductions*. University of Chicago Press, Chicago.

5 Plant population genetic structure and biological conservation

Anthony H. D. Brown and **Daniel J. Schoen**[a]
CSIRO Division of Plant Industry,
GPO Box 1600,
Canberra, A.C.T. 2601,
Australia

ABSTRACT

Isozyme surveys of population genetic structure, at first sight, seem to suggest that populations of the majority of plant species show little divergence. In outbreeding plants, the average single population appears to have over 80 % of the species genetic resources, as measured by the gene diversity index. This chapter reconsiders the apparent redundancy of plant populations at the morphological, isozyme and DNA sequence levels and rejects it as misleading. Plant species show a rich and varied structure with populations differing in the kind and amounts of genetic diversity they contain. The mating system is a key indicator of structure, and the pattern of heterozygosity a useful summary variable. Both should play a role in the choice of populations for any species nominated for conservation, and the choice of indicator species for monitoring an ecosystem.

[a] Permanent address:
Department of Biology, McGill University, Montreal, Quebec, Canada

INTRODUCTION

Sustainable development, in contrast to the unmitigated exploitative variety, must embrace biological conservation. Its aim should be the continued functioning of Earth's complex ecosystems, although any individual component species or variety may wax or wane. Genetic resources will be needed in the future, not only to incorporate new genes into the plants that humanity directly depends upon, but also for wild plants to persist and evolve in their changing environments.

Genetical parameters in biological conservation, as opposed to purely ecological ones, received more pointed attention after Frankel (1970) stressed that the continued existence of a species requires genetic variation to respond to future evolutionary pressures. Two recent perspectives, however, appear to challenge this emphasis on genetics. First the large body of genetic data, principally for isozyme markers, indicates that most if not all outbreeding plant and animal populations are very diverse genetically. In addition, many surveys that

Conservation of Biodiversity for Sustainable Development.
Edited by O. T. Sandlund, K. Hindar and A. H. D. Brown. 1992 Scandinavian University Press, Oslo

compare allozyme variation in several populations using Nei's (1973) G_{ST}-analysis seem to justify the conclusion that each population on average contains the bulk (usually over 80 %) of the species variation. Such a conclusion – that most of the genetic variation within the whole species resides in any single population – is profoundly challenging. It amounts to claiming that there is great redundancy of populations with respect to genetic variation. If the conclusion is generally true, it would appear to justify the conserving of very few populations of each species as adequate for further evolution. Does it indeed give a valid picture of the genetic structure of plant species?

Second, Simberloff (1988) has noted that conservation biology is now in a period of emphasis on population biology as seen in its current concepts of "minimum viable population size" and the "metapopulation". In the previous period of emphasis on island biogeography, ecological theory was directed at the conservation of whole ecosystems and the dynamics of total species numbers. Conservation strategies for individual species populations were based solely on genetic criteria such as the number of individuals needed to avoid inbreeding depression, or the number required to harbour sufficient quantitative genetic variation to allow future response to selection (Frankel and Soulé 1981). However Lande (1988) has claimed that the emphasis on genetics at the population level led to the relative neglect of basic population demography. He cites two bird species in which prescriptions for conservation based on blanket application of genetic criteria for minimum population size have failed. Does this mean that considerations of demography are entirely to displace those of genetic resources in conservation planning?

There is need to review studies of genetic structure of plant populations to see whether these perspectives are valid. In particular, they tend to minimize the extent and importance of genetic differences between populations. It will be our contention that plant population genetic structure is so rich and varied among species that it cannot be ignored in any long-term conservation planning. Indeed for many plant species, particularly inbreeders, it may well be the dominant consideration.

ALLOZYME FREQUENCY DIVERGENCE BETWEEN POPULATIONS

Let us look more closely at the claimed genetic similarity of plant populations. Current evidence on the extent and character of genetic structure in plant populations is mostly derived from isozyme studies. The technique has proved ideal for such studies. It combines the ability to look at variation close to the DNA level with the capacity to survey large numbers of individuals, populations and species for a moderate investment of resources.

The many published studies of allozyme divergence among plant populations were recently summarized by Hamrick and Godt (1989). The great majority of these studies are of natural populations of wild species. In order to reveal the broad determinants of genetic structure, they classified each of over 400 taxa for their taxonomic status, life form, geographic range, latitudinal distribution, breeding system, seed dispersal and successional stage. Table 1 is an extract of their results for the following genetic statistics – the average number of alleles per locus at the levels of either the population or the total species (A_P or A_S); and the average Nei gene diversity or expected panmictic heterozygosity (h_P or h_S); and the proportion of the total gene diversity found among populations ($G_{ST} = 1 - h_P/h_S$). This last statistic is a common measure of the extent of population divergence (Nei 1973).

Widespread species generally contain more total gene diversity (h_S) than do endemics or species of limited geographic distribution. Here is a clear and fundamental link between genetic variation and current success as measured by the capacity of species to range over wide areas and perforce cope with diverse environments. Of course there are many exceptions to this generality. Some aggressively colonizing, weedy species (e.g.*Hordeum murinum*) have achieved continental and sometimes global distributions without much allozyme polymorphism. Such plants usually have a uniparental mating system, like apomixis or autogamy (Brown and Burdon 1987).

Indeed, although many species characteristics affect the pattern of allozyme variation in Hamrick and Godt's analysis, the mating system emerges as the key indicator of gene diversity at the population level. The average gene diversity for self-pollinating plants is less than half those of wind pollinated outcrossers (Table 1). The averages for animal pollinated species are intermediate.

Table 1. The amount of allozyme variation at the level of species and population for species of higher plants with various attributes (extracted from Hamrick and Godt 1989).

Categories	N	Species		Population		Divergence
		A^a_S	h_S	A_P	h_P	G_{ST}
Breeding System – Pollination						
Selfing	78	1.69	0.12	1.31	0.07	0.51
Mixed						
– animal	60	1.68	0.12	1.43	0.09	0.22
– wind	9	2.18	0.19	1.99	0.20	0.10
Outcrossing						
– animal	124	1.99	0.17	1.54	0.12	0.20
– wind	102	2.40	0.16	1.80	0.15	0.10
Geographic Range						
Endemic	52	1.80	0.10	1.39	0.06	0.25
Narrow	82	1.83	0.14	1.45	0.11	0.24
Regional	180	1.94	0.15	1.55	0.12	0.22
Widespread	85	2.29	0.20	1.72	0.16	0.21
Life Form						
Annual	146	2.07	0.16	1.48	0.11	0.36
Perennial herbs	119	1.70	0.12	1.40	0.10	0.23
Long-lived woody	110	2.19	0.18	1.79	0.15	0.08
Seed Dispersal						
Gravity	161	1.81	0.14	1.45	0.10	0.28
Attached	52	2.96	0.20	1.68	0.14	0.26
Explosive	23	1.48	0.09	1.25	0.06	0.24
Ingested	39	1.69	0.18	1.48	0.13	0.22
Wind	105	2.10	0.14	1.70	0.12	0.14

[a]The measures of variation are A = allelic diversity or the number of alleles per locus, h = gene diversity or expected panmictic heterozygosity, G_{ST} = population divergence or proportion of the total gene diversity among populations, and N = the minimum number of taxa in the category. The subscript S refers to measures of variation within the species and subscript P to measures within the average population.

Turning to our basic question of differences among populations, allele frequency variation as measured by the G_{ST} statistic shows that more than half the allozyme variability in inbreeders accrues from population divergence. This figure is in stark contrast to outbreeders where 10 – 20 % can be discerned among populations.

How representative are isozymes of genetic variation? Clearly they afford only a limited and biased sample of the plant genome. Therefore we now proceed to seek evidence at other levels, to get a more comprehensive picture of plant population structure (see also Brown and Burdon 1987).

ADAPTIVE DIFFERENTIATION WITHIN SPECIES

Ecotypes

The notion that any single plant population contains on average 80 % of the species variation and, by inference, 80 % of its total evolutionary potential, is perhaps most discordant with the experience of the genecologist. Beginning with the work of Turesson, genecology has provided the classic evidence in plants of population divergence. Patient study first in common-garden experiments and later by reciprocal transplantation, has highlighted the extent of genotypic divergence between populations. The adaptive significance of this variation is evident in its correlation with habitat. Typical of the classic results from common-garden plantings are examples of variation patterns among and within nine races of *Campanula rotundifolia*, and *Solidago virgaurea* which are known or presumed outbreeders (Turesson 1925). From estimates of variance components within and between races, about 75 % of the total phenotypic variation for the trait, stem length, was between races. Such figures contrast with the low G_{ST} values estimated from allozymes.

Selective coefficients against alien genotypes, measured in reciprocal transplant experiments, are usually intense. "(But) it is already clear that different populations of a species have very different adaptations to different environments and that it is commonplace for a population in an alien environment to have a fitness of only half that of the native population" (Bradshaw 1984). The proportion of the total variation within a species that determines such strong selective coefficients is unknown and could be quite small. The more important alleles or allelic complexes involved are likely to be frequent within their local population, given the similar performance of individual plants from the same population. Once again the strength of population divergence appears to disagree with low G_{ST} values.

Scale is another important aspect of ecotypic divergence. The classic experiments of Clausen et al. (1940) were on the scale of hundreds of kilometres distance and thousands of metres elevation, well beyond the range of single generation migration. More recent studies have found adaptive differentiation on the scale of metres, often despite high levels of gene flow (Bradshaw 1984). Hence it is likely that the populations of a species inhabiting an area of great microenvironmental diversity (physical and/or biotic) are likely to contain higher levels of adaptively significant variation than would populations from more uniform habitats. Thus populations are likely to differ markedly for their evolutionary potential.

Life-history variation

In complete contrast to seeking evidence on genetic diversity at the protein or DNA level, are the current efforts to study "ecologically meaningful" phenotypic traits. Here are included all observable attributes which clearly contribute to a plant's survival and reproduction. Unlike DNA variation, life-history variation is said to be "directly interpretable ecologically" (Venable 1984). Variation among plants is evident *in situ*, and the main issues are the extent to which such variation is heritable, its genetic basis, and the extent to which differences between populations are genetic. Venable (1984) has summarized 24 *ex situ* and transplant studies of intraspecific variation in plants, either within or between populations, for traits affecting demography in over 20 species. From this list it is clear that much less is known about the heritability of interpopulation differences than that of intrapopulational variation, and about the dysgenic effects of crossing between populations.

Variation in biological interactions

Genes which regulate the interaction of plants with other organisms – such as competitors, pollinators, herbivores or pathogens – could show even more marked geographic patterns than the genes of abiotic responses. One reason for this is that spatial variation in environment will determine the prevalence of these interacting species, and hence the need to adapt to them. Burdon et al. (1989) stressed the inherent spatial patchiness of plant-pathogen interactions which arises from the mutual amplification of stochastic variation in host, pathogen and environment.

Clear examples of such strong patterns are evident in the frequencies of resistance to disease. Thus natural populations in Israel of the wild progenitor of barley, *Hordeum spontaneum*, show a pronounced geographic structure for major-gene resistance to the pathogen causing barley leaf scald, *Rhynchosporium secalis* (A. H. D. Brown, J. J. Burdon and D. C. Abbott, unpublished). The pattern (Table 2) is one where populations in the more mesic part of the country, where disease occurs, have much higher frequencies of resistance genes, than in dried (xeric) sites. Remarkably, resistance is common, despite the pathogen's well-known ability to evolve new pathotypes against single resistance sources.

Table 2. Resistance[a] in natural populations of *Hordeum spontaneum* in Israel to the disease barley leaf scald (Abbott, Brown and Burdon unpublished).

Region	Sites	Accessions	Resistance to *R. secalis*
Mesic:			
Mediterranean Coast	5	16	100 %
Galilee Mountains	7	16	94 %
Xeric:			
Jordan Valley & Judean Desert	5	19	32 %
Negev Desert	5	13	69 %

[a]Resistance was tested as seedling infection type to four isolates of *Rhynchosporium secalis*

MOLECULAR POLYMORPHISM

Investigation of the kinds and levels of molecular diversity in plant populations is only just beginning (Clegg 1989). Several major classes of DNA-sequence need to be studied, viz single-copy genes, multigene families, hypervariable sequences and organellar genomes. Surveys can range from the level of restriction site or fragment length differences, through physical maps of several restriction sites, to complete DNA sequences (Levin, this volume). Despite the few comparative data in plants at the population level, it is already clear from the studies outlined below that each class of sequence has different patterns of variation which reflect the varying importance of different evolutionary forces.

In the case of **nuclear sequences of low copy number**, the evidence comes from surveys with known probes (such as that by Gepts and Clegg (1989) in *Pennisetum glaucum* using cloned maize alcohol dehydrogenase (ADH) genes) or from mapping of restriction fragment length polymorphisms (RFLPs) using anonymous random genomic or cDNA clones in crop species. For this class of sequence the contrast in ease of finding polymorphism in the outbreeders (maize and pearl millet) compared with inbreeding species (lettuce, wheat, tomato and soybean) is striking. In tomato and soybean, interspecific crosses to wild related species were needed to detect polymorphism. It appears that small rearrangements (either insertion/deletions or inversions) are a common cause of polymorphism as well as the gain/loss of restriction sites.

At the level of complete DNA sequences, nothing yet is known about the partition of variation among and within natural plant populations. The recent polymerase chain reaction technique might now allow random samples of the required size to address this problem. So far only specific different alleles have been sequenced comparatively and not random samples of genes. The sequencing of selected different alleles at a locus (for example, the *Adh1* locus in maize, Osterman and Dennis 1989) has shown a variety of sequence changes, including rearrangements in the noncoding regions associated with transposable element activity. Thus in addition to the partition of variation between and within populations, the relative roles of tandem duplications, insertion/deletion events, silent base substitutions, amino acid replacements, in the various regions of the gene (introns, transcribed, and translated regions) are yet to be resolved.

Considerably more evidence is available on variation in the **nuclear multigene family coding for ribosomal RNA (rDNA)** (Dvorak 1989). The basic unit typically includes the 18S, 5.8S and 26S RNA sequences, with the internal and external transcribed spacers and an intergenic (nontranscribed) spacer. Part of the intergenic spacer comprises a number of subrepeats. Variation in the number of subrepeats gives rise to length variants in the basic unit. The unit is tandemly repeated in large blocks on one or more chromosomes. Thus an individual plant could be heterogeneous for its rDNA complement at up to four levels. Two or more types of units could occur within the one block, or could comprise two different blocks on nonhomologous chromosomes, on related or homoeologous chromosomes from the different genomes in polyploids, or in segregating alternative "alleles" in heterozygotes.

Two special modes of evolutionary change for spacer length variation are the generation of new kinds of units through unequal crossing over, and the rapid homogenization of the family within and between blocks of repeats through "concerted evolution". In addition, Dvorak (1989) has suggested that changes in the prevalence of the various blocks of the 18S-26S family in polyploids evolve jointly with changes in blocks in the 5S family. This would be an example of coadaptation in the Darwinian sense. In addition the number of repeats in a block can respond to stress making it one of the more labile elements in the plant genome (Walbot and Cullis 1985).

Studies by Sytsma and Schaal (1990) of the extent of population divergence for rDNA in the *Lisianthius skinneri* complex serve to emphasis another trend. Their data indicate a pattern of rDNA variation with almost total homogeneity within populations, but fixed for alternative repeats among populations. Thus it appears that the rDNA may show greater population divergence than that revealed by allozyme variation. As these genes are involved in the fundamental cellular process of protein synthesis, their variation patterns could be particularly important to differential adaptedness and thus to conservation strategies.

A hypervariable class of dispersed repeat sequences apparently occurs in plant genomes (Dallas 1988). They provide powerful genetic markers for studying individual genotypic identity but because of their excess variability, are unlikely to be useful as estimators of variation per se. Their potential lies more in tracing current or recent identity by descent among individuals or subpopulations, particularly in testing the isolation of fragmented populations.

Reports of intraspecific variation in **organellar DNA sequences** in plants have recently appeared but information on population structure of this variation is very limited. Yet levels and patterns appear to differ markedly from those of variation in nuclear sequences. The likely reasons for such contrasts are the predominantly uniparental inheritance of organelles, and the lack of regular recombination between variants. Presumably, other features like the high multiplicity of copy number per cell and specialized genome organization and function of organelles also regulate the nature and extent of changes accepted. Of the two types of organellar DNA, more is known of intraspecific variation in chloroplast DNA (cpDNA) than in mitochondrial DNA (mtDNA) so that cpDNA is considered further here.

Despite the slower rate of evolution of the chloroplast genes compared with nuclear genes, some cpDNA intraspecific variation has been found in several plants (e.g. *Hordeum vulgare, Lupinus texensis, Zea mays, Heuchera micrantha, Pinus banksiana* and *Glycine* spp.). Both changes to restriction sites and altered fragment lengths due to insertions or inversions have been detected. With uniparental inheritance and the lack of recombination, the population structure of this limited variation could resemble that for nuclear markers in selfing species and be highly differentiated between populations. Further, the restriction of migration of cpDNA variants to the seed dispersal component in maternally inherited systems would allow even greater population divergence than for nuclear markers under selfing.

A recent intensive survey of restriction site characters in the cpDNA's of four related diploid species of *Glycine* found considerable polymorphism (Doyle et al. 1990a). Table 3 summarizes the data as the number of accessions tested, the total number of plastomes (chloroplast genomes) distinguished within each species, the number unique to a species and the number of sites polymorphic within species. These figures show considerable divergence among some of the plastomes within species. In part this divergence was geographic. A cladistic analysis of site changes showed a limited congruence between plastome type and taxonomy based on morphology (Doyle et al. 1990a). In part, the limited concordance is probably due to introgression among the taxa and the maintenance of ancestral cpDNA polymorphism. It appears that in this group of species, morphological (nuclear DNA) divergence is only partially correlated with organellar DNA divergence. Such lack of correlation amounts to additional genetic discontinuities among populations.

Overall, the population structure of molecular variation is likely to show much greater divergence between populations than that revealed by allozymes. This is so

Table 3. Survey of 52 informative restriction site characters within related diploid species of *Glycine* (from Doyle et al. 1990a).

	Numbers			
Taxa	Accessions sampled	Total plastomes	Unique plastomes	Polymorphic sites
G. microphylla	26	8	6	21
G. tabacina (diploid)	11	5	4	9
G. latifolia	14	5	3	6
G. sp. aff. tabacina	9	3	3	2

not only because DNA variation is less prone to the problem of cryptic variation from the limits of electrophoretic resolution. It is also because these sequences reflect the action of additional modes of evolutionary change, such as unequal crossing-over, concerted evolution, and uniparental inheritance. Molecular analysis appears not to support the concept that the average population has over 80 % of the species variation.

OTHER MEASURES OF GENETIC STRUCTURE

Most recent studies of population divergence within plant species have employed the Nei gene diversity partition (e.g. Hamrick and Godt 1989) and related genetic distance computations. The diversity statistics are functions of allele frequencies, and are largely determined by the variation in frequency of the most common allele at each locus. The presence or absence of rare localized alleles has little effect on these statistics. More importantly, they are insensitive to variation in the level of polymorphism from one population to the next. Yet unevenness of diversity has emerged as an important problem in conservation planning (Brown 1989a).

Variation in gene diversity (h)

In order to define and test for such patterns we (D. J. Schoen and A. H. D. Brown, unpublished) analyzed the variation in gene diversities for allozymes among loci and across populations for several species with contrasting breeding system. The basic variable is h_{ij}, the gene diversity at the ith locus ($i = 1,..,I$) in the jth population ($j = 1,..,J$). Let us consider the average gene diversity ($h._j$) observed at the jth site. The variance over populations of this average can be split into two components, the gene diversity variance (GDV) and gene diversity covariance (GDC), as shown in the following formula. In this formula, the variances and covariances are over populations, and the summations are over single loci or pairs of single loci.

$$\text{var}\,[h._j] = \left\{ \sum_i \text{var}\,[h_{ij}] + 2 \sum_{i<k} \sum \text{cov}\,[h_{ij}\,,\,h_{kj}] \right\} / I^2$$

When the gene diversities at separate loci are independent, the gene diversity covariance (or GDC) component in the equation is zero. If however the diversities are correlated the GDC component is non-zero, as would be the case when populations highly diverse for one locus, tend to be diverse for others. The ratio of the covariance to the gene diversity variance (or GDV) component provides a measure of the extent of interdependence among h_{ij}'s within populations. The estimates in Table 4 show that selfing species exhibit large and significant positive covariance (GDC) between pairs of

Table 4. Gene diversity variance (GDV) and covariance (GDC) components of variation in mean population gene diversity

Species	GDV[a]	GDC[a]	Ratio
Self-fertilizing species:			
Arabidopsis thaliana	0.51	0.48	0.9**
Avena barbata (California)	0.12	0.31	2.5***
Avena barbata (Spain)	0.24	0.42	1.8***
Hordeum spontaneum	0.08	0.17	2.2***
Lycopersicon pimpinellifolium[b]	0.14	0.21	1.5***
Phlox cuspidata	0.23	0.10	0.5***
Outcrossing species:			
Echium plantagineum	0.10	0.00	0.0 ns
Helianthus debilis	0.34	-0.08	-0.2 ns
Phlox drummondii	0.52	0.28	0.5***
Phlox roemariana	0.49	0.00	0.0 ns
Picea abies	0.14	-0.10	-0.7 ns
Pinus sylvestris	0.03	0.01	0.4 ns
Plantago lanceolata	0.03	0.04	1.4***
Pseudotsuga menziesii	0.02	0.01	0.3 ns
Stephanomeria exigua	0.10	0.01	0.1 ns

[a]Values of GDV and GDC are all X 10^{-2}, e.g. 0.0051, 0.0048 etc.
[b]Populations with sample sizes of < 20 were excluded.
Significance testing was conducted by randomizing the entries within the rows of the data matrix of $\{h_{ij}\}$ containing *I* rows and *J* columns; 1000 randomizations were made for each species, and the computed variance and covariance components compared with those observed. **$P<0.01$, ***$P<0.001$, ns = non-significant. See Schoen and Brown (1991 a) for citations to sources of data.

single-locus gene diversity indices across populations. In contrast, the GDCs in populations of the outcrossing species differed from zero in only two out of nine cases. Also the variance component (GDV) was generally less in outbreeders than inbreeders.

Thus intraspecific variation in gene diversity possesses a pattern which is not obvious from the average diversity values alone. In particular, many selfing species consist of populations with both low and high levels of genetic variation. All other things being equal, the populations which merit priority in conservation programs are those with large gene diversity. This conclusion conflicts with strategies based solely on analysis of G_{ST}-statistics which have suggested that large numbers of populations are required to conserve the genetic diversity of selfing species (Hamrick 1983). If only few populations of such autogamous species are highly variable, it may well be more important to locate and conserve them, than to conserve the bulk of relatively uniform stands.

Variation in observed heterozygosity (*H*)

Variation in the observed levels of heterozygosity as revealed by surveys of allozyme diversity again illustrates that individual plant populations may differ considerably in genetic make-up. Table 5 summarizes results from 16 studies with estimates of the proportion of heterozygous loci per individual (*H*). The estimates of some species with a mixed mating system show more than a ten-fold range. In three studies with population data on the mating system, heterozygosity was significantly correlated with outcrossing rate (*L. pimpinellifolium*, *G. achilleifolia* and *C. tembloriensis*). Geographic patterns of variation in H in trees can occur on a continental scale (Moran et al. 1989). For example, in Norway spruce (*Picea abies*) populations in central Europe have lower H than those in USSR and Sweden (Lagercrantz and Ryman 1990).

Thus, in many plant species, conservation biologists may again be confronted with a potentially wide range of variation in yet another population genetic statistic, heterozygosity. This measure reflects both gene pool diversity and the capacity of the breeding system to assort and assemble new gene combinations. In addition, some evidence in plants indicates that adaptedness or productivity itself is related to this measure (see Mitton 1989). The measure is particularly important in species with intermediate or flexible mating systems, or in species where inbreeding depression may be

Table 5. Proportion of loci that are heterozygous per individual (H).

Species	No. pops.	H averaged across pops.	Range of H among populations	Reference
Selfing herbs:				
Hordeum spontaneum	28	0.003	0 – 0.02	1.
Phlox cuspidata	43	0.008	0 – 0.02	2.
Herbs with mixed mating systems:				
Amsinckia spectabilis	13	0.009	0 – 0.03	3.
Clarkia tembloriensis	8	0.050	0.01 – 0.15	4.
Eichhornia paniculata	6	0.08	0.01 – 0.12	5.
Gilia achilleifolia	7	0.106	0.02 – 0.18	6.
Impatiens capensis	11	0.039	0.01 – 0.10	7.
Lycopersicon pimpinellifolium	43	0.074	0 – 0.22	8.
Echium plantagineum	8	0.20	0.16 – 0.22	9.
Phlox drummondii	73	0.052	0.01 – 0.08	2.
Sarracenia purpurea	11	0.090	0.04 – 0.16	10.
Stephanomeria exigua	11	0.092	0.07 – 0.16	11.
Trees (predominantly outcrossing):				
Casuarina cunninghamiana	20	0.208	0.13 – 0.26	12.
Picea abies	70	0.115	0.08 – 0.17	13.
Picea mariana	21	0.107	0.05 – 0.16	14.
Pseudotsuga menziesii	11	0.155	0.14 – 0.18	15.

References: 1. Nevo et al. (1979). 2. Levin (1978). 3. Ganders et al. (1985). 4. Holtsford and Ellstrand (1989). 5. Glover and Barrett (1987). 6. Schoen (1982). 7. Knight and Waller (1987). 8. Rick et al. (1977). 9. Burdon and Brown (1986). 10. Schwaegerle and Schaal (1979). 11. Gottlieb (1975). 12. Moran et al. (1989). 13. Lagercrantz and Ryman (1990). 14. Yeh et al. (1986). 15. Yeh and O'Malley (1980).

substantial if forced to inbreed (Charlesworth and Charlesworth 1987). Knowledge of the mating system would again serve as a useful predictor.

Multilocus associations

Another dimension of genetic structure to plant populations is the extent of correlations in allelic state among loci. Comparisons among species for such associations are difficult to draw because, being second order effects, the measures cannot strictly be divorced from first order effects (single locus variability patterns). However, two overall themes seem to emerge from the data collected from plant populations.

First, mating system has a profound effect on the intensity of associations (Brown 1984). Hastings (1989) compiled data from nine studies of outbreeders and eleven of close inbreeders. The outbreeders showed very little or no linkage disequilibrium between loci unless they are very closely linked, whereas the close inbreeders often show intense associations. Even fewer studies have been reported of species with intermediate or flexible mating systems. In one such example, Waller and Knight (1989) reported a wide range for estimates of standardized relative disequilibria, from 0.09 to 1.0, in populations of *Impatiens capensis*, with some correlation between the intensity of association and the mean fixation index. This could reflect population differences in net outcrossing rate.

Second, populations of a highly inbreeding species can show a wide range in disequilibrium levels even without marked variation in the mating system. This is evident in the populations of *Hordeum spontaneum* in Israel (Brown et al. 1980). Such a pattern parallels that mentioned above in the variation among populations of inbreeders for genic diversity. Both patterns may stem from the same cause, namely variation in effective population size. If this pattern were general it would add yet another reason for priority to populations with high single-locus diversity, namely such populations are more likely to possess lower levels of disequilibrium and hence a greater array of multilocus genotypes, upon which selection can act.

MONITORING OF MATING SYSTEMS

The above-cited examples of correlation between population genetic variation in plants and the outcrossing rate illustrate the importance of mating system analysis as a useful tool in predicting genetic structure and discriminating among populations. In the first instance, the mating system can be measured as the fractions of apomixis (asexual reproduction), selfing and outcrossing occurring in a population, and there are many standard procedures for doing this (see Brown 1989b for review).

Yet there is scope for more detailed analysis at the individual plant level in estimating the variation in male or female fertility. Such variation can reduce effective population size substantially. For example, significant variation in male fertility was found among 33 distinct genotypes or clones of *Picea glauca* planted in a seed orchard (Schoen and Stewart 1987). Fewer than 20 % of the clones successfully contributed male gametes to over 50 % of the seed. In the second year, 20 % of the clones were the pollen parents of over 70 % of the seed. The male fertility variation in this seed orchard of white spruce has a marked effect on its effective population size (N_e). The proportionate reduction is given by the formula:

$$N_e/N = 1/[(1 + F)(s^2/z^2) + 1] = 0.19$$

where F is the fixation index in the parents

Table 6. Estimates of overall selfing rate, and its components from the whole- and part-flower selfing model in populations of *Glycine argyrea and G. clandestina* (from Schoen and Brown 1991 b).

Species	Population habitat	Whole[a]-flower selfing	Part-flower selfing	Overall selfing
G. argyrea	coastal,	0.02	0.50	0.53
(6 loci)	subtropical	(0.09)[b]	(0.07)	(0.05)
G. clandestina	mid elevation,	-0.05	0.56	0.53
(3 loci)	temperate	(0.07)	(0.08)	(0.10)
G. clandestina	subalpine,	0.58	0.69	0.89
(4 loci)	temperate	(0.13)	(0.07)	(0.02)

[a]Whole-flower selfing is the fraction of flowers that receive no outcross pollen, and part-flower selfing is the proportion of selfed seed from flowers that produce at least one outcrossed seed.
[b]Standard error in brackets

and s^2/z^2 is the variance in cone production divided by the mean squared cone production (Heywood 1986). The effective population size falls from 33 clones to six clones. Heywood (1986) gives several examples where female fertility variation leads to reduction in N_e.

Correlation of selfing events

Despite the manifold contrasting effects which different mating systems have on population genetic structure, many plant species possess mating systems of remarkable flexibility. For example, species of *Glycine* (Fabaceae) produce cleistogamous flowers in leaf axils specialized for selfing, and chasmogamous flowers specialized for insect attraction and cross-pollination. Using genetic markers and a model which partitions selfing into whole-flower and part-flower components, we (Schoen and Brown 1991 b) examined the mating system of the chasmogamous flowers in two closely related species in contrasting environments. The subalpine population, in conditions less favourable for the constant activity of pollinators, showed a marked increase in rate of whole-flower selfing compared with more benign conditions (Table 6.) Such effects not only illustrate the importance of conserving pollinators in decisions concerning the conservation of plant populations *in situ*, but also suggest levels of inbreeding might be controlled in *ex situ* conservation by deliberate fruit selection.

POLYPLOID ORIGINS

Over 30 % of angiosperm species are of polyploid origin (Stebbins 1971). In botanical texts, the point frequently made about polyploidy is that it affords speciation in a single step, in contrast to other modes of origin of reproductive isolation. However, this tenet does not imply that polyploidy is necessarily an isolated event in one lineage that acts as a drastic bottleneck on the sample of genes which pass from the parental diploid(s).

Recent evidence from chloroplast DNA variation in tetraploid *Glycine tabacina* shows that polyploid taxa may have multiple origins (Doyle et al 1990b). A survey of 65 accessions of this species (with genome formula BBB_2B_2) defined eight distinct plastome types based on cpDNA restriction site differences. The related diploid species, *G. microphylla*, *G. latifolia* and *G. tabacina* (diploid), share the genome formula *BB* as they can form fertile hybrids between each other despite considerable morphological, ecological and geographic divergence. These *BB* diploids are considerably polymorphic for cpDNA restriction sites which define at least 24 plastome types. Remarkably, six of the eight tetraploid plastome types are identical with corresponding plastome types in the *BB*-diploids for all restriction sites surveyed. The existence of pairs of diploid and polyploid accessions sharing cpDNA types with unique derived restriction sites suggests that the BBB_2B_2 polyploid *G. tabacina* has numerous independent origins. Thus a polyploid plastome shared more unique derived sites with a diploid plastome than it does with its nearest related tetraploid plastome. A few of the tetraploid plastome types are widespread geographically (Table 7).

Subsequent sporadic hybridization among such independent products, and segregation would be sufficient to generate the intermediate yet variable morphologies which tetraploid *G. tabacina* displays. Such a network of evolutionary relations among species could have a major effect on the population genetic structure in polyploid plants. In particular such networks might

Table 7. Multiple polyploid origin in *Glycine tabacina* – the number of accessions from each geographic region with various plastome groups in tetraploid *G. tabacina* and related diploid species (from Doyle et al. 1990b).

	Geographic distribution[a]				
Tetraploid/Diploid plastome pair	Vic. S.A.	N.S.W. A.C.T.	Queensland	Pacific	South Central Pacific
Identical plastomes:					
G. tabacina – Group 1	2	8	7	6	11
G. microphylla – Group 24	3	4	4	–	–
G. tabacina – Group 2	2	6	4	–	–
G. microphylla – Group 10	–	2	1	–	–
G. tabacina – Group 3	–	6	3	–	1
G. tabacina (2X) – Group 17	–	4	5	–	–
G. tabacina – Group 4	–	3	–	–	–
G. latifolia – Group 20	–	1	–	–	–
G. tabacina – Group 5	–	–	–	3	–
G. latifolia – Group 21	–	5	5	–	–
G. tabacina – Group 6	–	1	–	–	–
G. latifolia – Group 19	–	1	–	–	–
Unmatched:					
G. tabacina (4X) – 2 types	–	–	–	2	–
G. microphylla – 6 types	2	4	8	–	–
G. tabacina (2X) – 4 types	–	–	5	–	–
G. spp. – 7 types	–	2	8	–	–

[a]Vic. and S.A. = Victoria and South Australia, N.S.W. and A.C.T. = New South Wales and Australian Capital Territory

assist the survival of genetic resources present in contributing diploid species.

CONSERVATION IMPLICATIONS

We conclude from the above survey that population genetic structure varies dramatically among plant species. There are two components of this rich variety of structure. First, the several populations of a species differ in the kinds and amounts of genetic diversity they contain. Second, plant species differ markedly in their pattern of diversity.

It follows that we must reject as misleading the general notion that any single population has over 80 % of the species variation. Plant populations diverge to varying degrees, and are unequal in the amount of diversity they harbour. But does this inequality of populations matter to conservation planning?

We need to know much more about the factors determining population divergence to answer this question. Clearly genetic diversity itself must be studied at many levels, from DNA sequences to adaptive characters. The general aim should be to investigate the connection between genetic diversity and evolutionary potential; and between this potential and long-term sustainability of species. With the likely accelerated change in global climate, the capacity of populations to evolve is important to their future well-being. The need for evolutionary potential was argued a decade ago by Frankel and Soulé (1981), and climate change has underlined it. This is not to deny the point made by Lande (1988) on the need to seek sufficient habitat to meet the ecological needs for a minimum viable population size with specific duration and probability level.

With more knowledge of the relation between diversity and sustainability, our choices and conservation strategies can be more informed. For the conservation of a single species, the **selection of populations** is a crucial step. To compute the proportion of variance preserved from the number of populations conserved presuming they are equally variable, is only a first approximation. Variation is particularly uneven in inbreeding and apomictic species. Hence a check of the breeding system could be a helpful indicator. Isozyme surveys in plants could be perhaps more useful in testing the range of actual observed heterozygosity (such as in Table 4) than in partitioning gene diversity.

For conservation at the ecosystem level, the **selection of indicator species** could be improved by adding genetic criteria to those of mere size, connectedness of populations and ecological dominance. In plants, giant trees have sometimes been assigned this role as conspicuous organisms. Yet such species can show apparently low levels of population divergence (e.g. boreal forest trees have average G_{ST} values of about 0.05). Perhaps species that have flexible mating systems or that display wide ranges of actual heterozygosity among their populations could serve as better indicators of the genetic structure and well-being of many associated organisms in particular communities.

ACKNOWLEDGMENTS

The authors thank Drs J. J. Burdon, O. H. Frankel, O. Muona and T. Skrøppa for comments on previous drafts of this chapter. DJS thanks CSIRO Division of Plant Industry for support.

LITERATURE CITED

Bradshaw, A. D. 1984. Ecological significance of genetic variation between populations, pp. 213–228. In: R. Dirzo and J.

Sarukhan (eds) *Perspectives on Plant Population Ecology*. Sinauer, Sunderland, MA.

Brown, A. H. D. 1984. Multilocus organization of plant populations, pp. 159–169. In: K. Wöhrmann and V. Loeschcke (eds) *Population Biology and Evolution*. Springer-Verlag, Berlin.

Brown, A. H. D. 1989a. The case for core collections, pp. 136–156. In: A. H. D. Brown, O. H. Frankel, D. R. Marshall and J. T. Williams (eds) *The Use of Plant Genetic Resources*. Cambridge University Press, Cambridge.

Brown, A. H. D. 1989b. Genetic characterization of plant mating systems, pp. 145–162. In: A. H. D. Brown, M. T. Clegg, A. L. Kahler and B. S. Weir (eds) *Plant Population Genetics, Breeding and Genetic Resources*. Sinauer, Sunderland, MA.

Brown, A. H. D. and J. J. Burdon. 1987. Mating systems and colonizing success in plants, pp. 115–131. In: A. J. Gray, M. J. Crawley and P. J. Edwards (eds) *Colonization, Succession and Stability*. Blackwell, Oxford.

Brown, A. H. D., M. W. Feldman and E. Nevo. 1980. Multilocus structure of natural populations of *Hordeum spontaneum*. Genetics 96: 523–536.

Burdon, J. J. and A. H. D. Brown. 1986. Population genetics of *Echium plantagineum* L. – target weed for biological control. Aust. J. Biol. Sci. 39: 369–378.

Burdon, J. J., A. M. Jarosz and G. C. Kirby. 1989. Pattern and patchiness in plant-pathogen interactions – causes and consequences. Ann. Rev. Ecol. Syst. 20: 119–136.

Clausen, J., D. D. Keck and W. M. Hiesey. 1940. Experimental studies on the nature of species. 1. The effect of varied environments on western North American plants. Publ. Carnegie Instn. 520.

Clegg, M. T. 1989. Molecular diversity in plant populations, pp. 98–115. In: A. H. D. Brown, M. T. Clegg, A. L. Kahler and B. S. Weir (eds) *Plant Population Genetics, Breeding and Genetic Resources*. Sinauer, Sunderland, MA.

Charlesworth, D. and B. Charlesworth. 1987. Inbreeding depression and its evolutionary consequences. Ann. Rev. Ecol. Syst. 18: 237–268.

Dallas, J. F. 1988. Detection of DNA "fingerprints" of cultivated rice by hybridization with a human minisatellite DNA probe. Proc. Natl. Acad. Sci. USA 85: 6831–6835.

Doyle, J. J., J. L. Doyle and A. H. D. Brown. 1990a. Chloroplast DNA polymorphism and phylogeny in the B genome of *Glycine* subgenus *Glycine* Leguminosae. Amer. J. Bot. 77: 772–782.

Doyle, J. J., J. L. Doyle, A. H. D. Brown and J. P. Grace. 1990b. Multiple origin of polyploids in the *Glycine tabacina* complex inferred from chloroplast DNA polymorphism. Proc. Natl. Acad. Sci. USA 87: 714–717.

Dvorak, J. 1989. Evolution of multigene families: the ribosomal RNA loci of wheat and related species, pp. 83–97. In: A. H. D. Brown, M. T. Clegg, A. L. Kahler and B. S. Weir (eds) *Plant Population Genetics, Breeding and Genetic Resources*. Sinauer, Sunderland, MA.

Frankel, O. H. 1970. Variation – the essence of life. Proc. Linn. Soc. New South Wales 95: 158–169.

Frankel, O. H. and M. E. Soulé. 1981. *Conservation and Evolution*. Cambridge University Press, Cambridge.

Ganders, F. R., S. K. Denny and D. Tsai. 1985. Breeding system and genetic variation in *Amsinckia spectabilis* (Boraginaceae). Can. J. Bot. 63: 533–538.

Gepts, P. and M. T. Clegg. 1989. Genetic diversity in pearl millet (*Pennisetum glaucum* [L.] R. Br.) at the DNA sequence level. J. Hered. 80: 203–208.

Glover, D. E. and S. C. H. Barrett. 1987. Genetic variation in continental and island populations of *Eichhornia paniculata* (Pontederiaceae). Heredity 59: 7–17.

Gottlieb, L. D. 1975. Allelic diversity in the outcrossing annual plant *Stephanomeria exiguia* ssp *carotifera* (Compositae). Evolution 29: 213–225.

Hamrick, J. L. 1983. The distribution of genetic variation within and among natural plant populations, pp. 335–348. In: C. M. Schonewald-Cox, S. M. Chambers, B. MacBryde and W. L. Thomas (eds) *Genetics and Conservation: A Reference for Managing Wild Animal and Plant Populations*. Benjamin/Cummings, Menlo Park, CA.

Hamrick, J. L. and M. J. Godt. 1989. Allozyme diversity in plant species., pp. 43–63. In: A. H. D. Brown, M. T. Clegg, A. L. Kahler and B. S. Weir (eds) *Plant Population Genetics, Breeding and Genetic Resources*. Sinauer, Sunderland, MA.

Hastings, A. L. 1989. The interaction between selection and linkage in plant populations, pp. 163–180. In: A. H. D. Brown, M. T. Clegg, A. L. Kahler and B. S. Weir (eds) *Plant Population Genetics, Breeding and Genetic Resources*. Sinauer, Sunderland, MA.

Heywood, J. S. 1986. The effect of plant size variation on genetic drift in populations of annuals. Amer. Nat. 127: 851–861.

Holtsford, T. P. and N. C. Ellstrand. 1989. Variation in outcrossing rate and population genetic structure of *Clarkia tembloriensis* (Onagraceae). Theor. Appl. Genet. 78: 480–488.

Knight, S. E. and D. M. Waller. 1987. Genetic consequences of outcrossing in the cleistogamous annual *Impatiens capensis*. I. Population-genetic structure. Evolution 41: 969–978.

Lagercrantz, U. and N. Ryman. 1990. Genetic structure of Norway spruce (*Picea abies*): Concordance of morphological and allozymic variation. Evolution 44: 38–53.

Lande, R. 1988. Genetics and demography in biological conservation. Science 241: 1455–1460.

Levin, D. A. 1978. Genetic variation in annual Phlox: self-compatible versus self-incompatible species. Evolution 32: 245–263.

Levin, B. R. 1992. DNA technology and the release of genetically engineered organisms: some implications for the conservation of genetic resources. This volume.

Mitton, G. C. 1989. Physiological and demographic variation associated with allozyme variation, pp. 127–145. In: D. E. Soltis and P. S. Soltis (eds) *Isozymes in Plant Biology*. Dioscorides Press, Portland.

Moran, G. F., J. C. Bell and J. W. Turnbull. 1989. A cline in genetic diversity in river she-oak *Casuarina cunninghamiana*. Aust. J. Bot. 37: 160–180.

Nei, M. 1973. Analysis of gene diversity in

subdivided populations. Proc. Natl. Acad. Sci. USA 70: 3321–3323.

Nevo, E., D. Zohary, A. H. D. Brown and M. Haber. 1979. Genetic diversity and environmental association of wild barley, *Hordeum spontaneum* in Israel. Evolution 33: 815–833.

Osterman, J. C. and E. S. Dennis. 1989. Molecular analysis of the ADH1-C^{m} allele of maize. Plant Mol. Biol. 13: 203–212.

Rick, C. M., J. F. Robes and M. Holle. 1977. Genetic variation in *Lycopersicon pimpinellifolium*: evidence of evolutionary change in mating systems. Plant Syst. Evol. 127: 139–170.

Schoen, D. J. 1982. Genetic variation and the breeding system of *Gilia achilleifolia*. Evolution 36: 361–370.

Schoen, D. J. and A. H. D. Brown. 1991 a. Intraspecific variation in population gene diversity and effective population size correlates with the mating system in plants. Proc. Natl. Acad. Sci. USA. 88: 4494–4497.

Schoen, D. J. and A. H. D. Brown. 1991 b. Whole- and part-flower self-pollination in *Glycine clandestina* and *G. argyrea* and the evolution of autogamy. Evolution 45: 1651–1664.

Schoen, D. J. and S. C. Stewart. 1987. Variation in male fertilities and pairwise mating probabilities in *Picea glauca*. Genetics 116: 141–152.

Schwaegerle, K. E. and B. A. Schaal. 1979. Genetic variability and founder effect in the pitcher plant *Sarracenia purpurea* L. Evolution 33: 1210–1218.

Simberloff, D. 1988. The contribution of population and community biology to conservation science. Ann. Rev. Ecol. Syst. 19: 473–511.

Stebbins, G. L. 1971. *Chromosomal Evolution in Higher Plants*. Edward Arnold, London.

Sytsma, K. J. and B. A. Schaal. 1990. Ribosomal DNA variation within and among individuals of *Lisianthius* (Gentianaceae) populations. Plant Syst. Evol. 170: 97–106.

Turesson, G. 1925. The plant species in relation to habitat and climate. Hereditas 6: 147–236.

Venable, D. L. 1984. Using intraspecific variation to study the ecological significance and evolution of plant life-histories, pp. 166–187. In: R. Dirzo and J. Sarukhan (eds) *Perspectives on Plant Population Ecology*. Sinauer, Sunderland.

Walbot, V. and C. A. Cullis. 1985. Rapid genomic change in higher plants. Ann. Rev. Plant Physiol. 36: 367–396.

Waller, D. M. and S. E. Knight. 1989. Genetic consequences of outcrossing in the cleistogamous annual, *Impatiens capensis*. III. Interlocus associations. Heredity 63: 1–9.

Yeh, F. C., M. A. K. Khalil, Y. A. El-Kassaby and D. C. Trust. 1986. Allozyme variation in *Picea mariana* from Newfoundland – genetic diversity, population structure and analysis of differentiation. Can. J. For. Res. 16: 713–720.

Yeh, F. C. and D. O'Malley. 1980. Enzyme variation in natural populations of Douglas fir, *Pseudotsuga menziesii* [Mirb.] Franco, from British Columbia. 1. Genetic variation patterns in coastal populations. Silv. Genet. 29: 83–92.

SECTION 3

The ecological context of genetic erosion

In this section, the conservation of biodiversity is viewed from various perspectives within the life sciences. Berry (Chapter 6) describes how concepts of ecological genetics may be applied to conservation genetics. Nee, Harvey and Cotgreave (Chapter 7) outline how new knowledge about population abundance patterns can help understand population persistence, and Stenseth (Chapter 8) reviews the use of models for predicting ecological change. Two case studies conclude the section; in the first, Parsons (Chapter 9) addresses the effects of climatic change on biodiversity, exemplified by Australian *Drosophila*, in the second, Hindar (Chapter 10) outlines how wild populations may be affected by releases of captive stocks, using Atlantic salmon in Norway as an example.

The four cornerstones of ecological genetics are variation, natural selection, island biogeography, and the ecological niche. Berry (Chapter 6) emphasizes the dynamic nature of these factors and concepts, and challenges the unity of conservation biology by arguing that human-caused reduction of genetic diversity is *not* necessarily a threat to natural populations. Sufficient variation for adaptation exists, he maintains, even in populations undergoing severe bottlenecks, and extinction and immigration are the norm rather than the special case in the history of Earth.

An understanding of the factors influencing species abundances will significantly improve the estimation of the minimum viable sizes for their long-term persistence. Such an understanding does not exist for any natural community. Nee, Harvey and Cotgreave (Chapter 7) present new knowledge about *how* natural communities are structured, which, even if not fully understood, may provide better guidelines for conservation. Specifically, the authors outline some relationships between species body sizes and abundances. The species abundance patterns may indicate how large areas must be to protect viable populations of species within different body size categories.

To what extent can models improve the understanding of the environmental changes leading to losses of biodiversity? Stenseth (Chapter 8) emphasizes that a pluralistic approach to modelling is needed, and exemplifies this by address-

ing two issues with major implications for biodiversity conservation, (1) the interaction between economy and ecology, and (2) landscape ecology. He advises that ecological understanding be sharpened by the application of theoretical and empirical models to generate and test hypotheses. Such work requires collaboration between ecologists, mathematicians and statisticians, and it must be integrated with the social scientists' insight about socio-economic processes in order to develop guidelines for sustainable use of natural resources.

Climate plays an important role in determining the abundance and distribution of species, and there is considerable concern that climatic change will stress natural communities to the extent that a reduction in biodiversity is inevitable. Parsons (Chapter 9) discusses stress as an environmental factor, and how genetic variation for stress response is a prerequisite for adaptations in situations where organisms cannot escape stressful conditions by moving into more favorable habitats. Studying *Drosophila* in Australian rainforests, Parsons demonstrates how extinction and replacement of stress-sensitive *Drosophila* are likely following quite small increases in stress caused by temperature changes and/or desiccation. Knowledge about stress response traits can help to predict whether or not a particular species can adapt to changing conditions.

How should species be managed that contain wild populations alongside captive stocks at various stages of domestication? This question is addressed by Hindar (Chapter 10) with reference to Atlantic salmon in Norway, where the escape of vast numbers of fish from a growing aquaculture industry challenges the genetic and ecological integrity of wild populations. A breakdown of local adaptations and of the genetic structure of the species, and losses of entire populations are likely outcomes following interbreeding, disease transmission, and/or competition between escaped and native fish. Hindar concludes that the cost of complete containment of reared fish must be borne for the benefit of the industry as well as for the protection of wild populations – and the necessary steps must be taken before there is nothing left to protect.

6 The role of ecological genetics in biological conservation

Robert J. Berry
Department of Biology,
University College London,
London WC1E 6BT,
UK

ABSTRACT

Conservation genetics is an application of ecological genetics, a subject that has come of age in the last 25 years through the recognition of large amounts of variation (revealed by electrophoresis), significant strengths of natural selection, the developments of island biogeography, and the application of game and control theory to refine concepts of ecological niches and habitats. It still lacks an adequate understanding of genetic architecture and the role of stress in natural systems. We have a good working knowledge of genetical processes in natural systems, although more data are needed, particularly on long-term series, the flexibility of life histories, and the vulnerability of ecosystems and habitats to disturbance or invasion. But the application of needed conservation measures depends ultimately on moral commitment, building upon as good as possible guidance from science.

BACKGROUND

The science of conservation genetics began in the 1960s, largely as a result of the recognition that concentration by agriculturists on a few favoured varieties was leading to a probable loss of alleles in the unimproved "land-races" whence the high-yielding forms came. One of the aims of the International Biological Programme was to collect and preserve seed from a wide variety of threatened varieties (Frankel and Bennett 1970). A parallel but less energetic response came from zoologists, and zoos began to adopt a conservation as well as an educational role. But the theory and urgency of conservation genetics were slow to develop.

In 1970, the British Ecological Society organized as their 11th Annual Symposium a meeting on the "Scientific Management of Animal and Plant Communities for Conservation" (Duffey and Watt 1971). At that meeting, I presented a paper on "Conservation aspects of the genetical constitution of populations", arguing from first principles that reduction in population size and fragmentation of habitat will cause genetical

Conservation of Biodiversity for Sustainable Development.
Edited by O. T. Sandlund, K. Hindar and A. H. D. Brown. 1992. Scandinavian University Press, Oslo.

pressures on species but that the evidence for adaptation and factors maintaining variation was such that genetical problems were unlikely to produce acute conservation problems (Berry 1971). Michael Soulé subsequently criticized this conclusion on the grounds that **any** loss of genetic variation would reduce the potential for persistence and adaptation of a population or species; he identified me as a "phyletic optimist" (Frankel and Soulé 1981: p. 84). In retrospect I believe we were both right. The available scientific evidence pointed to my "optimistic" judgement, but my stated conclusion was an open invitation to policy-makers and managers to ignore the complex of factors which affect long-term viability and survival. We need constantly to be aware of the difference between the interpretation and application of scientific facts.

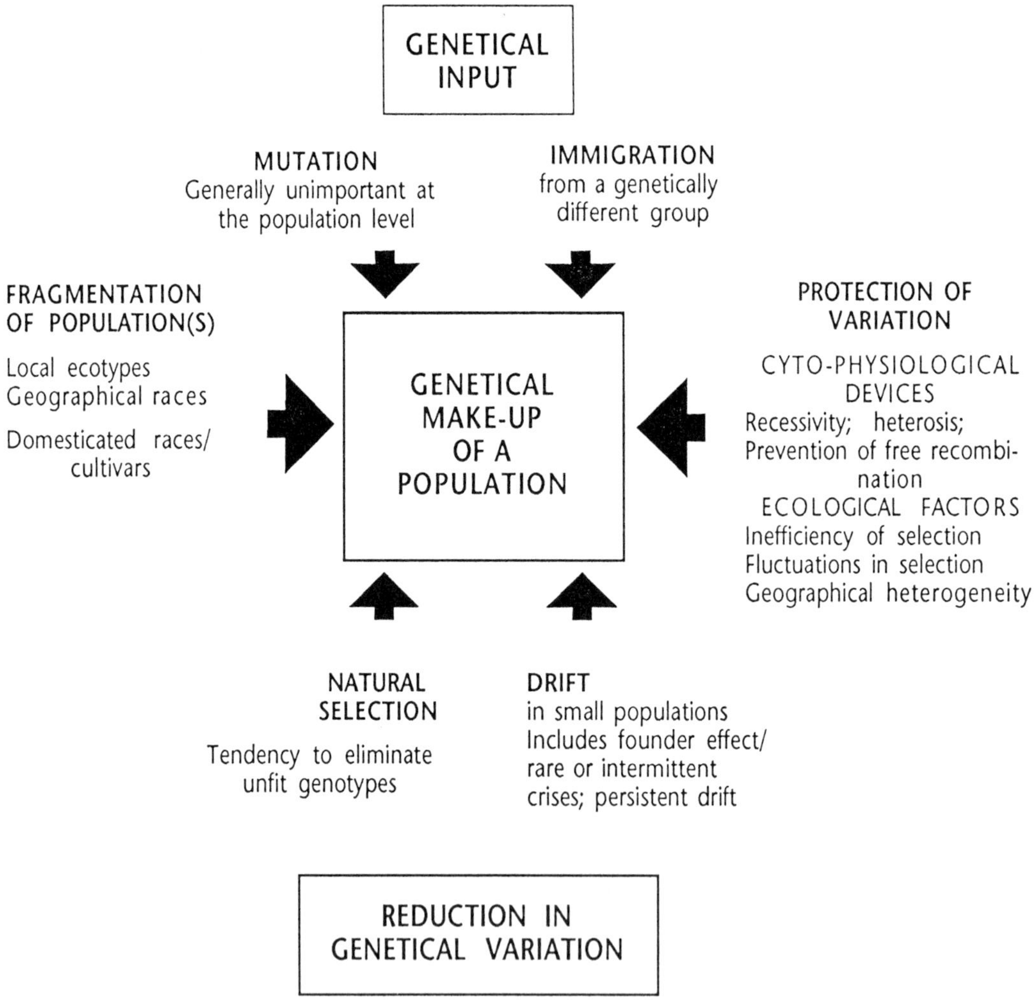

Figure 1. Factors contributing to the genetical make-up of a population or group of populations. Note that all the factors affecting genetical constitution are influenced by the environment (from Berry 1974).

Problems arise when management decisions are extrapolated too far from their justifying data so that the underlying science is doubted and falls into disrepute (Ashby 1978, 1980; Berry 1990a). The purpose of this essay is to examine the interface between ecology and genetics in order to provide an understanding of the state and direction of conservation genetics.

THE GENETICS – ECOLOGY INTERFACE

Ecological genetics can be regarded as having its roots in the work of the biometricians around the turn of the century, but its development lay in the hands of Turesson and the genecologists on the botanical side and Huxley and Ford on the animal side. Unfortunately the melding of genetics and ecology has never really taken place. Population genetic theory is well-developed: two agents (mutation, and immigration from a genetically different population) increase variation within populations, and two (natural selection and drift) decrease it (a summary of these interacting forces is given by Crow 1986) (Figure 1). In contrast, ecology has many key concepts, but no core theoretical structure (Berry 1989a). Its central question is to determine "why is what where", and this can be answered in a number of ways and with different emphases. As a result, the practice of ecological genetics has been influenced and perhaps distorted by techniques and the idiosyncracies of its leading exponents.

During the 1950s, population genetics came under a tight theoretical thrall, epitomized in two seminal papers, on genetic load by Müller (1950) and on the cost of natural selection by Haldane (1957); ironically for conservation geneticists, both used quantitative arguments to show the deleteriousness of too much variation. The spuriousness of this theoretical straitjacket was exposed by the discovery (through the application of protein electrophoresis to population samples, beginning in 1966) that highly successful species were carrying far too much variation for their survival on the prevalent genetic load dogma; yet showed no signs of becoming extinct. Further electrophoretic work has shown that virtually every species (plant and animal) is extremely variable; the old assumptions about population structure and evolutionary possibilities were radically wrong.

The use of electrophoresis has revolutionized ecological genetics over the same period that conservation genetics has become established as a discipline in its own right. It seems, therefore, worth reviewing the successes and failures of ecological genetics over the past 25 years or so as a basis for understanding the tasks facing conservation geneticists. There have been four areas of major importance at the genetics –ecology interface, and two further topics which have not received the attention they merit.

Extensive variation

The realisation that enormous amounts of variation segregate in natural populations of animals and plants has forced a reappraisal of its effects and maintenance. Although both theoreticians and breeders traditionally assumed that variation was readily available for evolutionary adjustment or artificial selection, ecologists tended to regard inherited variation as the exception rather than the rule; in the 1950s there was a debate about "classical" and "balanced" population structures. The revelations from electrophoresis precipitated the selection – neutralism debate of the 1970s and 1980s. The neutralist predictions of mutant accumulation and the accuracy of the molecular clock have not been borne out by subsequent research, and it now seems indisputable that variation is maintained in the traditional neo-Darwinian manner, by selection leading to adaptation in an environment varying in space and time (Berry 1989b; although see Kimura 1983; Crow 1986).

The big disappointment of efforts to discover the "causes" of particular variants is that most studies have shown a statistical correlation only between genetical and environmental variables. Although there is a clear relationship between overall measures of variation, such as heterozygosity, and measures of ecological heterogeneity, only about 40 per cent of the genetic diversity can be explained in this way (Nevo et al. 1984). Adaptive consequences of individual alleles have been identified, e.g. at the *Hbb* locus in house mice (Berry 1978) and the *Lap* locus in mussels (Koehn and Hilbish 1987), and the cumulative effects of heterozygotes in particular metabolic pathways have been dissected (Koehn et al. 1988), but the most promising relationship of heterozygosity to any extrinsic variable is with community species diversity (Valentine 1976; Berry 1979).

One clear lesson from the neutralism debate is the folly of assuming a direct causal link between genetical variety and any environmental trait (Hedrick and Miller, this volume). For example, the mean heterozygosity of different populations of a species may be very different for no apparent reason or defect. An example of this is *Spartina anglica*, a grass species which arose in southern England about 1890 from the tetraploidization of a hybrid between a European species and a colonizer from America. *S. anglica* now occupies an estimated 10,000 ha in Britain and 36,000 ha in China whither it was introduced in 1963 to stabilize mud flats, but both the tetraploid and its parental species seem to be allozymically monomorphic (Raybould et al. 1991); success on conventional measures (except, for obvious reasons, long-term persistence) is clearly unrelated in this case to genetical variation.

Significant selection

Endler (1986) has listed published examples of natural selection in 141 wild species (i.e. omitting laboratory experiments), most (199/314) involving morphological traits, but also affecting physiological and biochemical traits. Observed selection coefficients or differentials range "from weak to remarkably strong" (Endler's description). It is important to recognize that proven examples of selection are both common and biologically non-trivial; prior to field investigations of selection (first appreciated by J. B. S. Haldane, and pursued by E. B. Ford and his colleagues), it was commonly assumed that selection coefficients were usually so weak as to be frequently undetectable. There are now many examples of both directed and stabilizing selection where the selection coefficients are of the order of 0.3 to 0.5 or even more (Endler 1986: p. 211). This means that adaptive response will be rapid and precise, and conversely that random genetic changes are likely to be rather rare. Indeed, I am not aware of any established example of allele change through genetic drift under natural conditions, with the important exception of founder effects (which are frequent and important, but a special case) and fluctuations affecting the third base of codon triplets (which are almost certainly trivial in effect) (Gillespie 1986).

However, the neutralist debates of the past two decades have led to a much more ecological understanding of natural selection. Although the fact that selection coefficients vary in time and space is a basic tenet of population genetics, the heterogeneity of habitats has not always been appreciated to the full. Examples of variable selection pressures are: Berry and Crothers's (1968) demonstration that selection coefficients recognized in the inter-tidal snail *Nucella lapillus* by exposure to wave action varied from nought to over 0.90, with a correlation between selection coefficient and environmental harshness on different (and nearby) shores of more than 70 %; and the findings that selection shown by genotypic changes on the haemoglobin ß locus in house mice only takes place in cold (i.e. stressful) winters (Berry 1978). Geneticists and ecologists are now conceptually closer than formerly, but there is still much mutual confusion (Travis and Mueller 1989).

Island biogeography

The models of MacArthur and Wilson (1967) have been a major stimulus to ecologists in general and conservation biologists in particular over the past two decades (e.g. Wilcox 1980; Diamond and May 1981). The original MacArthur and Wilson work was mainly concerned with diversity rather than differentiation, although conservationists are usually (and probably wrongly) more anxious about endemics and rare forms, than the overall species composition of particular localities. The balance between immigration and extinction of species on islands is a logical necessity (Williamson 1981); the interesting questions about biodiversity which can usefully be approached on islands (because of their relatively depauperate biota), concern the factors affecting species turn-over, extinction, and the conditions for successful establishment.

Indeed a major dis-service that island biogeography has rendered to conservation biology is to strengthen the assumption that ecological communities are in equilibrium or deterministic progression towards a successional climax. This is ironical, because a major lesson from island studies is the importance of opportunism and stochasticism. For example, the house mouse population of Skokholm, a small Welsh island has a very high frequency of spina bifida occulta. This is almost certainly due to the idiosyncratic chance constitution of the original founding population, which came from an area where the frequency of the trait was c. 0.10 as opposed to less than 0.01 in mice generally (Berry 1964). Well documented similar cases in humans are the high local incidences of Huntington's chorea in the USA, porphyria variegata in South Africa, retinitis pigmentosa on the island of Tristan da Cunha, and – almost certainly – multiple sclerosis in the Orkney and Shetland archipelagoes (Berry 1972).

Four points need to be made about island populations. Firstly, unless they are relicts (i.e. survivors which have been isolated from a previously widespread population by the severance of a land-link), their genetic composition will be largely determined by the variants carried by the initial founders and not, as the chapter describing island differentiation in MacArthur and Wilson (1967) is called, "evolutionary changes following colonization". There are three species and 14 subspecies of the long-tailed field mouse *Apodemus* described on the islands to the north and west of Britain, but there is no pattern in their relationship except that all are more like Norwegian than Scottish mice; they are probably the descendants of inadvertent passengers on Viking boats during the colonial expansions of the 8th and 9th centuries, with a few individuals landing in seed or bedding and then being distributed randomly to other islands (Berry 1969). We need to recognize and respect the amount of genetic differences which can result immediately and unimportantly from founder effects. There is a cautionary tale in the house mouse population on the Faroe group. In his magisterial *Evolution – the Modern Synthesis*, Huxley (1942) reviewed evidence which implied that subspeciation usually takes about 5000 years, but "the facts concerning rats and mice show that subspeciation may sometimes occur much more quickly. In particular the Faroe house mouse, which was introduced into the islands not much more than 250 years ago, is now so distinct that certain modern authorities have assigned full specific status to it". In fact, the situation is more complicated. Taxonomic status has been accorded to the mice of the islands of Mykines and Nolsøy, while animals from the main island of Streymøy have been regarded as indistinguishable from western European mice. However, Mykines was colonized by mice sometime after 1800, and another island (Hestur) acquired its population during the 1939–45 war – yet the latter is as distinct (and "unique") as that of Mykines or Nolsøy (Berry et al. 1978). The only viable explanation seems to be that these island populations are differentiated through the chance characteristics of their founders. A similar conclusion was reached by Patton et al. (1975) studying *Rattus rattus* on the Galapagos; and there are many other similar examples.

A second lesson from islands is that selection is opportunistic, not perfecting. Adaptation depends upon the variation present in a population, not on progress towards an ideal goal. Island populations may become extinct when stressed because of the non-availability of genes for appropriate traits; or they may survive because they are not exposed to the harshness of competition or abiotic stress experienced by mainland relatives. The high frequency of spina bifida in Skokholm house mice has already been mentioned; an even more extraordinary finding is that the frequency of individuals with this condition increases by a third every summer, from about one in four to one in three, presumably as a consequence of selection on linked or associated loci; the reciprocal change takes place each winter. Generalising from this example, we must regard all species as no more than contemporary survivors, potentially susceptible to a change of conditions on a micro or macro-scale, such as the effect of pollution for cryptically coloured moths or the mass extinction of South American marsupials when invaded by eutherians.

Interestingly, there is no direct evidence that island biota suffer from the reduction in variation as a result of the founder bottleneck, although it would be difficult to collect critical evidence (Table 1). Moreover, a bottleneck has less effect on variation if numbers expand rapidly afterwards (Nei et al. 1975; Bryant et al. 1986).

Thirdly, we must avoid assuming Panglossian adaptation (Gould and Lewontin 1979; Berry 1985). As Rothschild and Clay (1952) have pointed out about parasitism, a situation which has been treated as a paradigm of adaptation or coevolution: "Parasitism can develop suddenly or gradually. It can be the outcome of complicated interactions or the result of isolated accidents which occured a million years ago or only this morning....There is only one vital factor in the genesis of a parasitic relationship, and that is opportunity". Conservation biology is too full of attempts to rescue species or ecosystems which are transients in the dynamic procession of life, such as the enormous efforts that go into preserving island endemics; conservation biologists are prone to preservationism whereas their science demands a practice of dynamic management which has to accept that extinction is a natural process.

And fourthly, island ecology should teach us that our common assumption of equilibrium and the related concept of determinism are unjustified; the "balance of nature" is a fruit of mediaeval theology not confirmed by modern science (Egerton 1973). Connell (1978, 1979) has experimentally examined temperate sea shores and tropical forests, which are commonly taken to typify communities controlled by climax equilibria. His conclusion is that both are

Table 1. Allozymic variation (heterozygosity) in island and mainland populations of the same species (from Berry 1986, after Kilpatrick 1981). *N* = number of populations.

	N	**Mainland** Mean heterozygosity per locus (range)	*N*	**Island** Mean heterozygosity per locus (range)	Reduction in heterozygosity (%)
Macrotus waterhousii	3	0.021(0–0.043)	1	0.040	–
Macaca fuscata	13	0.019(0–0.035)	5	0.013 (0.03–0.018)	31.6
Spermophilus spilosoma	12	0.090(0.049–0.160)	1	0.009	90.0
Peromyscus eremicus	44	0.040(0.006–0.079)	2	0.009(0–0.022)	77.5
Peromyscus leucopus	3	0.080(0.076–0.084)	3	0.071(0.052–0.078)	11.3
Peromyscus maniculatus	22	0.088(0.054–0.124)	11	0.068(0.010–0.131)	22.7
Peromyscus polionotus	26	0.063(0.050–0.086)	4	0.052(0.018–0.086)	17.5
Sigmodon hispidus	4	0.022(0.017–0.025)	1	0.021	4.6
Microtus pennsylvanicus	4	0.142(0.120–0.171)	9	0.056(0.023–0.114)	60.6
Mus domesticus	16	0.091(0.032–0.114)	20	0.041(0–0.079)	55.0
Rattus fuscipes	3	0.047(0.020–0.100)	9	0.011(0–0.040)	76.6
Rattus rattus	1	0.031	11	0.026(0.008–0.056)	16.1

dependent on recurrent disturbances and local responses; it parallels the recognition that the apparently simple balance between colonization and extinction on islands is a trivial rather than a fundamental relationship (Williamson 1981; Haila et al. 1982). Indeed the much-studied and undoubted phenomenon of ecological succession should be regarded as nothing more than a particular and general example of island biogeography: as habitats are modified and niches alter, species decline and are subject to replacement by available – not inevitable – colonists.

Niches and trade-offs

The genetic constitution of any species is determined by the sum of past environments experienced by that species and to that extent species are prisoners of their evolutionary history (Wilson 1975). This means that organisms are restricted in their response to any influence, whether biotic or abiotic; in more conventional language, organisms have definable niches. The interactions of genetic limits, possible responses and environmental instability and heterogeneity have been usefully combined by Southwood (1988). He lists the problem areas in the life of any organism, pointing out that each results in "combinations of tactics determined by the potential of the genome...for which selection can act. Each of these involves a different amount of risk (i.e. a "trade-off") between alternative courses of actions; each of them involves investment in a trait" (Figure 2).

A good example of a trade-off is the behaviour of black rabbits (Berry 1977: p. 271). An apparently pleiotropic effect of the colour is an increase in placidity over the normal. Black rabbits usually occur on small islands or headlands where population density is high and food is limiting. When a feeding group is disturbed, the

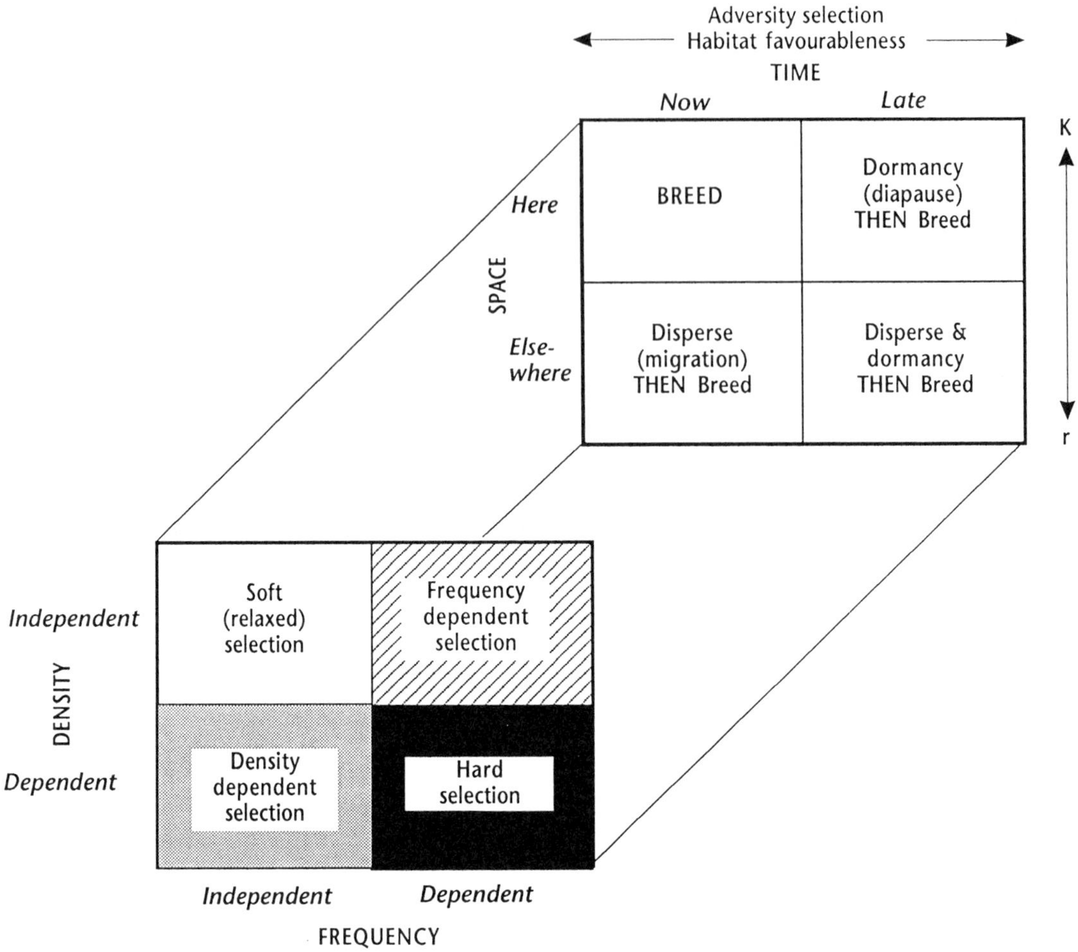

Figure 2. Wallace's (1975) classification of selection complements Southwood's (1977) habitat templet, putting the adaptive adjustability of genetical constitution alongside environmental heterogeneity and introducing the variable stresses experienced by organisms. The links between the two models are not causal: although allele frequencies are commonly regarded as changing with time and density in space, frequency may change in space (clinally) and density with time (through recruitment and/or mortality), i.e. selection modes interact with each other and with the environment.

blacks are last to take refuge and the first to reappear. Consequently they have more opportunity to find food, and this partially offsets their cryptic disadvantage (their life-span is only one third of that of agouti-coloured animals, presumably as a result of visual predation).

Wallace (1968, 1975) has distinguished between "hard" and "soft" selection in a way which complements Southwood's habitat templet. Wallace began from the unsatisfactoriness of the genetic load concept, and argued that hard selection acts on alleles which produce their effects in all known conditions (for example, lethal genes are inevitably fatal to their carriers), whereas soft selection involves varying survival (or death) of individuals as conditions change; soft selection is both frequency- and density-dependent, and varies with other population parameters (Figure 2).

The four topics reviewed above have been the major factors influencing and transforming ecology in the last 25 years or so. Two other topics have not had the impact they should have had.

Genetic architecture

In my 1971 conservation genetics paper, I pointed out that genetic architecture produced by the linkage of genes on chromosomes and by the need to maintain biochemically complementary mechanisms meant that both developmentally necessary gene combinations and "super-genes" will be protected; consequently it must be a significant factor in retarding genetic erosion. This point has been almost entirely neglected since (e.g. Lande 1988; although see Templeton 1986); virtually all discussions have been conducted on the straightforward probabilistic axioms of "bean-bag genetics" (Haldane 1964). The most likely reason for this has been lack of strong evidence for gene associations in wild populations, apart from a limited number of special cases of linkage disequilibria (such as the mammalian histocompatibility complex, incompatibility controls in flowering plants, mimicry patterns in butterflies, colour and shell banding in *Cepaea*) (e.g. Barton and Charlesworth 1984). However, failure to find evidence does not mean that genetic architecture is unimportant. Recent interest has concentrated on testing Mayr's (1954) hypothesis of "genetic revolutions", which is only a particular case of genetic architecture. As has been the situation with neglecting the evidence for strong selection when attention concentrated on neutralism, so memories of the early work of Mather (1943, 1974), Clausen and Hiesey (1958), Thoday (1961), and others have been displaced by recent observations with allozyme associations. We must not forget that species are determined by gene associations rather than genes per se, and that theoretical considerations indicate that complex traits are likely to be controlled by many loci, with selection favouring the formation of "balanced" chromosomes with positive and negative alleles intermingled. Experimental results (mainly from selection experiments in *Drosophila*) confirm this expectation (Bodmer and Parsons 1962).

I recently examined the genes which control aggressive behaviour in the house mouse (Berry 1989c). About one in six of the 1300 identified gene loci in mice can be regarded as affecting behaviour, and they are distributed over 19 out of the 20 chromosomes in the species.

The truth about genetic architecture is that we know far too little about its importance and resilience. This is well illustrated by an experiment my colleagues and I have recently completed. In 1982, we released 77 house mice from Orkney into an established population of about two thousand mice on an uninhabited 57 ha island in the Firth of Forth, Scotland, and monitored their spread and hybridization using genetic markers (Berry et al. 1990). Within four years, the frequency of introduced variants had more or less stabilized and thereafter remained fairly constant but at different frequencies from those in the receiving population, the ancestors of the introduced animals, and the introduced animals themselves (Figure 3); after a period of rapid change following the introduction, all the new alleles were apparently stabilized at new frequencies by forces undetectable by the techniques available to us, but presumably related to intra-genomic associations, i.e. to genetic architecture.

Stress

The second neglected topic in evolutionary ecology is stress, possibly because of argu-

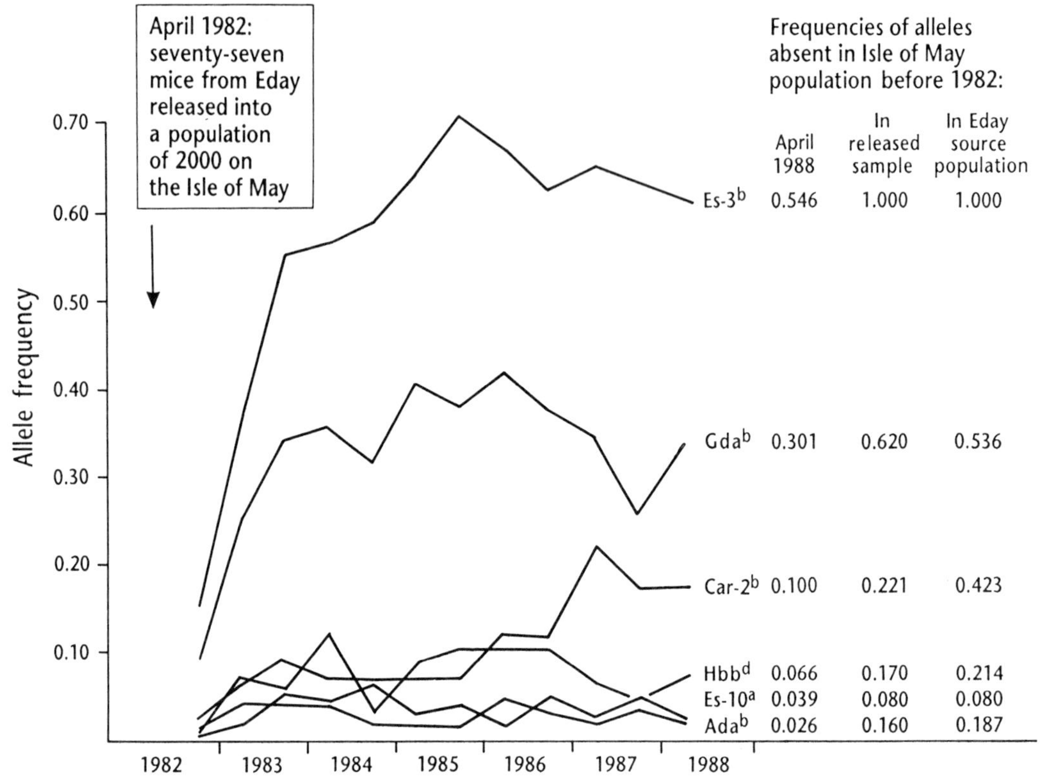

Figure 3. Changes in allele frequencies in the house mouse population of the Isle of May following release of mice from Eday (Orkney) carrying alleles not previously present in the May mice (based on Berry et al. 1991).

ments over its correct use (Calow and Berry 1989). A useful working definition for ecologists is that stress "is any environmental influence that impairs the structure and functioning of organisms such that their neo-Darwinian fitness is reduced" (Calow 1989). This incorporates survival probability, developmental rate and fecundity, and hence links responses at the level of individuals to parameters that influence the density of their populations and their future contribution to the gene pool.

The advantage of this definition is that it invites tests in a variety of situations (e.g. Parsons 1989, this volume). A particular topic of interest for conservation biology is the link between genetic variation and life-history traits (Allendorf and Leary 1986). The most intensive investigations on this have been carried out by Koehn, Bayne and their collaborators on the physiological responses of marine molluscs with different genotypes to varying temperature and food stresses. They have shown that the association between heterozygosity and growth (or productivity) found for various traits is between particular phenotypes, rather than heterozygosity *per se*, with homozygous individuals liable to suffer more than more heterozygous ones from a negative energy balance in environments that deviate from the optimum for the phenotype concerned (Koehn and Bayne 1989; Koehn 1991).

The importance of an adequate concept of stress is that it provides a framework for recognizing the dynamic complex of interactions between phenotype and environment. Survival is a property of these interactions, not of either phenotype or environment by themselves (Figure 2); the same genotype will respond differently to a range of conditions through life, and different genotypes will respond differently to a particular environmental stress (Berry et al. 1981). For example, Sikes (1968) found a high incidence of arterial disease in African elephants living at high densities or in disturbed habitats, but virtually none in low-density populations in their natural habitat. Clearly harsh environmental pressures cause problems for living beings, but to evaluate these problems, we need to study organism – environment interactions (or levels of stress) and not merely extrapolate from a knowledge of either genetics or environment.

IMPLICATIONS

Ecology is a synthesis of disciplines, concerned with explaining why is what where. Evolutionary biology is even more obviously a synthesis of different subjects, and it is salutary to recall the repeated trauma it has suffered by allowing its components to fall apart. As conservation geneticists drawing upon both ecology and evolutionary genetics, we need to be particularly aware of the dangers of distortion through over-simplistic interpretation, and the consequent hazards as we attempt to apply our results as conservation managers. In particular

1. We need data:
    a. Almost all biological systems are special cases, so we must collect information on many food webs, breeding systems, ecosystem stability and inertia as a deterrent to generalising from inadequate information (Ralls et al. 1986; Ledig 1986; Lawton 1989).
    b. On a long-term basis (Taylor 1988). Gray (1989) has reviewed the effect of pollution on marine benthic macrofauna, using abundance – biomass comparisons (ABC curves) to show a characteristic sequence of response involving a rapid increase of biomass, followed by a fall in species number, then dominance by a few, small-sized opportunist species, until both species number and biomass decreases. He points out that these changes are paralleled in the intertidal zone; they are also comparable to those in colonizing and successional stages generally.
    c. On life-history variation and its causes, since this is a convenient and important coordinating corrective to ecological reductionism (Caswell 1989; Berry and Bronson in press).
    d. Tropical ecosystems are species-rich in comparison with temperate ones. Too little is known about the status of tropical species, particularly whether temperate morphs are equivalent to tropical species (Hubbell and Foster 1986).
2. The effect of stress on ecosystems can often be detected by changes in energetics, community structure or system characteristics (such as openness, disease incidence, mutualism, etc.) (Rapport 1989). These effects in turn frequently reflect life history changes for individual species or populations. As we accumulate data (particularly under (b) and (c) above), we will be better able to predict and hence manage biological systems (Spotila et al. 1989).
3. There is a habit among conservationists to assume that change is necessarily detrimental and that there exists an innate "balance of nature". Both these assumptions must be resisted; the no-

tion of an unchanging world where only man is vile, owes more to Greek philosophy than modern biology (Egerton 1973). Anthropogenic changes in the natural world are now proceeding at a frighteningly dangerous rate, but it is no help to aim for the status quo. Although it is wrong wantonly to exterminate species, yet some species are more "important" (in terms of their interactions, biomass, distribution, etc.), and habitat conservation is more important than species conservation. And heretical though it may seem in 1990, DNA techniques are likely to be able to help us to maintain gene combinations more easily in culture than to preserve whole species.

4. Linked to the notion of a "balance of nature" is that of "saturated habitats". Virtually any habitat is susceptible to invasion, although the characteristics of invaded habitats or the invaders themselves have so far proved impossible to define (Groves and Burdon 1986; Kornberg and Williamson 1986; Gray 1986). The common assumptions of ecologists and geographers about such systems are too static to explain observed invasions, and depend too much on equilibrium-type theories (Hengeveld 1989).
5. We need to be much clearer than hitherto about our motives for conservation. Much of the utilitarian rhetoric about the value of undescribed species for mankind is unconvincing. We can almost certainly preserve most genetic variety *sensu stricto* in gene and embryo banks if that is all we want to do. But we want – and need – to do more. Norton (1987) has argued for the importance of species conservation on "weak anthropocentric grounds", that we are transformed by our contacts with natural variety in a powerful and effectively humanizing way; the instrumental value of living animals and plants (i.e. "strong" anthropocentricism) is relatively unimportant in comparison.

This brings us full circle, and back to my designation by Frankel and Soulé (1981) as a "phyletic optimist" for believing that variation loss may not be as common or catastrophic as is sometimes claimed. There can be little doubt that human activities threaten our continued existence, but I believe that our efficiency in eroding the world's overall genetic sources is probably not too great. We may lose species, but that does not necessarily mean that the genetic variation carried by them is also lost; many gene combinations and chromosomal segments are common to different species. Notwithstanding our **attitude** to our environment is crucial; species conservation is both symbolic and symptomatic of our commitment to an environmental stewardship on which the biological future and quality of life of the planet depends (Berry 1990b). Introducing the (American) National Forum on BioDiversity in 1986, E. O. Wilson (1988) concluded "In the end, I suspect it will all come down to a question of ethics – how we value the natural worlds in which we evolved and now, increasingly, how we regard our status as individuals....The drive towards perpetual expansion – or personal freedom – is basic to the human spirit. But to sustain it we need the most delicate, knowing stewardship of the living world that can be devised". Conservation genetics involves moral choice as well as the best support from its parent science.

ACKNOWLEDGMENTS

My thanks to the Norwegian Institute for Nature Research (NINA) for their stimulus to write this essay.

LITERATURE CITED

Allendorf, F. W. and R. F. Leary. 1986. Heterozygosity and fitness in natural populations of animals, pp. 57–76. In: M. E. Soulé (ed) *Conservation Biology: The Science of Scarcity and Diversity*. Sinauer, Sunderland, MA.

Ashby, E. 1978. *Reconciling Man with the Environment*. Oxford University Press, London.

Ashby, E. 1980. *The Search for an Environmental Ethic*. The Tanner Lecture on Human Values. Oxford University Press, Oxford.

Barton, N. H. and B. Charlesworth. 1984. Genetic revolutions, founder effects and speciation. Ann. Rev. Ecol. Syst. 15: 133–164.

Berry, R. J. 1964. The evolution of an island population of the house mouse. Evolution 18: 468–483.

Berry, R. J. 1969. History in the evolution of *Apodemus sylvaticus* (Mammalia) at one edge of its range. J. Zool. 159: 311–328.

Berry, R. J. 1971. Conservation aspects of the genetical constitution of populations, pp. 177–206. In: E. Duffey and A. Watt (eds) *The Scientific Management of Animal and Plant Communities for Conservation*. Blackwell, Oxford.

Berry, R. J. 1972. Genetical approaches to taxonomy. Proc. R. Soc. Med. 65: 853–854.

Berry, R. J. 1974. Conserving genetical variety, pp. 99–115. In: A. Warren and F. B. Goldsmith (eds) *Conservation in Practice*. Wiley, London.

Berry, R. J. 1977. *Inheritance and Natural History*. Collins New Naturalist, London.

Berry, R. J. 1978. Genetic variation in wild house mice: where natural selection and history meet. Amer. Scient. 66: 52–60.

Berry, R. J. 1979. Genetical factors in animal population dynamics, pp. 53–80. In: R. M. Anderson, B. D. Turner and L. R. Taylor (eds) *Population Dynamics*. Blackwell, Oxford.

Berry, R. J. 1985. Natura non facit saltum. Biol. J. Linn. Soc. 26: 301–305.

Berry, R. J. 1986. Genetics of insular populations of mammals, with particular reference to differentiation and founder effects in British small mammals. Biol. J. Linn. Soc. 28: 205–230.

Berry, R. J. 1989a. Ecology: where genes and geography meet. J. Anim. Ecol. 58: 733–759.

Berry, R. J. 1989b. Electrophoretic studies: perspectives for population biology, pp. 7–16. In: H. D. Loxdale and J. Den Hollander (eds) *Electrophoretic Studies on Agricultural Pests*. Clarendon, Oxford.

Berry, R. J. 1989c. Genes, behaviour and fitness in mice: concepts and confusions, pp. 23–48. In: P. T. Brain, D. Mainardi and S. Parmigiani (eds) *House Mouse Aggression: A Model for Understanding the Evolution of Social Behaviour*. Harwood Academic, Chur.

Berry, R. J. 1990a. Science, mankind and ethics, pp. 289–304. In: P. H. Bourdeau, P. M. Fasella and A. Teller (eds) *Environmental Ethics*. Commission of the European Communities, Luxembourg.

Berry, R. J. 1990b. Environmental knowledge, attitudes and action: a code of practice. Sci. Publ. Affairs 5: 13–23.

Berry, R. J. and F. H. Bronson (in press). Life history and bioeconomy of the house mouse. Biol. Rev.

Berry, R. J. and J. H. Crothers. 1968. Stabilizing selection in the Dog whelk (*Nucella lapillus*). J. Zool. 155: 5–17.

Berry, R. J., M. E. Jakobson and J. Peters. 1978. The house mice of the Faroe Islands: a study in microdifferentiation. J. Zool. 185: 73–92.

Berry, R. J., R. D. Sage, W. Z. Lidicker, and W. B. Jackson. 1981. Genetic variation in three Pacific House mouse populations. J. Zool. 193: 391–404.

Berry, R. J., G. S. Triggs, V. Bauchau, C. S. Jones and P. Scriven. 1990. Gene flow and hybridization follow introduction of *Mus domesticus* into an established population. Biol. J. Linn. Soc. 41: 279–283.

Berry, R. J., G. S. Triggs, P. King, H. R. Nash and L. R. Noble. 1991. Hybridisation and gene flow in House mice introduced into an existing population. J. Zool. 225: 615–632.

Bodmer, W. F. and P. A. Parsons. 1962. Linkage and recombination in evolution. Adv. Genet. 11: 1–100.

Bryant, E. H., S. A. McCommas and L. M. Combs. 1986. The effect of an experimental bottleneck upon quantitative genetic variation in the housefly. Genetics 114: 1191–1211.

Calow, P. 1989. Proximate and ultimate responses to stress in biological systems. Biol. J. Linn. Soc. 37: 173–181.

Calow, P. and R. J. Berry (eds) 1989. *Evolution, Ecology and Environmental Stress*. Academic Press, London.

Caswell, H. 1989. Life-history strategies, pp. 285–307. In: J. M. Cherrett (ed) *Ecological Concepts*. Blackwell, Oxford.

Clausen, J. and W. M. Hiesey. 1958. Experimental studies on the nature of species. Publ. Carnegie Instn. 615: 1–312.

Connell, J. H. 1978. Diversity in tropical rain forests and coral reefs. Science 199: 1302–1310.

Connell, J. H. 1979. Tropical rain forests and coral reefs as open non-equilibrium systems, pp. 141–163. In: R. M. Anderson, L. R. Taylor and B. Turner (eds) *Population Dynamics*. Blackwell, Oxford.

Crow, J. F. 1986. *Basic Concepts in Population, Quantitative, and Evolutionary Genetics*. Freeman, New York.

Diamond, J. M. and R. M. May. 1981. Island biogeography and the design of nature reserves, pp. 163–186. In: R. M. May (ed) *Theoretical Ecology: Principles and Applications*. Blackwell, Oxford.

Duffey, E. and A. S. Watt (eds) 1971. *The Scientific Management of Animal and Plant Communities for Conservation*. British Ecological Society Symposium No. 11. Blackwell, Oxford.

Egerton, F. N. 1973. Changing concepts in the balance of nature. Quart. Rev. Biol. 48: 322–350.

Endler, J. A. 1986. *Natural Selection in the Wild*. Princeton University Press, Princeton, NJ.

Frankel, O. H. and E. Bennett (eds) 1970. *Genetic Resources in Plants*. IBP Handbook No. 11. Blackwell, Oxford.

Frankel, O. H. and M. E. Soulé. 1981. *Conservation and Evolution*. Cambridge University Press, Cambridge.

Gillespie, J. H. 1986. Rates of molecular evolution. Ann. Rev. Ecol. Syst. 17: 637–665.

Gould, S. J. and R. C. Lewontin. 1979. The spandrels of San Marco and the Panglossian paradigm: a critique of the adaptationist programme. Proc. R. Soc. Lond. B 205: 581–598.

Gray, A. J. 1986. Do invading species have definable genetic characteristics? Phil. Trans. R. Soc. Lond. B 314: 655–674.

Gray, J. S. 1989. Effects of environmental stress on species rich assemblages. Biol. J. Linn. Soc. 37: 19–32.

Groves, R. H. and J. J. Burdon (eds) 1986. *Ecology of Biological Invasions: An Australian Perspective*. Australian Academy of Science, Canberra.

Haila, Y., I. Hanski, O. Järvinen and E. Ranta. 1982. Insular biogeography: a Northern European perspective. Acta Oecol. Gen. 3: 303–318.

Haldane, J. B. S. 1957. The cost of natural selection. J. Genet. 55: 511–524.

Haldane, J. B. S. 1964. A defense of beanbag genetics. Persp. Biol. Med. 7: 343–360.

Hedrick, P. W. and P. Miller. 1992. Conservation genetics: theory and management of captive populations. This volume.

Hengeveld, R. 1989. *Dynamics of Biological Invasions*. Chapman and Hall, London.

Hubbell, S. P. and R. B. Foster. 1986. Commonness and rarity in a neotropical forest: implications for tropical tree conservation, pp. 205–231. In: M. E. Soulé (ed) *Conservation Biology: The Science of Scarcity and Diversity*. Sinauer, Sunderland, MA.

Huxley, J. S. 1942. *Evolution, the Modern Synthesis*. Allen and Unwin, London.

Kilpatrick, C. W. 1981. Genetic structure of insular populations, pp. 28–59. In: M. H. Smith and J. Joule (eds) *Mammalian Population Genetics*. University of Georgia Press, Athens, GA.

Kimura, M. 1983. *The Neutral Theory of Molecular Evolution*. Cambridge University Press, Cambridge.

Koehn, R. K. 1991. The cost of enzyme synthesis in the genetics of energy balance and physiological performance. Biol. J. Linn. Soc. 44: 231–247.

Koehn, R. K. and B. L. Bayne. 1989. Towards a physiological and genetical understanding of the energetics of the stress response. Biol. J. Linn. Soc. 37: 157–171.

Koehn, R. K., W. J. Diehl and T. M. Scott. 1988. The differential contribution by individual enzymes of glycolysis and protein catabolism to the relationship between heterozygosity and growth rate in the coot clam, *Mulinia lateralis*. Genetics 118: 121–130.

Koehn, R. K. and T. J. Hilbish. 1987. The adaptive importance of genetic variation. Amer. Scient. 75: 134–141.

Kornberg, H. and M. H. Williamson (eds) 1986. Quantitative aspects of the ecology of biological invasions. Phil. Trans. R. Soc. Lond. B 314: 501 – 742.

Lande, R. 1988. Genetics and demography in biological conservation. Science 241: 1455–1459.

Lawton, J. H. 1989. Food webs, pp. 43–78. In: M. J. Cherrett (ed) *Ecological Concepts*. Blackwell, Oxford.

Ledig, F. T. 1986. Heterozygosity, heterosis, and fitness in outbreeding plants, pp. 77–104. In: M. E. Soulé (ed) *Conservation Biology: The Science of Scarcity and Diversity*. Sinauer, Sunderland, MA.

MacArthur, R. H. and E. O. Wilson. 1967. *The Theory of Island Biogeography*. Princeton University Press, Princeton, NJ.

Mather, K. 1943. Polygenic inheritance and natural selection. Biol. Rev. 18: 32–64.

Mather, K. 1974. *Genetical Structure of Populations*. Chapman and Hall, London.

Mayr, E. 1954. Change of genetic environment and evolution, pp. 157–180. In: J. Huxley, A. C. Hardy and E. B. Ford (eds) *Evolution as a Process*. Allen and Unwin, London.

Müller, H. J. 1950. Our load of mutations. Amer. J. Hum. Genet. 2: 111–176.

Nei, M., T. Maruyama and R. Chakraborty. 1975. The bottleneck effect and genetic variability in populations. Evolution 29: 1–10.

Nevo, E., A. Beiles and R. Ben-Shlomo. 1984. The evolutionary significance of genetic diversity: ecological, demographic and life history correlates. Lect. Notes Biomath. 53: 13–213.

Norton, B. G. 1987. *Why Preserve Natural Variety?* Princeton University Press, Princeton, NJ.

Parsons, P. A. 1989. Environmental stresses and conservation of natural populations. Ann. Rev. Ecol. Syst. 20: 29–49.

Parsons, P. A. 1992. Biodiversity and climatic change. This volume.

Patton, J. L., S. Y. Yang and P. Myers. 1975. Genetic and morphologic divergence among introduced rat populations (*Rattus rattus*) of the Galapagos Archipelago, Ecuador. Syst. Zool. 24: 296–310.

Ralls, K., P. H. Harvey and A. M. Lyles. 1986. Inbreeding in natural populations of birds and mammals, pp. 35–56. In: M. E. Soulé (ed) *Conservation Biology: The Science of Scarcity and Diversity*. Sinauer, Sunderland, MA.

Rapport, D. J. 1989. Symptoms of pathology in the Gulf of Bothnia (Baltic Sea): ecosystem response to stress from human activity. Biol. J. Linn. Soc. 37: 33–49.

Raybould, A. F., A. J. Gray, M. J. Lawrence and D. F. Marshall. 1991. The evolution of *Spartina anglica* (Gramineae): genetic variation and status of the parental species in Britain. Biol. J. Linn. Soc. 44: 369–380.

Rothschild, M. and T. Clay. 1952. *Fleas, Flukes and Cuckoos*. Collins New Naturalist, London.

Sikes, S. K. 1968. Observations on the ecology of arterial disease in the African elephant (*Loxodonta africana*) in Kenya and Uganda. Symp. Zool. Soc. Lond. 21: 251–269.

Southwood, T. R. E. 1977. Habitat, the templet for ecological strategies. J. Anim. Ecol. 46: 337–365.

Southwood, T. R. E. 1988. Tactics, strategies and templets. Oikos 52: 3–18.

Spotila, J. R., E. A. Standora, D. P. Easton and P. S. Rutledge. 1989. Bioenergetics, behavior and resource partitioning in stressed habitats: biophysical and molecular approaches. Physiol. Zool. 62: 253–285.

Taylor, L. R. 1988. Objective and experiment in long-term research, pp. 20–70. In: G. E. Likens (ed) *Long-term Studies in Ecology*. Springer-Verlag, New York.

Templeton, A. R. 1986. Coadaptation and outbreeding depression, pp. 105–116. In: M. E. Soulé (ed) *Conservation Biology: The Science of Scarcity and Diversity*. Sinauer, Sunderland, MA.

Thoday, J. M. 1961. The location of polygenes. Nature 191: 368–370.

Travis, J. and L. D. Mueller. 1989. Blending ecology and genetics: progress towards a unified population biology, pp. 101–124. In: J. Roughgarden, R. M. May and S. A. Levin (eds) *Perspectives in Ecological Theory*. Princeton University Press, Princeton, NJ.

Valentine, J. W. 1976. Genetic strategies of adaptation, pp. 78–94. In: F. J. Ayala (ed) *Molecular Evolution*. Sinauer, Sunderland, MA.

Wallace, B. 1968. *Topics in Population Genetics*. Norton, New York.

Wallace, B. 1975. Hard and soft selection revisited. Evolution 29: 465–473.

Wilcox, B. A. 1980. Insular ecology and conservation, pp. 95–117. In: M. E. Soulé and B. A. Wilcox (eds) *Conservation Biology: An Evolutionary – Ecological Perspective*. Sinauer, Sunderland, MA.

Williamson, M. H. 1981. *Island Populations*. Oxford University Press, Oxford.

Wilson, E. O. 1975. *Sociobiology. The New Synthesis*. Belknap Press of Harvard University Press, Cambridge, MA.

Wilson, E. O. 1988. The current study of biological diversity, pp. 3–18. In: E. O. Wilson (ed) *Biodiversity*. National Academy Press, Washington, DC.

7 Population persistence and the natural relationships between body size and abundance

Sean Nee, Paul H. Harvey and **Peter Cotgreave**
A.F.R.C. Unit of Ecology and Behaviour,
Department of Zoology,
University of Oxford,
South Parks Road,
Oxford OX1 3PS,
UK

ABSTRACT

Conservation and management decisions should be based on more than guesswork. It is always preferable to understand **why** natural communities are structured in the way that they are, and to base decisions on a proper scientific understanding of the various factors influencing species abundances. When such understanding does not exist, and that is usually the case, decisions must be underpinned instead by our knowledge of **how** communities are structured. In fact such knowledge, based on the description of observed patterns within and between natural communities, can sometimes be used for hypothesis generation and to test among alternatives. In this chapter, we describe several perspectives on the description of the natural relationships that exist between body size and abundance. The fact that relationships **do** exist, some of them surprising, supports the hope that gross biological features such as body size may provide useful information for the evaluation of species minimum viable population sizes and other statistics of interest to conservationists. Only conservation managers themselves can decide what use to make of the information ecologists extract about the structure of natural communities.

INTRODUCTION

Although conservation management decisions are ideally based on a complete understanding of ecological processes, pragmatism demands that we exploit observed patterns even if we currently lack a good understanding of those patterns. Our general approach is motivated by the fact that environmental change and conservation policies do not only affect the handful of species in which we may be interested for particular reasons, but affect thousands of species simultaneously. This has prompted a broad-brush approach to questions about minimum viable population size in a search for general statements relating gross features of species, such as body size, to their requirements for long-term persistence. It

Conservation of Biodiversity for Sustainable Development.
Edited by O. T. Sandlund, K. Hindar and A. H. D. Brown. 1992. Scandinavian University Press, Oslo.

has been suggested, for example, that organisms with smaller body sizes must maintain greater abundances than bigger organisms or they may be more likely to go extinct (Belovsky 1987; Pimm et al. 1988; Lawton 1989).

If this is true, then we should expect to observe a dearth of species in the wild which are small bodied and of low abundance. So, are small-bodied/low-abundance species uncommon? There is no simple answer to this question because of a suite of problems arising from the interpretation of imperfect data. In this contribution, we discuss these problems and consider two new lines of evidence, one inferential and one empirical, that pertain to the question and suggest the answer is "yes".

A broad-brush approach to conservation biology can only be successful if we have a good understanding of the natural relationships that exist between gross features of species and their abundances. This understanding is a long way off. We describe two novel approaches to the exploration of the relationship between body size and abundance which have revealed new patterns that have yet to be fully understood.

DAMUTH'S HOLES

Over a large range of body sizes, species abundance typically declines quite steeply with increasing body size according to a power relationship of the form ABUNDANCE = K(BODY SIZE)^{P}, where P is a negative number. For example, Figure 1 shows the relationship, plotted on logarithmic coordinates, for Damuth's (1987) compilation of organisms ranging from plankton to elephants. The slope of this relationship is approximately –0.75 and Damuth interpreted this result as being consistent with the energetic equivalence hypothesis. This hypothesis claims that roughly equal amounts of energy flow through species of different weights. Since metabolic rate increases with body weight to the 0.75 power, the hypothesis requires abundance to decrease with the –0.75 power of body weight.

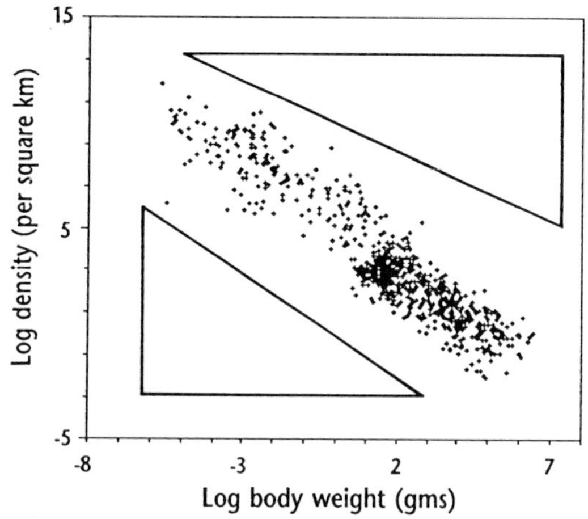

Figure 1. Relationship between population density and body weight among organisms ranging in size from plankton to elephants (after Damuth 1987). The reasons for no species occurring in the triangles at the top right and bottom left (Damuth's top and bottom holes) of the figure are discussed in the text. If all organisms are cubes, have a specific gravity of one, and do not live on top of each other, i.e. they live in a two–dimensional world, then no species can occupy the space above the hypotenuse of the upper triangle.

But the data can be interpreted in other ways. Imagine that the data points were initially uniformly scattered throughout the square box in Figure 1. Now we need to explain Damuth's triangular holes. Damuth's top hole has two types of explanation. The first is one of physical packing. Only so many organisms can be packed into a given space. And the same amount of space can hold fewer large than small ones. Most large organisms, such as terrestrial mammals, live in a two dimensional world. That is, while they live side by side, they do not actually live on top of each other. If organisms had the specific gravity of water and were all the same shape (e.g. cubes), no population density could be high enough to enter the top triangle on Figure 1. If we are considering terrestrial organisms, if one species did reach the lower limit of the triangle, no other species could coexist with it. In fact, many smaller organisms such as plankton actually live in three dimensions, and this may be part of the reason why smaller organisms approach closer to the top hole than larger organisms.

The second explanation for Damuth's top hole is one of resource limitation. Larger organisms use more resources per capita per unit time, and so a given amount of resource can support more small than large organisms. Inasmuch as population density is the reciprocal of territory size, Damuth's top hole has received considerable attention, and it has been found that larger organisms and those which live on scarcer resources (such as predators and carnivores) occupy larger territories (e.g. McNab 1963; Schoener 1968; and many studies since reviewed by Harvey and Pagel 1991).

So Damuth's top hole is not problematic. His bottom hole is. In fact, it may be just an artefact and contain large numbers of species. Damuth used ecological density as a measure of abundance, which is the density recorded in the literature from places where the species was actually censused, as opposed to population size in some arbitrary, large area such as Norway. But this procedure of culling data from the literature may misrepresent the distribution of population densities in natural communities. In particular, it may overestimate density, as naturalists are probably only going to have censused species in areas where those species are thought to commonly occur. This is less of a problem for animals such as elephants, which are hard to miss, than it is for animals such as *Daphnia*, which would be very easy to miss if they were not abundant. Furthermore, very small-bodied species, such as diatom species, that exist at low abundance everywhere will be completely unknown to us; after all, how would we discover them? Who would record a density of one diatom per square kilometre? So Damuth's bottom hole may be an observational bias whose effects decrease as the species get larger. In the next section we will present an argument that suggests that, for diatoms at least, Damuth's bottom hole is real, i.e. all diatom species exist at very large abundances.

Strangely, studies of the abundances of species from single taxa, such as birds, which span a narrower range of body sizes have revealed either a much shallower relationship between abundance and body size, or a **triangular** relationship (Brown and Maurer 1987; Morse et al. 1988). Furthermore, the triangular relationships observed may be sampling artefacts and are consistent with there being no actual underlying relationship between abundance and body weight (Blackburn et al. 1990). So how are we to reconcile Damuth's relationship with these narrow-range non-relationships? It may be that within a single taxon spanning a narrow size range there **is** no relationship between body size and abundance. Or, on the other hand, it may be that poor quality data cannot reveal a steep negative relationship between abundance and body size over a narrow range of sizes. In the third section, we will look at a new data set of unusually high quality, British birds, which, in fact, reproduces in miniature many of the important features of Damuth's relationship.

We will now present an argument suggesting that, indeed, very small-bodied species do not occur at low abundances.

PRESTON'S OBSERVATIONS

For some years, species abundance data were represented as in Figure 2, plotting the number of species in a sample represented by 1, 2, 3 etc. individuals (e.g. Williams 1964). These plots are typically L-shaped, with the number in each abundance class declining precipitously from the class of minimum abundance, which is one. Then Preston (1948) had the innovative idea of plotting **frequency histograms** of the numbers of species in **logarithmic** abundance classes. In many cases, this transforms the L-shape into a humped distribution, truncated on the left at what Preston called the "veil line" (see Figure 3). He suggested that enlarging the sample would shift the veil line to the left and reveal more of the underlying distribution, leaving unchanged the shape of the distribution already revealed. There is some evidence for this (e.g. Williams 1964).

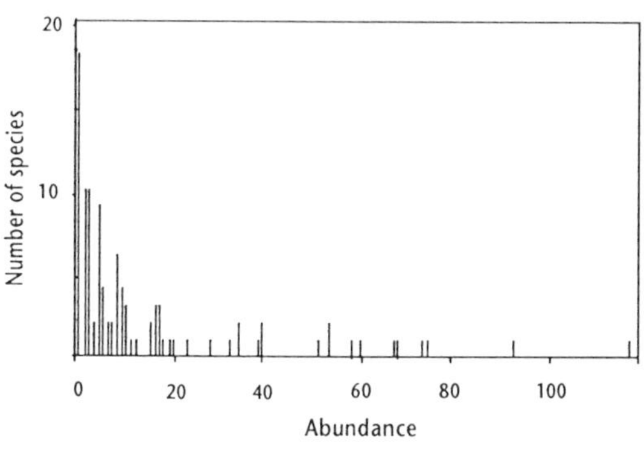

Figure 2. The number of diatom species represented by one individual, two individuals, three individuals, and so on, in a sample from a natural community. Data from Patrick (1968: p. 176, Box 8). For clarity, we have not included those species represented by more than 120 individuals, which would produce a long tail extending to 1,446.

Preston called the humped distribution "lognormal". This is not a precise statistical statement: it simply suggests that if we could see beyond the veil line the entire community would display a roughly symmetrical, bell-shaped distribution (Figure 3). From the few data sets available to him at the time, he generalised certain numerical relationships among the various parameters of the distribution and called distributions exhibiting these relationships "canonical lognormal". These relationships can be expressed in a variety of ways (see May 1975), but the simplest statement of the canonical hypothesis, and the one most relevant to what follows, is given by Preston (1962: p. 168): if there are more than 100 species in the community, then the standard deviation of abundances (measured on a $\log_2$ scale) is roughly 4. Data compiled by Sugihara (1980) suggests that the standard deviations are, indeed, roughly 4 (Figure 4).

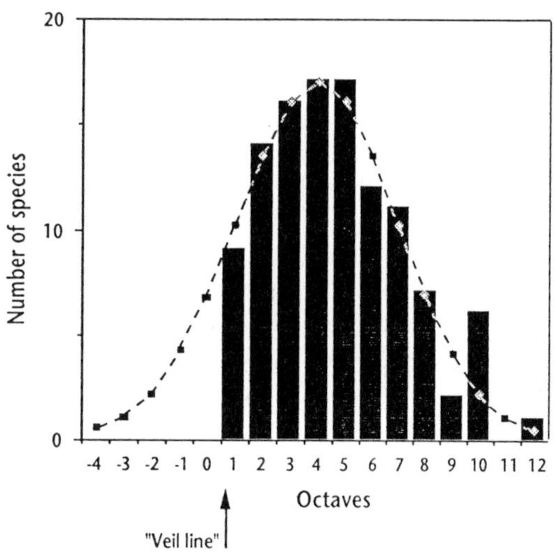

Figure 3. The same data as Figure 2 represented using Preston's (1948) technique. An octave represents abundances between 2^n and 2^{n+1}. The frequency histogram is constructed according to the convention that species with exactly 2^n individuals are evenly divided between the two adjacent octaves. The numbering of the octaves is arbitrary. The curve is a fitted lognormal, extrapolated beyond the veil line to describe those species in the community represented by less than one (i.e. zero) individuals in the sample.

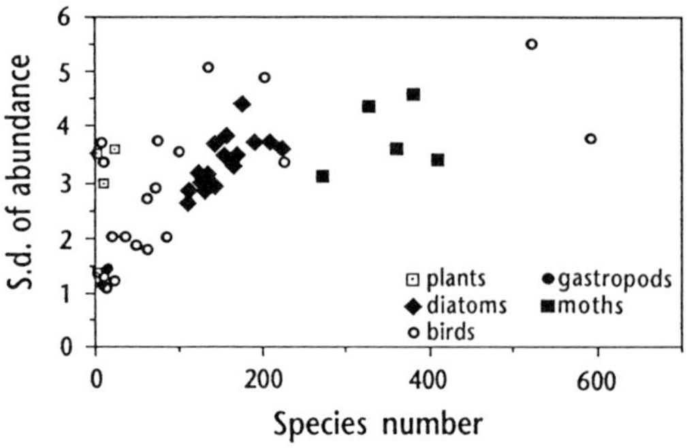

Figure 4. Relating the standard deviation in abundance to the number of species in the community. Redrawn from Sugihara (1980). Except for very small communities, the species number and standard deviation (s.d.) are actually inferred from the sample data by a variety of techniques for fitting lognormal parameters.

As May (1975) suggested, it is not surprising, for macroscopic organisms such as birds at least, that the standard deviation is roughly 4, because of the relationship between the standard deviation and **range** of humped distributions. We can see this as follows (the subject is discussed more fully in Nee et al. in press).

Suppose that the standard deviation (s.d.) of a lognormally distributed community was, in fact, 7.5. As 99 % of the species are expected to occur ± 3 s.d. of the mean, we may take 6 s.d. (= 45) as the range of abundances. Since we are on a $\log_2$ scale, this means the abundances will vary over a range of 2^{45} (= $3.5*10^{13}$). If the rarest species in the community has 1 individual, then the commonest species has $3.5*10^{13}$ = 35,000,000,000,000 individuals. This is certainly an absurd figure for species such as birds.

More plausible ranges are generated by s.d.'s in the range of 2.5 to 5.5. For example, if the s.d. = 3, then there is a range of 2^{18} in abundance. So if the rarest species has 1 individual, the most abundant has 260,000 individuals. If the s.d. = 5.5, then there is a range of 2^{33} in abundance. So if the rarest species has 1 individual, the most abundant has 8,500,000,000 individuals.

These arguments are neither sensitive to the precise shape of the distribution (it suffices for it to be humped and vaguely symmetrical), nor to the choice of a range corresponding to 99 % of the area under the normal density curve. One way to convince oneself of this is to compute the standard deviations of humped distributions from fields far removed from community ecology. For example, the s.d. of the populations of the nations of the world, measured on a $\log_2$ scale, is 4.3. The s.d. of the populations of the capital cities is 3.2. Such points would not look out of place in Figure 4.

But what prevents the s.d.'s being even smaller? Why is the most abundant species not just twice as common as the least abundant species? Again, it is obvious why not in communities of hundreds of bird species: basic biology, such as differences in trophic levels occupied, and niche requirements produce substantial abundance differences between, for example, eagles and sparrows.

Interestingly, diatom communities also have standard deviations of abundance of roughly 4 (Figure 4). As we will now see, this is another manifestation of Damuth's bottom hole. Hutchinson (1967: p. 373) noted that it is very difficult for diatoms not to be extremely abundant. A species that exists in a 1 km^2 lake at the undetectable density of 1 individual per litre will have a total population size of 10^9 even if it is confined to a layer 1 metre thick. But if the minimum diatom population size in a community really is only 1, as we supposed for birds, then an abundance s.d. of 7.5 is not as impossible for a diatom community as it would be for birds since the most abundant diatom species can be very abundant indeed.

But what do we mean by minimum population size? Diatom population sizes change enormously from the spring bloom to the winter collapse. If we are able to sample the community at all, we are doing so at a time when at least some species are highly abundant, so it is the minimum abundance at this time that we are interested in (see Figure 5). Figure 4 shows that the diatom community s.d., like everything else, is about 4. This implies that if the abundance of the most abundant species is some colossal number, which it certainly is, then the abundance of the least abundant species must also be quite enormous, i.e. the community looks like C in Figure 5. **Hence, Damuth's bottom hole manifests itself once again.**

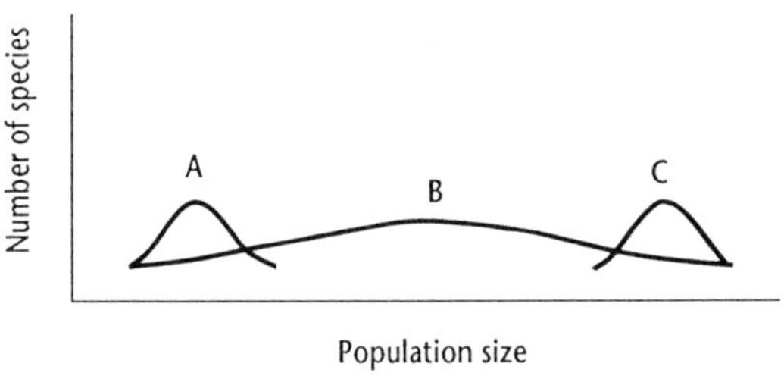

Figure 5. Two ways in which a community containing few individuals (A) may expand. In case B, the least abundant species do not change much in population size, so the range of sizes increases. In case C, the range does not change, so the least abundant species becomes very abundant. We are able to sample the community when it is in either states B or C.

We can now adopt one of two attitudes. First, we may decide to reject Preston's theory that our sample data are from a veiled lognormal distribution and suppose, instead, that there is some massive non-normality lurking beyond the veil line which gives the community a much larger range of abundances than we infer. This idea is illustrated in Figure 6. Or we may suppose that the abundance of the least abundant species actually **is** enormous.

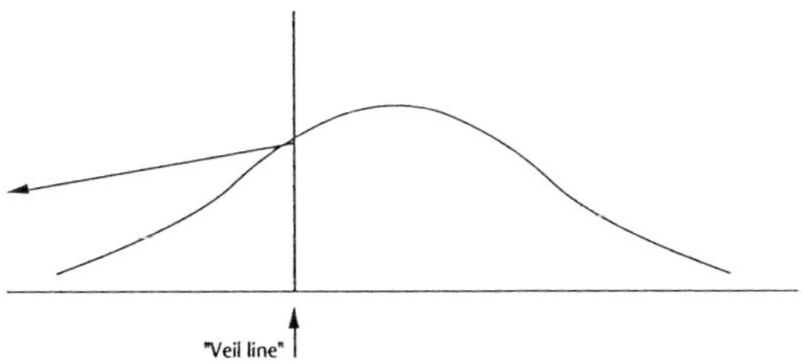

Figure 6. We must infer the shape of the community abundance pattern beyond the veil line – see Figure 3 for axis information. If the distribution is normal for diatom communities, the least abundant species must be very abundant. If there is a large departure from normality, then the range of abundances in the community can be much larger.

BRITISH BIRDS

The abundance estimates of British birds are of unusually high quality because of the strength and popularity of ornithology in Britain. The abundances of rare birds are particularly well known because of their attractiveness to ornithologists. As a result, this data set can illuminate many of the issues we have come up against.

First of all, as Figure 7 shows, the distribution of abundances in British birds, a completely unveiled community, is indeed lognormal. The distribution exhibits the slight left-skewness that was suspected by MacArthur to lie beyond the veil line and is a prediction of Sugihara's (1980) sequential stick breaking model of abundances (Nee et al. 1991). This skewness is insufficient to substantially inflate the standard deviation of abundances or compromise the argument relating abundance and range which we used above. So there was no substantial non-normality lurking beyond the veil line in this community at least.

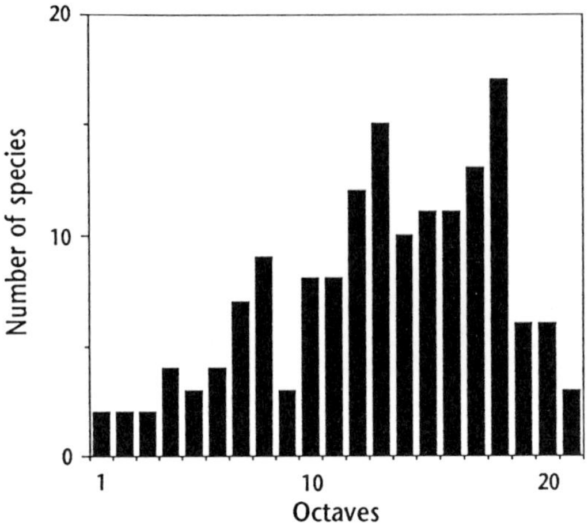

Figure 7. The distribution of abundances of British inland breeding birds measured on a log_2 scale. The measure of abundance is the number of breeding pairs published by the British Trust for Ornithology (Marchant et al. 1990) for 146 species of British inland breeding birds. The BTO lists 228 species in total, and we have excluded occasional breeders, seabirds, species which have recently colonised Britain (such as the collared dove, Streptopelia decaocto*), escapees and birds which are stocked (such as the red-legged partridge,* Alectoris rufa). *The standard deviation and skewness are 4.7 and –0.4.*

Figure 8 illustrates the relationship between abundance and body size. The slope of the relationship is –0.75, in remarkable agreement with Damuth's slope and the energetic hypothesis, and there is a dearth of species that are both small bodied and of low abundance. This suggests that Damuth's bottom hole is neither an artefact of the use of ecological density as a measure of abundance (Figure 8 uses actual population size in Britain) nor a phenomenon that only exists over very large body size ranges. Although there is no significant relationship between body size and abundance within the two groups distinguished in Figure 8, passerines and non-passerines, which occupy the lower and upper half of the body size range respectively, there is still a dearth of small rare species **within** each group.

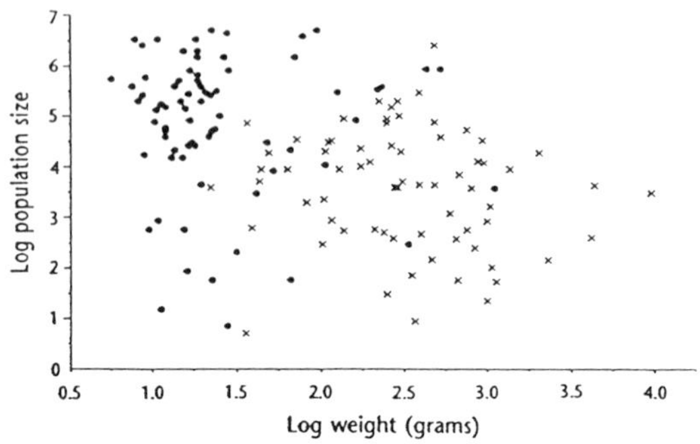

Figure 8. The relationship between population size and body weight in British inland breeding birds. ● = *passerines, x = all other orders.*

Instead of concluding with a synthetic resolution of the puzzles concerning the relationship between body size and abundance, we will now describe two new approaches to the analysis of the relationship between abundance and body size, one phylogenetic, the other comparative, which have revealed new patterns. These new perspectives cast a very different light on the relationship between body size and abundance.

THE EVOLUTIONARY ECOLOGY OF BRITISH BIRDS

We looked at the relationship between size and abundance within tribes in British birds. We followed the taxonomy of Sibley et al. (1988). This is based on DNA hybridisation data and we infer phylogeny from taxonomy. Surprisingly, within 18 of the 25 British tribes there is a **positive** relationship between abundance and body size, more than expected by chance alone (binomial p = 0.043), and the *z*-transformed Spearman rank correlation coefficients are more positive than expected by chance alone ($t_{[24]}$ = 3.05, p = 0.006). Thus, across closely related species, increases in body size are associated with **increased** population size, in contrast to the negative association across all species.

How distantly related must species be before negative relationships are detected more often than expected by chance alone? Relationships across and within taxa can be analysed by considering correlations across species within genera, across generic means (calculated from constituent log-transformed species points) within tribes, across tribal means (calculated from constituent generic means) within subfamilies and so on. Overall there is no consistent within-taxon association between abundance and body size (39 positive and 39 negative associations), but the relationships across higher taxa are opposite to those found within lower taxa. Across species within genera, and generic means within tribes, there are almost twice as many positive as negative associations between size and abundance (binomial p = 0.072). Across constituent taxa means within middle-ranked taxa (subfamilies, families and superfamilies), there are six positive and eleven negative associations, as expected by chance alone (binomial p = 0.33). Across taxa within higher taxa (parvorder and above) there are twelve negative associations and four positive associations (binomial p = 0.077). The pattern within higher taxa is significantly different to that within lower taxa ($X^2_{[1]}$ = 7.40, p = 0.007).

In summary, there are positive associations between body weight and population size across closely related species and taxa, but when more distantly related taxa are compared, those characterised by larger bodied species have lower population densities. In this data set, then, the negative relationship between body size and abundance is a higher-taxon phenomenon.

As we will now describe, phylogeny is implicated in the relationship between body size and abundance in a very different way which may be relevant to understanding the positive relationship that exists at low taxonomic levels. We examined whether or not the degree of relatedness of each tribe to other tribes in Britain has any effect on the relationship between body size and abundance within the tribe. Taxonomic levels were numbered 1 to 10 from the subfamily to the infraclass level. A tribe was denoted as joining the phylogenetic tree of British birds at a particular level on the criterion that it joins a taxon already containing more than half as many species as the original tribe. So, for example, the woodpecker tribe joins the tree at level 10 as it has no British sister taxon below the infraclass level, whereas the two warbler tribes join the tree at levels 2 and 3, respectively. Tribes with a positive relationship between abundance and body size are more likely to join the tree at higher taxonomic levels (Mann-Whitney U = 30.5, p = 0.045). That is, the more unrelated a tribe is to other tribes in Britain, the more likely it is to show a positive relationship between size and abundance. In fact, the six tribes which join the phylogenetic tree above the order level all have positive relationships.

It may often be the case that larger bodied species have an advantage in interspecific competition and that the effects of such competition on resource access, either now or **in the evolutionary past**, adversely affect the abundance of the smaller-bodied species. If this is the case, then we would be more likely to observe a positive relationship between abundance and body size within tribes that encompass entire guilds. Phylogenetically distinct tribes, such as woodpeckers, owls and pigeons, probably include all the species playing a distinct ecological role, whereas tribes with many close relatives, such as the warbler tribes, may not. Thus, insofar as phylogenetic divergence represents ecological divergence, tribes which are more likely to represent guilds are more likely to have a positive relationship between body size and abundance.

This analysis suggests that phylogenetic relatedness will eventually allow deeper insight into patterns of species abundance, for British birds at least. Whether phylogenetic relatedness stands in for some ecological similarity remains to be seen. Closely related species will share many characteristics and adaptations in addition to those considered by any single analysis. And they are more likely to compete for limiting resources than are distantly related species.

COMPARATIVE ECOLOGY

While it is of scientific interest that particular structures occur within the composite community of British birds, there is no guarantee that those patterns will be replicated in different countries. Furthermore, conservation and management decisions have often to be focused on smaller scales to encompass specific habitats. Both of these considerations should lead to the search for repeated patterns among local communities. That is where the comparative approach becomes important. To what extent can the findings from one community in a specific habitat be extrapolated to other communities in similar habitats? And what patterns can be replicated in different types of communities? Can a land manager working in Norway base some of his decisions on the experiences of Scottish Estate workers?

To answer such questions, one of us (P. Cotgreave, in preparation) has compiled species abundance relationships **within** many bird communities to see if the distributions describing the form of those relationships differ **among** communities in ways that correlate with biological differences between the communities. For example, do species abundance curves vary among habitat types or as a function of average bird body size in the community? The results we shall describe are preliminary but they reveal interesting patterns that require explanation.

Ninety one bird community counts were extracted from the literature on the basis of meeting the following criteria: each comprises a complete census of a relatively clearly defined area containing at least fifteen species. As a way of summarising species abundance relationships, species were ranked from most to least abundant on the X axis, and logarithmically transformed abundance values were plotted on the Y axis. By definition, the points describe a monotonically decreasing functional relationship. If all bird species had similar abundances the slope of the relationship would be shallow but negative. If the different species had widely varying abundances, the slope would be negative but steep. Most relationships were fairly linear and the slopes of best fit regression lines were used as measures of equability (we use equability which means "evenness" instead of the more usual equitability which means "reasonableness"). Across the 91 communities, slopes vary from just below zero to about minus 0.73. The latter values describe bird communities with low equability (different species have very different population densities).

One of the most obvious sources of variation among organisms from any taxonomic group is body size. Bird species from different communities tend to have different sizes: some communities consist of larger-bodied birds than others. It turns out that communities in which the birds have larger average body sizes have low equability.

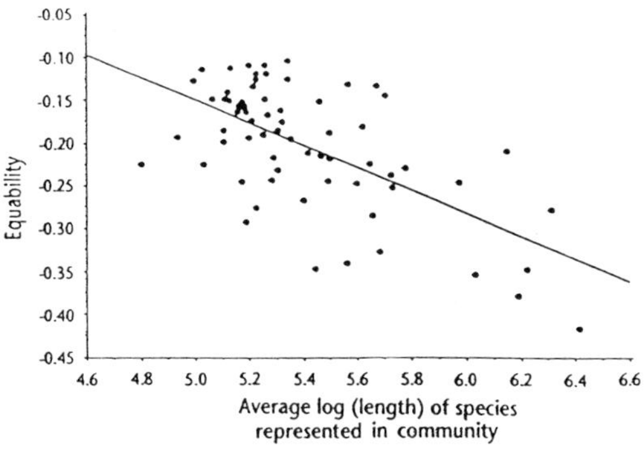

Figure 9. Bird communities in which species tend to be larger have less equable species abundances. Each point represents a different bird community. The X axis is the average logarithmically transformed length of the different species represented in the community. The Y axis is the slope of the graph in which species logarithmically transformed abundances are plotted against rank abundance. Only the data for the 15 most abundant species in each community are used for this analysis to remove potentially confounding effects of the number of species in the community.

Could this size related equability pattern be a biologically uninteresting statistical artefact of some sort? We do not think so. The phenomenon is remarkably robust:

Number of species in sample

The pattern holds for the fifteen most common bird species from the different communities (thus controlling for the number of species in the sample).

Body size per se

The pattern holds across the same communities for subsamples of middle-sized birds: when only the commonest five species between 250 mm and 450 mm in length are used for analysis, those drawn from communities of generally big birds have less equable abundance distributions than birds of the same size which come from assemblages of birds which are on average small.

It seems, then, that the average body size of birds in a community is a surrogate for some other difference causing differences in equability.

Sample area

It might be possible that large species, say top predators, could only maintain viable populations in communities that cover large areas. Two lines of evidence suggest that the pattern is not dependent on the area of community sampled. First, variation in the size of the birds in the assemblage does not correlate with equability. Second, communities containing more feeding guilds do not have less equable distributions.

Coloniality

Preston (1962) suggested that in small areas, colonial versus territorial nesting might influence species abundance relationships. He is correct: the larger the proportion of species in the assemblage which is colonial, the less equable is the species abundance curve (Figure 10).

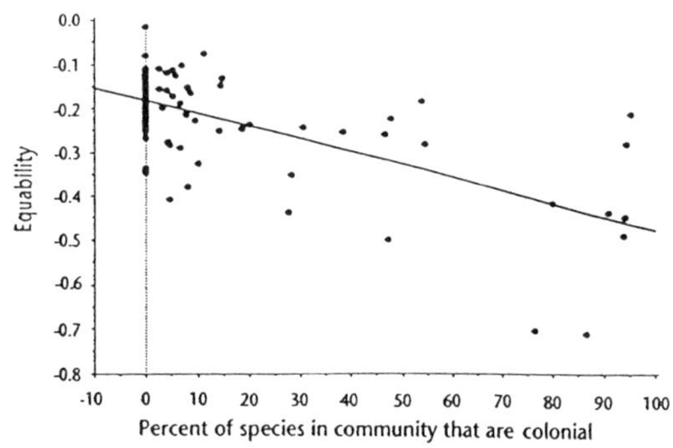

Figure 10. Bird communities containing a higher proportion of colonial species have less equable species abundances. The same data as Figure 9, but here the X axis denotes the percentage of bird species in the community which is colonial.

Latitude

Temperate and polar communities are less equable than tropical ones (Figure 11), although there is no relationship between the average size of species and latitude – this is therefore a second relationship, which in fact turns out to be independent of but not as strong as the body size one.

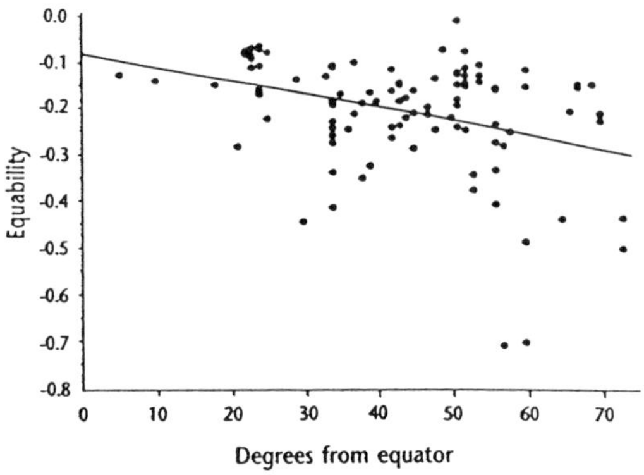

Figure 11. Tropical bird communities have more equable species abundances. The data from Figures 9 and 10 with the same Y axis, but with each community plotted according to distance from the equator. The average body size of species in the communities (as in Figure 9) is not significantly correlated with distance from the equator.

Habitat

There are differences in equability for communities from different habitat types, even when latitude is taken into account (Figure 12). The pattern revealed here may well be related to habitat complexity or number of niches available within the habitat. We might argue that more complex habitats (like forests, woodlands, parks and gardens) provide more niches for specialist species, none of which can be relatively common. In contrast, simple communities (like small islands which provide nesting places for marine birds) can provide one or two major niches, with the rest of the species being relegated to rarity, hence communities with low equability. For example, small rocky islands provide nesting habitats for a few species of very common colonial seabirds, but a few small passerines and other birds manage to maintain relatively small populations, feeding in the terrestrial habitat itself.

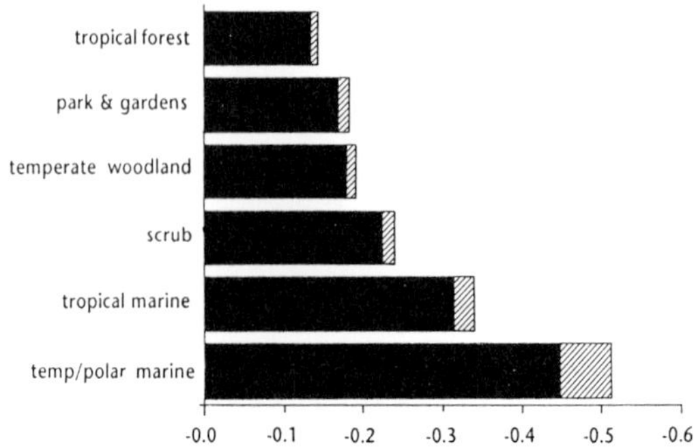

Figure 12. Equability of bird communities from different habitat types. The same data are used as in Figures 9 to 11. Hatched bars denote a single standard error above the mean.

CONCLUSION

Conservation is not only about saving individual species; it's about saving entire communities. But until we know and understand the rules governing community structure, any conservation action will be carried out on an *ad hoc* basis. Ecology has had a tradition of attempting to understand how communities are structured through the analysis of species abundance relationships. It is now that more, and more accurate, data sets are available and our knowledge is such that we can examine it from new perspectives, that a new and more realistic range of hypotheses can be entertained and tested.

Conservation policy would benefit from an understanding of the relationship between body size and minimum viable population sizes. As we have seen in this chapter, we are still in the process of describing the relationships between body size and abundance that now exist in Nature. The fact that our understanding is not sufficiently advanced to assist conservation managers in their day-to-day decisions must not be allowed to disguise the fundamental, long-term importance of pure research in ecology. Ultimately, political and policy discussion must be informed by a knowledge and understanding of natural patterns. So if this research is still in a very early stage, this should encourage a greater investment of effort, not disregard.

LITERATURE CITED

Belovsky, G. E. 1987. Extinction models and mammalian persistence, pp. 35–58. In: M. Soulé (ed) *Viable Populations for Conservation.* Cambridge University Press, Cambridge.

Blackburn, T. M., P. H. Harvey and M. Pagel. 1990. Species number, population density and body size relationships in natural communities. J. Anim. Ecol. 59: 335–345.

Brown, J. H. and B. A. Maurer. 1987. Evolution of species assemblages: effects of energetic constraints and species dynamics on the diversification of the North American avifauna. Amer. Nat. 130: 1–17.

Damuth, J. 1987. Interspecific allometry of population density in mammals and other animals: the independence of body mass and population energy use. Biol. J. Linn. Soc. 31: 193–246.

Harvey, P. H. and M. Pagel. 1991. *The Comparative Method in Evolutionary Biology.* Oxford University Press, Oxford.

Hutchinson, G. E. 1967. *A Treatise on Limnology.* Vol. 2. Wiley, New York.

Lawton, J. H. 1989. What is the relationship between population density and body size in animals. Oikos 55: 429–434.

Marchant, J. H., R. Hudson, S. P. Carter and P. Whittington. 1990. *Population Trends in British Breeding Birds.* British Trust for Ornithology, Tring, Hertfordshire.

May, R. M. 1975. Patterns of species abundance and diversity, pp. 81–120. In: M. Cody and J. M. Diamond (eds) *Ecology and Evolution of Communities.* Harvard University Press, Cambridge, MA.

McNab, B. K. 1963. Bioenergetics and the determination of home range size. Amer. Nat. 97: 133–140.

Morse, D. R., N. E. Stork and J. H. Lawton. 1988. Species number, species abundance and body length relationships of arboreal beetles in Bornean lowland rainforest trees. Ecol. Entomol. 13: 25–37.

Nee, S., P. H. Harvey and P. Cotgreave. (in press). Species abundances. Proceedings of the Oji International Ecology Congress.

Nee, S., P. H. Harvey and R. M. May. 1991. Lifting the veil on abundance patterns. Proc. R. Soc. Lond. B 243: 161–163.

Patrick, R. 1968. The structure of diatom communities in similar ecological conditions. Amer. Nat. 102: 173–183.

Pimm, S. L., H. L. Jones and J. M. Diamond. 1988. On the risk of extinction. Amer. Nat. 132: 757–785.

Preston, F. W. 1948. The commonness, and rarity, of species. Ecology 29: 254–283.

Preston, F. W. 1962. The canonical distribution of commonness and rarity. Ecology 43: 185–215, 410–432.

Schoener, T. W. 1968. Sizes of feeding territories among birds. Ecology 49: 123–141.

Sibley, C. G., J. E. Ahlquist and B. L. Munroe. 1988. A classification of the living birds of the world based on DNA-DNA hybridisation studies. Auk 105: 409–423.

Sugihara, G. 1980. Minimal community structure: an explanation of species abundance patterns. Amer. Nat. 116: 770–787.

Williams, C. B. 1964. *Patterns in the Balance of Nature*. Academic Press, New York.

8 Models for predicting ecological change

Nils Chr. Stenseth
Division of Zoology, Department of Biology,
University of Oslo,
P.O. Box 1050, Blindern,
N-0316 Oslo 3,
Norway

and

Norwegian Institute for Nature Research (NINA),
Tungasletta 2,
N-7005 Trondheim,
Norway

ABSTRACT

We need a pluralistic approach to modelling if we are to understand the environmental changes occurring today. In this paper I use models to address two main issues of the interaction between humans and their environment: 1) Ecology – economy interaction, and 2) landscape ecology. The first relates to temporal dynamics whereas the second relates to temporal as well as spatial dynamics of ecological processes.

Models developed for improving our understanding of economy – ecology interaction should contain a) interactions between the participating actors, b) consideration of future versus present rewards, c) dynamic processes within and between the various components of the system, and d) constraints such as institutional designs and incentive structures. The "tragedy of the **un**managed commons" is in this connection discussed both with and without a time dimension.

Theoretical models in landscape ecology should be combined with observational studies and experiments on empirical model systems. Here I show how an experimental set-up using voles as experimental organisms can be used for obtaining an understanding of ecological processes such as space use under various degrees of fragmentation.

INTRODUCTION

Habitats and landscapes have always been undergoing changes due to the influence of biotic as well as abiotic factors; extensive pest outbreaks and landslides are such examples. Similarly, species have always gone extinct; the majority of species that have existed on this earth have long ago gone

Conservation of Biodiversity for Sustainable Development.
Edited by O. T. Sandlund, K. Hindar and A. H. D. Brown. 1992. Scandinavian University Press, Oslo.

extinct. However, today the rate of such environmental changes and species extinction has accelerated (e.g. Fisher et al. 1969) due to the direct or indirect human influence – influences which continuously are becoming stronger.

Scientists as well as politicians and laymen agree about one thing: Human destruction of nature needs to be minimized and indeed reduced to a far lower level than it is today. In order to achieve such a goal, we need to understand the causes and the dynamics of the ecological changes taking place around us. That is, we need to understand the dynamic interaction between social, economic and ecological processes. Each of these processes is rather complicated. Equally complicated are also the interactions between them. We therefore need to develop a variety of models – both theoretical (i.e. mathematical and conceptual) and empirical (e.g. Stenseth 1985a; Ims and Stenseth 1989).

Besides giving some general perspectives on the issue of the human-ecosystem interaction, I will discuss what characterizes useful models – i.e. models which may help us understand and moderate destructive ecological changes. I will detail on one particular mathematical model and an experimental set-up, an experimental model system (EMS; sensu Ims and Stenseth 1989) using small rodents as experimental organisms, for improving our understanding of landscape ecological processes and changes. This latter section focuses on the ecological effects of habitat fragmentation.

Throughout, my approach will be a reductionistic one (e.g. Łomnicki 1988). That is, I approach the building of an ecological theory from processes operating at the level of the individual organism. This may seem an ambitious and difficult goal to realize, partly because individual behaviour must be seen in a very broad ecological context, incorporating the entire food web and a landscape ecological context. Properties and features at higher levels of organization are then seen as emerging properties of processes at the level of the individual. I do so because, as far as I can see, very little progress has been achieved through a more holistic ecosystem approach (sensu Odum 1969).

We must remember that in all scientific enterprise, we have to build our understanding and our models of the reality on things we reasonably understand. Building ecosystem models of the kind that was done in, for instance, the International Biological Programme (IBP) (Watt 1966; Swartzman and Van Dyne 1972) did not follow this precept. Although much was learnt from that programme, particularly about how to model ecosystem processes, we did so very inefficiently. Specifically we learnt that a more reductionistic approach is a far more efficient one than a more holistic one. However, we also learnt to see the individual organisms in an ecosystem context.

Fairly often those taking a reductionistic approach to ecological dynamics, forget that the environment in which individuals live, is itself a dynamic system (Stenseth 1986). Many of the single species experiments are examples of this fallacy (Gause 1934; Gause et al. 1936; but see, e.g. Huffaker 1958; Luckinbill 1974; Kareiva 1987), particularly when results are extrapolated to natural situations.

On this basis, I will illustrate some of the ecological effects of humans' use of their environment. The first major part is rather general and presents perspectives needed for protecting biodiversity. The other major part of this chapter is much more specific and focuses on empirical and theoretical modelling in landscape ecology.

ECOLOGY – ECONOMY INTERACTION

Preamble

Today economic forces control human exploitation of natural resources (both living and non-living). Hence, in order to understand this exploitation, we must at least understand the economic – ecological interaction. Typically, this is an interaction between processes occurring at rather different rates (or time scales): most of the economic processes are much faster than most of the ecological processes. This sort of dynamical interaction between fast and slow processes resembles the interaction between ecology and evolution (e.g. the Red Queen hypothesis in evolutionary biology; Stenseth and Maynard Smith 1984; Stenseth 1985c). Rather complicated and unexpected dynamic behaviour may emerge from such interactions (Stenseth and Maynard Smith 1984).

One contrast in particular complicates the study of economy – ecology interactions. Ecologists generally focus more on the study of dynamic changes resulting from perturbations around an equilibrium (e.g. Maynard Smith 1974). Economists, however, often aim to find equilibrium conditions rather than study the dynamical changes after some perturbation from an equilibrium (e.g. Arrow et al. 1960; Lancaster 1968).

Exploitation of unmanaged common resources

Common resources are those available to all within a further specified group, if they are available to one member of that group. Pastures in larger parts of the Norwegian mountains – and elsewhere (e.g. Sahel) – are such resources. Other examples are fish in the sea, blueberries in the forest, clean water and clean air.

Pasture herbage, fish and blueberries are living resources having their own pronounced intrinsic dynamics: individuals give birth to new individuals. The water and air have much less of an intrinsic dynamic: specifically, these resources have no birth process. In addition, living resources may actively exploit the resources in their environment – just like human beings. It is important not to forget about these differences when developing models (e.g. McKelvey 1989), and when extrapolating the conclusions from them. Unfortunately economists often talk about economy – ecology interactions in general terms when in fact they are talking about **non**-living resources.

Some common resources are common resources by their nature: one cannot put a fence around a piece of clean air. Other resources are common resources for more practical and economic reasons: it does not pay to put a fence around the relatively unproductive and highly seasonal land pastures in the Norwegian mountains – nor in the arid and seasonal Sahelian savannas. Such resources are characterized by low renewal rates. In this respect, social scientists and economists may have much to learn from the ecological literature on territorial defence (e.g. Davies and Houston 1986; Carpenter 1987). Testing the predictions derived from these territoriality-based models on socio-economic and social anthropological data might indeed be a rewarding enterprise. Many of the resources on this earth are, in fact, common resources.

Part of our understanding of the human – nature interaction follows from an understanding of the tragedy of **un**managed commons (Hardin 1968), a result and an argument which Garett Hardin made well known in the sixties (but see Hardin and Baden 1977). Today the "tragedy of the commons"-argument is much debated both among social scientists as well as among environmental scientists and mathematicians (e.g. Hardin and Baden 1977; Berkes 1989; Berkes et al. 1989). Unfortunately, much misunderstanding prevail in this field.

The tragedy of commons goes like this: Consider, as an example, a herdsman grazing his herd in some common property land. He is free to choose how many animals he wants in his herd. He asks himself: "Should I add more animals to my herd?" If he does so, he will obtain almost all the profit from this extra animal when selling it on the market. This is so, because by letting his animals graze the common pastures, he in effect transfers a piece of common property (the living vegetation) to some private property (the meat of his animal). The cost of letting an additional animal graze on the pasture, will, however, be shared by all herdsmen using the pasture. It follows, as an obvious consequence, that the herdsman finds it economically rewarding and rational at the moment of taking the decision to add one more animal to his herd. By the same logic it also follows that it is rational to add yet another one, a third one, etc. It also follows that all herdsmen will behave in such a way.

To increase the herd this way may have negligible environmental impact if there are not too many herdsmen exploiting the same pasture. However, if there are too many herdsmen, each having too many animals, this economically based reasoning will lead to an ecological catastrophe – the pasture will become overutilized or overgrazed. And that is the tragedy of the commons – the **un**managed commons – which Hardin made familiar to us some 20 years ago. That is, the transfer of common property (the pastures) into private property (the animals) leads to a tragedy if too many individuals try to behave this way.

It follows that the unmanaged common must be turned into a managed one. That is, there need to be rules and restrictions (either enforced by law or by traditions) on each herdsman's use of the common and renewable resource. Up until fairly recently, each locally adapted society very often had its own set of rules (e.g. McEvoy 1986). Now these traditional rules seem to have broken down in several communities, either due to some passive process or due to an active centralizing policy (e.g. Agarwal, this volume).

It also follows that the number of herdsmen must be restricted: Population growth may therefore be a problem when trying to manage the Earth's resources (Ehrlich and Ehrlich 1970; Ehrlich et al. 1973). Many people have said so before me. Unfortunately, too many – particularly politicians – tend to overlook this conclusion.

Restrictions, institutional design and the commons

Through rules and institutional design we may be able to exploit the renewable resources in a sustainable manner (sensu WCED 1987). That is, by such a scheme we may design environmental and developmental strategies so that coming generations have the option of exploiting the same resources in a similar way. Certainly, this could be easier **if** living resources were owned and inherited as a whole by a son or a daughter (as, e.g. forests and agricultural land are inherited in Norway). However, as already pointed out, many common resources cannot be privatized for economic reasons. Notice, however, that the problem may not be solved by privatization alone. Indeed many examples demonstrate that private property land also may be badly mismanaged (e.g. Berge 1990).

Although rules are needed, it is not clear who should set them. If a subset of the group exploiting the common resource sets the rules, the larger group is unlikely to agree to them. If everybody participates in the decision process, the rules are unlikely to be effective. Reasoning along the line of the "tragedy of the commons"-argument leads to such a conclusion (think, e.g. about the "free rider" who benefits from belonging to the group without contributing anything to the welfare of the group; e.g. Taylor 1987). That is, a democratic – laissez faire – society may not be able to manage common property resources in a sustainable manner.

Many of these difficulties rest on poor communication between scientists, politicians, managers and laymen. Modern computer-based techniques (e.g. expert systems, and various psychology-based techniques designed for improving communication and mutual understanding among participants in a group) may be of some use when trying to avoid these difficulties. Unfortunately, such techniques are not used as often as they ought to be (but see Jeffers (1988), for an example of how such techniques can be used in connection with environmental management).

Discounting the future

Sustainable development dictates that we consider future resource levels – and expected rewards – whenever deciding about the exploitation strategy to be used **now** (e.g. Moxnes 1989): we compare present and future rewards. Hence, we need to incorporate temporal dynamics into the "tragedy of the commons"-argument.

To evaluate present fairly **certain** rewards against **expected** (i.e. more or less **uncertain**) future rewards either to myself and/or to my descendants, interest rates and discount rates need to be included in the arguments and the models. From the above argument relating to the "tragedy of the commons", the rational behaviour of a herdsman considering whether or not he should add an animal to his herd, is to do so even if the environment deteriorates as a consequence of such a decision. He does so because **he** himself wants to obtain the reward or profit rather than give it to somebody else. This type of "short-sighted" strategy becomes even more profitable – or economically rational – if the discount rate is high. Then, it may even pay to harvest the last fish or whale and invest the proceeds into a bank-account or some more profitable enterprise (e.g. Clark 1976, 1989). This is especially so if the biological resources being exploited have a low renewal rate (i.e. a long generation time)[1]. Unfortunatcly, several of the common renewable resources on Earth do indeed have a low growth (or renewal) rate (e.g. the Sahelian savannas).

What sort of models are needed?

"All models are wrong but some are useful" (Box 1979; quoted from Kanji 1986).

Any model – or argument – which helps us understand a particular issue better, is useful both in science and in management. Furthermore, the simpler a model is, the more useful it often is when we try to improve our understanding of the world around us (e.g. Maynard Smith 1974; Stenseth 1977a, c, 1985a, b).

However, we need a multitude of models, partly because we do not know which approach is the better one. In ecology, for instance, we do not know which variables are the most appropriate ones to use when developing mathematical models (e.g. Maynard Smith 1974; Stenseth 1986). Presumably the same situation exists for economic modelling, especially when integrating na-

[1] So far I assumed there is a real choice. However, if I am poor, I cannot afford to think of the future. To be responsible then, takes on a different meaning. It is rather discomforting to realize that in most of the economically poor societies and countries the discount rate is often very high. This certainly brings us into a vicious circle. It begs the question of how it is possible to manage resources of this planet – many of which are **un**managed common resources – in a sustainable manner.

tural and social processes (Stenseth 1985d).

A pluralistic approach to modelling enables different types of models to focus on different aspects of the issue; hence they can complement each other. Many well known modellers do indeed favour such a strategy (e.g. Maynard Smith 1974).

Finally, if the model is to mimic some real system, it may be of the curve-fitting type of regression (e.g. Stenseth 1977a). However, if the model is to provide insight, it often must be as simple as possible. To quote Maynard Smith (1974) (discussing ecology, but his reasoning applies much wider):

> "A theory of ecology must make statements about ecosystems as a whole, as well as about particular species at particular times, and it must make statements which are true for many different species and not just for one. Any actual ecosystem contains far too many species, which interact in far too many ways, for simulation to be a practical approach. The better a simulation is for its own purposes, by the inclusion of all relevant details, the more difficult it is to generalize its conclusions to other species. For the discovery of general ideas in ecology, therefore, different kinds of mathematical description, which may be called (theoretical) models, are called for. Whereas a good simulation should include as much details as possible, a good (theoretical) model should include as little as possible."

Both deterministic and stochastic approaches are needed. Certainly, a proper treatment of uncertainty and stochasticity is of paramount importance both in ecology – economy modelling and in other types of modelling. With uncertainty inherent in the processes under study, the designing of sustainable development strategies requires decisions without being able to predict the outcome precisely.

Statistical models are distinct from **mathematical (and conceptual) models**. Statistical models will, through data from the real world, pinpoint which of a set of potentially important factors may be the more decisive ones in producing a particular pattern. They represent a compact description of the system from which the data derive. Such models generate hypotheses as well as test other hypotheses, and deduce quantitative predictions.

Mathematical models are meant to be caricatures of the real world. The development of such models requires knowledge about the processes occurring at the level of the individual alone in its environment and together with other individuals in some dynamic environment; this applies to ecological processes as well as to social- and political science processes. Analysis of such models may then lead to qualitative – and in rare cases, quantitative – predictions about specified situations in the real world. These predictions may then be tested against real life data using statistical models and statistical techniques.

Mathematical models may be either deterministic or stochastic, or a combination thereof. They may assume continuous or discrete time; which one to choose depends on the issue under investigation.

In order to understand the real world around us better, a strong and continuous interplay between theoretical modelling work and empirical work is needed. One provides input to the other in a mutualistic manner. To quote Maynard Smith (1982):

> "Mathematics without natural history is sterile, but natural history without mathematics is muddled."

Although this refers to ecology and natural history, it certainly applies much more widely.

Models integrating socio-economic and ecological processes

Although the above general observations, or arguments, do not help politicians and managers to design **how** they should manage resources better than is done today, the arguments and the insights are of great value: For instance, the insight dictating the necessity of rules and restrictions on the use of resources is crucial. That is, institutional design becomes important when trying to achieve sustainable development. The processes operating at the market-place – or the free market – will not lead to sustainable development. If anything, the free-market processes will take us further away from a sustainable situation.

The design of more specific strategies for managing the world's resources rests on detailed models. A diversity of mathematical models is needed to account for the complexity of the interacting processes. Such models must integrate social, economic, ecological (biotic as well as abiotic) and physical processes. As a consequence, these models must also integrate fast and slow processes.

Bio-economic models do, of course, exist. Most of them do, however, treat rather primitively either the economic part or the ecological part. This problem has to do with the difficulty of interdisciplinary work. There are two ways out of this. Either, the scientist carrying out the research is trained in both – or all – subject-fields involved (i.e. **both** ecology and economics), or a group of scientists working closely together is established. Such groups should not, as unfortunately often is the case, be biased to one or the other field. Natural and social scientists should collaborate on an equal basis. Good interdisciplinary work requires first rate scientists as well as equality of the participants. Interdisciplinary projects violating one or both of these requirements often turn into low quality projects, as judged both by pure scientific and applied standards.

Prime examples of models integrating respectable economics and ecology are those reported by Clark (1976, 1989). Unfortunately, this kind of all-encompassing and high quality effort is rare. But it is precisely what is required for developing strategies for sustainable management of natural resources.

I close this discussion of modelling the human-environment interaction, with a few recommendations. Models for predicting ecological changes caused by humans should contain the following dynamic features:

- Interaction between the participating actors (or members in general, countries, companies, etc.), for instance, by using game theory (e.g. Neuman and Morgenstern 1947; Axelrod 1984; Taylor 1987; Axelrod and Dion 1988).
- Consideration of future versus present benefits; i.e. incorporating the discount rates and hence economic dynamics. Here it is important to consider both the more narrowly defined discount rates as used by economists (e.g. Clark 1989) as well as the more general discount rate commonly defined by socio-anthropologists (Pearce et al. 1990 1966; Moxnes 1989).
- Dynamic processes within the ecological components of the system (e.g. population dynamics, Maynard Smith 1974) and in the interaction between the ecological and socio-cultural processes of the system (e.g. frequency dependent resource exploitation, Clark 1976).
- Constraints caused by, for instance, institutional design and incentive structures in the particular society we are dealing with (e.g. Bromley 1989; Hovi 1990). Here the historical and cultural setting as well as the law system of the particular society being studied, enter the models (e.g. McEvoy 1986).

LANDSCAPE ECOLOGY

Preamble

The remainder of this chapter deals with the spatio-temporal ecological aspect of resource management[2]. Such aspects are particularly pertinent to sustainable development. They belong to the branch of ecology called landscape ecology (sensu Forman and Gordon 1986; Urban et al. 1987; Turner and Gardner 1990). Specifically, I will turn to one of the most urgent issues of landscape ecology, namely that of the ecological effects of habitat fragmentation (for reviews, see Wiens 1976; Hansson 1977a, b; Forman 1985; Forman and Gardner 1986; Urban et al. 1987; Nahev 1987; Turner 1989; Turner and Gardner 1990; Carpenter 1990).

Habitat fragmentation probably represents a serious ecological threat to biodiversity on both a local and global scale. The effects of this primarily human driven process can be seen almost everywhere (e.g. Forman and Gordon 1986). It is, for instance, rather conspicuous when travelling through the Eastern Valley to and from Røros (Figure 1): Thirty years ago, huge areas of old boreal forests were broken with small patches of clearcuts and young regrowth areas. Today huge areas of clearcuts have only some few and smaller patches of pristine forests left.

Figure 1. An example of extensive man-made landscape ecological changes in the Eastern Valley of Norway. In the hillside huge areas of clearcuts are seen. In the bottom of the valley, the experimental set-up for the Evenstad Empirical Model System described later in the text is seen. (Photo: R. A. Ims)

A theoretical view on habitat patchiness

Answers to questions about the ecological consequences of habitat fragmentation must incorporate the fact that different species use the environment and habitat in different ways and on different spatial scales. Figure 2 illustrates various spatial ways of utilizing the habitat during an organism's lifetime. In this figure, the inner rectangles represents the livable segments of the environment (which for simplicity is assumed to coincide for two of the species).

In this idealized world, I have depicted caricatures of three life-types, a **plant**-type

[2] Although I disregard incorporation of economic arguments in this section, I do add the spatial component to our dynamic scheme: ecology in general and landscape ecology in particular are concerned with spatial dynamics.

of organism, a **mouse**-type of organism, and a **weasel**-type of organism. In Figure 2, the plant-type disperses only once in its lifetime as seeds (or immature propagule). It may disperse to another site within the same patch, to another similar patch, or to a less favorable (possibly a non-livable) habitat type. The mouse-type may disperse (or shift home range) several times during its lifetime, partly inside the same patch and partly to another patch (by moving through some transition habitat; sensu Stenseth 1983). Finally, the weasel-type exploits the entire habitat complex (i.e. more than one patch) during its daily search for resources. The differences between these modes of exploiting the environment relate to the graininess of the habitat for the particular species (MacArthur 1972; Pianka 1988). Notice, in particular, that different organisms may view the same environment differently with respect to graininess. The following two subsections discuss how the features embodied in Figure 2 may be taken into consideration when modelling habitat fragmentation.

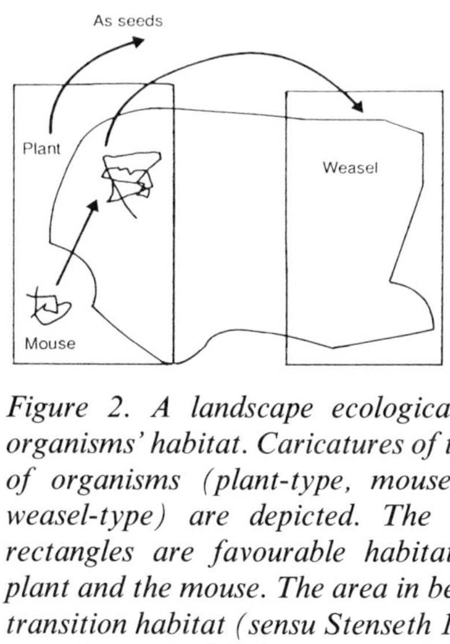

Figure 2. A landscape ecological view on organisms' habitat. Caricatures of three types of organisms (plant-type, mouse-type and weasel-type) are depicted. The two inner rectangles are favourable habitats for the plant and the mouse. The area in between is a transition habitat (sensu Stenseth 1983). The weasel uses both habitat types. See the main text for details.

Experimental model systems, EMS

Background

Both theoretical models (e.g. Pulliam 1988; Burkey 1989) and empirical studies (e.g. Lovejoy et al. 1986; Rolstad and Wegge 1989; Rolstad 1991) suggest that habitat fragmentation may have strong negative effects on population viability. Yet the demographic and behavioural processes interacting with the spatial patterns of the environment are poorly understood (Kareiva 1989). Consequently, the characteristics of species that might be particularly vulnerable to habitat fragmentation are largely unknown. Moreover, it is still unclear which of the spatial parameters characterizing a given habitat mosaic (e.g. patch size, shape and connectivity) that are most crucial for the viability of specific populations.

Observational studies conducted on natural populations in fragmented landscapes, are seldom capable of illuminating the **processes** behind patterns of distribution and abundance (Simberloff 1988; Wiens and Milne 1989; Turner 1989; but see Johnston 1989). This has had the unfortunate consequence of making theoreticians base their models more on speculations and guesses than on observations (Kareiva 1989).

The gaps in knowledge about the effect of habitat fragmentation are unlikely to be filled through the more conventional approaches for studying the effects of habitat fragmentation (i.e. theoretical modelling and observational studies; see Turner 1989; Wiens and Milne 1989; Ims and Stenseth 1989). These approaches need to be complemented by carefully designed manipulative experiments exposing the causal mechanisms behind individual organisms' response to changes in the spatial patterns (e.g. Ims and Stenseth 1989).

This subsection discusses the use of **experimental model systems** (EMS, sensu Ims and Stenseth 1989); some of the more theoretical aspects being of interest in this context are discussed below. In either case, it is important that we derive and extend our insight about the functioning of the ecological system from the processes relating to the individual organism (cf. Stenseth 1985c).

The use of Experimental Model Systems (EMS)

Ims and Stenseth (1989), Wiens and Milne (1989) and J. A. Wiens et al. (in preparation) have argued that crucial information about the causal mechanisms behind population responses to habitat fragmentation may be obtained from studies of EMS. In essence, EMS constitute experimentally "tractable" species-systems studied under more controlled conditions and on a smaller absolute spatial scale than, for instance, wildlife species currently being threatened by habitat fragmentation. When designing such empirical model systems, care must be exercised so as to achieve an appropriate scaling of the experimental habitat mosaic (e.g. R. A. Ims et al., in preparation). Otherwise the EMS may not mimic the spatial processes occurring in connection with the habitat fragmentation experienced by species (a wildlife species, say) on macro-scale (Morris 1987; Wiens 1989; Wiens and Milne 1989).

The use of empirical model systems is fairly common in science: For instance, in medical science, such model systems are commonly used when testing medicines to be used by humans. Furthermore, physical models are commonly used by engineers when building oil platforms, ships, etc. In such cases there will always be uncertainty whether, in fact, the conclusions drawn about the model system really apply to the real-life system. This applies to the kind of physical systems mentioned here, as well as to social experiments (Elster 1979).

To extrapolate from empirical model systems to the real system, mathematical modelling together with comparative studies (on the real system and on the model system) will be of great value. I will return to this issue later in this section.

An EMS experiment

We have recently initiated a project using voles (*Microtus oeconomus*) in artificially fragmented 'landscapes' as an EMS to explore how fragmentation affects various individual and population parameters (R. A. Ims et al. in preparation). We manipulate the habitat on an appropriate spatial scale equivalent to that experienced by several wildlife species in the boreal forest of Scandinavia.

Of four half hectare mouse-proof fenced grids, two of the grids had continuous optimal habitat, whereas two had fragmented habitats (Figure 3). All four populations (one in each of the grids) were initiated by animals born and reared in the laboratory (R. A. Ims et al. in preparation).

Each pair of continuous and fragmented grids was colonized (by release) by pure populations of either Southern or Northern voles. The populations in the four grids

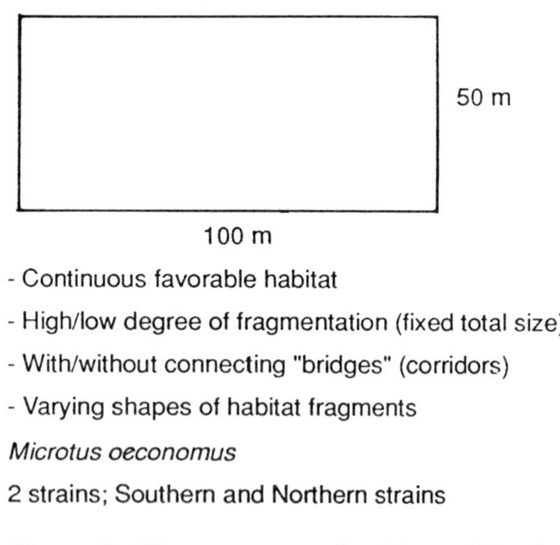

Figure 3. The mouse-proof grids used in the experimental model system (EMS) discussed in the main text (R. A. Ims et al. in preparation). Altogether there are seven 0.5 ha grids; only four were used in the 1989-experiments.

differed only in the origin of animals (i.e. genetic strain), and had the same treatment in the laboratory, social structure, and population structure and size.

In the 1989-season, space use responses to habitat fragmentation were found to be both strain and sex specific: Northern male home ranges overlapped little on the continuous plot, whereas both overlapped extensively on the fragmented plot. Home ranges of Southern males overlapped extensively on both plots. Home range size of males seemed not to respond to habitat fragmentation.

In the EMS-experiment, Southern and Northern females were both of the mouse-type (although they showed intrinsic differences). Northern males exploited the habitat complex in a weasel-type manner, whereas the Southern males exploited the habitat complex in a mouse-type manner.

Southern females had a response similar to that of Northern males both with respect to home range overlap (which was more common and extensive on the fragmented than on the continuous plot) and with respect to home range size (which did not differ significantly among the plots). Northern females responded differently to habitat fragmentation than the other strain- and sex categories: Home range overlap, which in Northern females was less than in other animal categories (i.e. strain and sex categories), was also less extensive on the fragmented plot than on the continuous plot. Northern females were the only category that differed with respect to home range size between the fragmented and continuous plots.

These results are summarized in Figure 4 comparing the space use on the continuous habitat grids with that for the two sexes and the two strains on the fragmented habitat grids.

The main result from this study was that the response to habitat fragmentation is phenotype- (i.e. sex-) and genotype-specific (i.e. strain-specific) (cf. R. A. Ims et al. in preparation). The fact that genetic diversity is today being reduced and changed, makes such results very important for predicting the ecological effects of human induced changes in the environment.

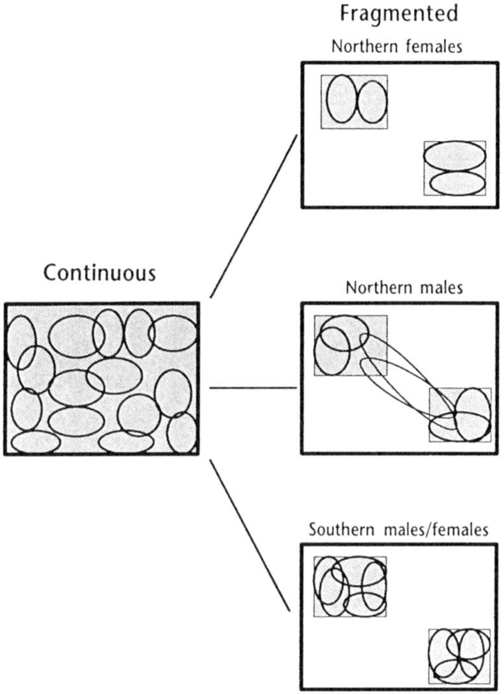

Figure 4. Summary of the results from the 1989 EMS-experiments (R. A. Ims et al. in preparation). Northern females stayed within patches without extensive overlap; Northern males often used more than one patch and exhibited extensive home range overlap; Southern males and females compressed their home range into single patches resulting in an extensive home range overlap.

Experiments on empirical model systems (EMS) can further seek the causes behind the genotypic (strain) and phenotypic (sex) differences being observed. In fact, the observed behavioural differences relate to behavioural differences observed in the laboratory (see R. A. Ims et al. in preparation).

Thus, conclusions about expected behavioural responses of individuals in an ecological setting, may be deduced from observation on individuals in a rather benign and sterile laboratory situation. However, in order to draw conclusions on a higher level of organization (e.g. the population), mathematical modelling is often necessary, to which we now return.

Mathematical models: an example of a new model with new predictions

Mathematical models based on experimental data generated from such experimental studies are important when trying to deduce conclusions about the expected **dynamic** consequences of fragmentation. For general reviews of theoretical models see, for instance, Levin (1976), Stenseth (1977b, 1981, 1983), Hanski (1989), and May and Southwood (1990). Here I will briefly refer some results of a new model specifically designed to help us predict the population dynamic consequences to be expected on the basis of the EMS-experiment.

Recently Tormod Burkey and I (N. C. Stenseth and T. V. Burkey, in preparation) have developed a mathematical model for the weasel-type of strategy depicted in Figure 2. This model complements those developed for the mouse-type strategy (cf. Hastings 1990; see also Stenseth 1977b, 1981, 1983). Altogether this will enable us to predict expected differences in the population dynamic responses of animals from Northern Norway (the weasel-type) and Southern Norway (the mouse-type). The model incorporates seasonality and habitat fragmentation; both the degree of fragmentation and the length of the reproductive season may be altered in the analysis. The model is based upon the ideas of Łomnicki (1988) for the resource utilization within a population. Reproduction is assumed only to take place during the summer; these reproductive periods may be short or long, depending on the values of the parameters of the model.

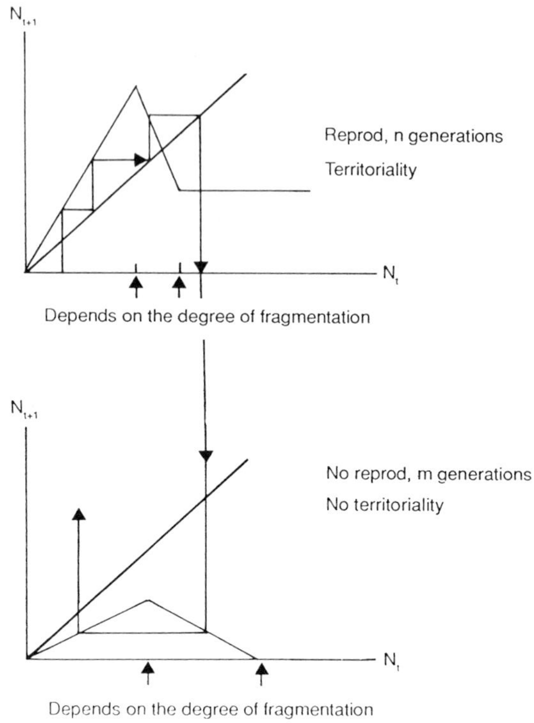

Figure 5. The Stenseth-Burkey model (N. C. Stenseth and T. V. Burkey, in preparation) incorporates separate demographies during summer (A) and winter (B). Depending upon the length of the summer and the winter, the model "turns" around a different number of times within one season before changing to the other season. The effect of incorporating summer-winter conditions into the demographies is of a similar kind to incorporating fragmentation into the demographies. Other details of the model also depends, as indicated in the diagrams, on the degree of fragmentation.

Figure 5 depicts the basic ideas of the model, including how the degree of habitat fragmentation and how the length of the reproductive period are incorporated into the model. The resources exploited by the animal species are assumed not to have its own dynamics.

Analysis of this model leads to two predictions (both of which relate to the weasel-type organisms; cf. Figure 2): First, the population dynamics becomes destabilized as the length of the summer becomes shorter; this has been shown earlier (e.g. Stenseth 1985b). Second, with increasing degree of fragmentation (and hence more moving between patches), the population dynamics becomes increasingly less stable. This latter result contrasts with that emerging from other models (e.g. Stenseth 1983; Hastings 1990), most of which assume a mouse-type of organism, and study the effect of dispersal on population dynamics. It follows from this contrast that it becomes very important to be specific about the way animals exploit patches (i.e. their space use strategy) in heterogeneous landscapes. This means, essentially, that the spatial scale relative to the organism's movement distance becomes very important.

Although further theoretical analysis is needed, the current results suggest that the Northern animals (exploiting the habitat patches in a weasel-type manner; cf. Figure 2) may respond to habitat fragmentation with destabilized population dynamics, whereas the Southern animals (exploiting the habitat patches in a mouse-type manner; cf. Figure 2) may respond to the same manipulation by becoming more stable. This study therefore shows the necessity of combining empirical results with those from theoretical models. Its conclusions are important both to basic and applied ecology.

Conclusion

Clearly we need both theoretical and empirical models to understand the dynamics resulting from the environmental changes caused by humans today. Such work requires the expertise and collaboration of ecologists, mathematicians and statisticians. That way our ecological understanding may be sharpened so as to integrate ecological insight with the social scientists' insight about the socio-economic processes taking place in our society.

CONCLUSION

Processes relating to the human-environment interaction are complex and difficult to investigate either theoretically or empirically. Furthermore, generalizing from studies undertaken under particular circumstances requires much care and some degree of certainty that the situations involved are really comparable.

Essentially, this means that there are no easy shortcuts. Taking such shortcuts can have irreversible and unfortunate consequences with even more environmental problems than those we currently are faced with. Guidelines are needed on how to develop extrapolation rules, and I contend that combining experiments, empirical modelling and theoretical analysis helps to develop such guidelines.

On this basis I make two recommendations:

- To the scientists advising managers and politicians: Be careful and do the nitty-gritty detailed work and take no unjustified shortcuts when advising politicians and managers.
- To the politicians and managers: Make strong efforts to obtain the most updated scientific insight – be it based on social, economic or natural sciences – before introducing new management practices. Scientists are generally the most able ones to advice on that, and on whether or not research is needed before some new management strategy can be developed and introduced. Involve the very best scientists in research aimed at improving nature management. Allow the scientists to do the

detailed work they claim is necessary before they advise politicians and managers. And do remember that what we search for is reliable scientific results which may be applied for managing Earth's resources better than we do today.

ACKNOWLEDGMENTS

I wish to acknowledge generous long-term funding from the Norwegian Research Council for Science and the Humanities (NAVF) for ecological studies. In particular I like to acknowledge the "landscape ecology"-grant from NAVF. Discussions with several people have influenced my thinking on the issues discussed in this chapter, in particular my collaborator Rolf Anker Ims who is also thanked for commenting on an earlier version of this chapter. An anonymous referee also made valuable comments on an earlier version of the manuscript.

LITERATURE CITED

Agarwal, A. K. 1992. Sociological and political constraints to biodiversity conservation: a case study from India. This volume.

Arrow, K., S. Karlin and P. Suppers (eds) 1960. *Mathematical Methods in the Social Sciences*. Stanford University Press, Stanford.

Axelrod, R. 1984. *The Evolution of Cooperation*. Basic Books, New York.

Axelrod, R. and D. Dion. 1988. The further evolution of cooperation. Science 242: 1385–1390.

Berge, E. 1990. Some notes toward a property rights perspective on institutional change in the welfare state. NAS-notat 1990: 9, Inst. Appl. Social Research, Oslo.

Berkes, F. (ed) 1989. *Community Property Resources. Ecology and Community-Based Sustainable Development*. Belhaven, London.

Berkes, F., D. J. Feeny, B. J. McLay and J. M. Acheson. 1989. The benefits of the commons. Nature 349: 91–93.

Bromley, D. W. 1989. *Economic Interests and Institutions*. Basic Books, Oxford.

Burkey, T. V. 1989. Extinction in nature reserves: the effect of fragmentation and the importance of migration between reserve fragments. Oikos 55: 75–81.

Carpenter, F. L. (ed) 1987. Territoriality: conceptual advances in field and theoretical studies. Amer. Zool. 27: 223–409.

Carpenter, S. R. 1990. Large scale perturbations: opportunities for innovation. Ecology 71: 2038–2043.

Clark, C. W. 1976. *Mathematical Bioeconomics: the Optimal Management of Renewable Resources*. Wiley Interscience, New York.

Clark, C. W. 1989. Bioeconomic modelling and resource management, pp. 11–57. In: S. A. Levin, T. G. Hallam and L. J. Gross (eds) *Applied Mathematical Ecology*. Springer-Verlag, Berlin.

Davies, N. B. and A. I. Houston. 1986. Territory economics, pp. 148–169. In: J. R. Krebs and N. B. Davies. *Behavioural Ecology. An Evolutionary Approach*, 2nd Edition. Blackwell, Oxford.

Ehrlich, P. R. and A. H. Ehrlich. 1970. *Population, Resources, Environment: Issues in Human Ecology*. Freeman, San Francisco.

Ehrlich, P. R., A. H. Ehrlich, and J. P. Holdren. 1973. *Human Ecology*. Freeman, San Francisco.

Elster, J. 1979. Skeptiske tanker om samfunnsplanlegging. (Sceptical thoughts on community planning). Forskning og framtid, No 1 (1984): 27–36. (In Norwegian).

Fisher, J., N. Simm and J. Vincent. 1969. *Wildlife in Danger*. Viking Press, New York.

Forman, R. T. T. 1983. An ecology of the landscape. BioScience 33: 535.

Forman, R. T. T. and M. Gordon. 1986. *Landscape Ecology*. Wiley, New York.

Gause, G. F. 1934. *The Struggle for Existence*. Williams & Wilkins, Baltimore.

Gause, G. F., N. P. Smogdarova and A. A. Witt. 1936. Further studies of interactions between predator and prey. J. Anim. Ecol. 5: 1–18.

Hanski, I. 1989. Metapopulation dynamics: Does it help to have more of the same? Trends Ecol. Evol. 4: 113–114.

Hansson, L. 1977a. Landscape ecology and the stability of populations. Landscape Planning 4: 85–93.

Hansson, L. 1977b. Spatial dynamics of the field vole *Microtus agrestis* in heterogeneous landscapes. Oikos 29: 539–544.

Hardin, G. 1968. The tragedy of the commons. Science 162: 1243–1248.

Hardin, G. and J. Baden. 1977. *Managing the Commons*. Freeman, San Francisco.

Hastings, A. 1990. Spatial heterogeneity and ecological models. Ecology 71: 421–423.

Hovi, J. 1990. Promoting cooperation through institutional design. Eur. J. Pol. Economy 6: 397–406.

Huffaker, C. 1958. Experimental studies on predation: Dispersion factors and predator – prey oscillations. Hilgardia 27: 343–383.

Ims, R. A. and N. C. Stenseth. 1989. Divided the fruitflies fall. Nature 342: 21–22.

Jeffers, J. 1988. Decision-thinking about land use. Land Use Policy 1988: 75–78.

Johnston, C. A. (ed) 1979. Spatial and temporal analysis using geographic information systems. Landscape Ecol. 4: 1–80.

Kanji, G. K. 1986. Editorial: "All models are wrong but some are useful." The Statistician 35: 95.

Kareiva, P. 1987. Habitat fragmentation and the stability of predator – prey interactions. Nature 326: 388–390.

Kareiva, P. 1989. Renewing the dialogue between theory and experiments in ecology, pp. 68–88. In: J. Roughgarden, R. M. May and S. A. Levin (eds) *Perspectives in Ecological Theory*. Princeton University Press, Princeton, NJ.

Lancaster, K. 1968. *Mathematical Economics*. MacMillan, New York.

Levin, S. A. 1976. Population dynamic models in heterogeneous environments. Ann. Rev. Ecol. Syst. 7: 287–310.

Łomnicki, A. 1988. *Population Ecology of Individuals*. Princeton University Press, Princeton, NJ.

Lovejoy, T. E., R. O. Bierregaard Jr., A. B. Rylands, J. R. Malcolm, C. E. Quintela, L. H. Harper, K. S. Brown Jr., A. H. Powell,

G. V. N. Powell, H. O. R. Schubart and M. B. Hays. 1986. Edge and other effects of isolation on Amazonian forest fragments, pp. 257–285. In: M. E. Soulé (ed) *Conservation Biology: The Science of Scarcity and Diversity*. Sinauer, Sunderland, MA.

Luckinbill, L. S. 1974. Coexistence in laboratory populations of *Paramecium aurelia* and its predator *Didivium nasutum*. Ecology 55: 1142–1147.

MacArthur, R. H. 1972. *Geographical Ecology: Patterns in the Distribution of Species*. Harper & Row, New York.

May, R. M. and T. R. E. Southwood. 1990. Introduction, pp. 1–22. In: B. Shorrocks and I. R. Swingland (eds) *Living in Patchy Environments*. Oxford University Press, Oxford.

Maynard Smith, J. 1974. *Models in Ecology*. Cambridge University Press, Cambridge.

Maynard Smith, J. 1982. Storming the fortress. New York Rev., 13 May: 41–42.

McEvoy, A. F. 1986. *The Fisherman's Problem. Ecology and Law in the California Fisheries*. Cambridge University Press, Cambridge.

McKelvey, R. 1989. Common property and the conservation of natural resources, pp. 58–80. In: S. A. Levin, T. G. Hallam and L. J. Gross (eds) *Applied Mathematical Ecology*. Springer- Verlag, Berlin.

Morris, D. W. 1987. Ecological scale and habitat use. Ecology 68: 262–269.

Moxnes, E. 1989. Sustainable Development. Chr. Michelsens Institute, Bergen: 44 pp.

Nahev, Z. 1982. Landscape ecology as an emerging branch of human ecosystem science. Adv. Ecol. Res. 12: 189–237.

Neumann, J. von, and O. Morgenstern. 1947. *Theory of Games and Economic Behavior*. Princeton University Press, Princeton, NJ.

Odum, E. P. 1969. The strategy of ecosystem development. Science 164: 262–270.

Pearce, D., E. Barbier and A. Markandya. 1990. *Sustainable Development. Economics and Environment in the Third World*. Edward Elgar Publishing Co., Brookfield, USA.

Pianka, E. R. 1988. *Evolutionary Ecology*. Harper & Row, New York.

Pulliam, H. R. 1988. Sources, sinks and population regulation. Amer. Nat. 132: 652–661.

Rolstad, J. 1991. Consequences of forest fragmentation for the dynamics of bird populations: conceptual issues and the evidence. Biol. J. Linn. Soc. 42: 149–163.

Rolstad, J. and P. Wegge. 1989. Capercaillie *Tetrao urogallus* populations and modern forestry – a case for landscape ecological studies. Finnish Game Res. 46: 43–52.

Simberloff, D. 1988. The contribution of population and community biology to conservation science. Ann. Rev. Ecol. Syst. 19: 473–511.

Stenseth, N. C. 1977a. Forecasting of rodent outbreaks: Model and the real world. EPPO Bull. 7: 303–315.

Stenseth, N. C. 1977b. On the importance of spatio-temporal heterogeneity for the population dynamics of rodents: towards a theoretical foundation of rodent control. Oikos 29: 545–452.

Stenseth, N. C. 1977c. Modelling the population dynamics of voles: models as research tools. Oikos 29: 449–456.

Stenseth, N. C. 1981. Spatial heterogeneity and population stability: Some evolutionary consequences. Oikos 35: 165–184.

Stenseth, N. C. 1983. Causes and consequences of dispersal in small mammals, pp. 61–101. In: I. Swingland and P. Greenwood (eds) *The Ecology of Animal Movement*. Oxford University Press, Oxford.

Stenseth, N. C. 1985a. Why mathematical models in evolutionary ecology? In: J. H. Cooley and F. B. Golley (eds) *Trends in Ecological Research for the 1980s*. Plenum, New York.

Stenseth, N. C. 1985b. Models of bank vole and wood mouse populations, pp. 339–376. In: J. R. Flowerdew, J. Gurnell and J. H. W. Gipps (eds) *The Ecology of Woodland Rodents*. Clarendon Press, Oxford.

Stenseth, N. C. 1985c. Darwinian evolution in ecosystems: the Red Queen view, pp. 55–72. In: P. Greenwood, P. Harvey, and M. Slatkin (eds) *Evolution – Essays in Honour of John Maynard Smith*. Cambridge University Press, Cambridge.

Stenseth, N. C. 1985d. A coevolutionary circuit model for cultural and biological evolution, pp. 268–279. In: J. P. Aubin, D. Saavi and K. Sigmund (eds) *Dynamics of Macrosystems*. Lecture Notes in Economics and Mathematical Systems. Springer-Verlag, Berlin.

Stenseth, N. C. 1986. Darwinian evolution in ecosystems: a survey of some ideas and difficulties together with some possible solutions, pp. 145–185. In: J. L. Casti and A. Karlquist (eds) *Complexity, Language and Life: Mathematical Approaches*. Biomathematics, Springer-Verlag, Berlin.

Stenseth, N. C. and J. Maynard Smith. 1984. Coevolution in ecosystems: Red Queen evolution or stasis? Evolution 38: 870–880.

Swartzman, G. L. and G. M. Van Dyne. 1972. An ecologically based simulation-optimization approach to natural resource planning. Ann. Rev. Ecol. Syst. 3: 347–398.

Taylor, M. 1987. *The Problem of Cooperation*. Cambridge University Press, Cambridge.

Turner, M. G. 1989. Landscape ecology: The effect of patterns on processes. Ann. Rev. Ecol. Syst. 20: 171–197.

Turner, M. G. and R. H. Gardner. 1990. *Quantitative Methods in Landscape Ecology. The Analysis and Interpretation of Landscape Heterogeneity*. Springer-Verlag, New York.

Urban, D. L., R. V. O'Neill and H. H. Shugart. 1987. Landscape ecology: A hierarchial perspective can help scientists to understand spatial patterns. BioScience 37: 119–127.

Watt, K. E. F. (ed) 1966. *Systems Analysis in Ecology*. Academic Press, New York.

Wiens, J. A. 1976. Population responses to patchy environments. Ann. Rev. Ecol. Syst. 7: 81–120.

Wiens, J. A. 1989. Spatial scaling in ecology. Functional Ecol. 3: 385–397.

Wiens, J. A. and B. Milne. 1989. Scaling of landscapes in landscape ecology, or landscape ecology from a beetle's perspective. Landscape Ecol. 3: 87–96.

World Commission on Environment and Development. 1987. *Our Common Future*. WCED. Oxford University Press, Oxford.

9 Biodiversity and climatic change

Peter A. Parsons
Waite Campus,
University of Adelaide,
Glen Osmond, S.A. 5064,
Australia

ABSTRACT

Postulated global climatic changes and localized habitat disturbances focus attention on the ability of organisms to adapt to stressful conditions. Studies of climatically stressed outlier populations and those at forest edges indicate that extinctions and replacements of rare stress-sensitive *Drosophila* species are likely following quite small increases in temperature/desiccation stress. In this way biodiversity would fall by a combination of the extinction of such species and their replacement by stress-resistant generalist species.

At species boundaries, physiological constraints occur whereby the amount of energy required to counteract stress becomes excessive for survival. Metabolic costs may therefore preclude spread into climatically more extreme habitats, especially if there are synergistic interactions among stresses. This means that conservation biology needs to incorporate the phenotypic and genotypic analysis of specific stress response and related traits in order to predict whether particular species can adapt to changing conditions.

INTRODUCTION

The prediction of future climatic trends is uncertain, however, substantial increases in world temperature have been postulated within the next century which may be exceedingly rapid compared with prehistoric changes of a similar magnitude (Schneider 1989). In any case, a consequence of human activities is the destruction of natural habitats, especially rain forests, and this is associated with an increase in environmental stress as habitats become increasingly marginal so converging towards the ecological situation of species borders (Lovejoy et al. 1986; Parsons 1989). Consequently evolutionary change under stress is an important primary area of investigation which needs to be incorporated into conservation strategies. Genetic variation for stress response traits is a prerequisite for adaptation in situations where organisms cannot escape stressful conditions by moving into more favourable habitats. The conservation of such variation is therefore important in ensuring long-term survival of species.

Conservation of Biodiversity for Sustainable Development.
Edited by O. T. Sandlund, K. Hindar and A. H. D. Brown. 1992. Scandinavian University Press, Oslo.

Following Hoffmann and Parsons (1991) stress is defined as an environmental factor causing a potentially injurious change in a biological system. The environmental factors are mainly physical rather than biological in agreement with most recent discussions of stress in biology. Defined in this way at the extreme end of a gradient at the limits of resistance, stress has a major impact on many evolutionary and ecological processes since there will be nonreversible damage and mortality irrespective of the density of organisms. Its importance in natural populations has frequently been underestimated because stressfull periods are often assumed to be relatively unimportant or to predominate at species borders.

Darwin (1859) considered stress in The Origin of Species when discussing extreme external conditions, principally climate: "Climate plays an important role in determining the average numbers of a species, and periodical seasons of extreme cold or drought seem to be the most effective of all checks". "When we reach the Arctic region, or snow-capped summits, or absolute deserts, the struggle for life is almost exclusively with the elements". Even so, Darwin regarded evolution as a mainly gradual process involving intricate and complex interactions among species. He greatly emphasized the importance of competition in the process of natural selection, and hence in evolutionary change. The overall impression is a continuum of organism – environment interactions ranging from competition being exceedingly important in benign environments, to being of minor importance in extreme (and rare) environments where the effects of environmental stresses predominate.

Ecologists have often emphasized quantitative traits (especially those related to physical features of the environment) as important in determining the distribution and abundance of organisms based upon the assumption that the organism is the unit of selection (Andrewartha and Birch 1954). In *D. melanogaster* such ecological phenotypes vary clinally in a manner predictable from *a priori* considerations of geographic variation in climate (Stanley and Parsons 1981). Periods of short-term extreme temperature and desiccation stress are effective in determining climatic races. This has led to the recent increasing emphasis upon genetic variation for resistance to extreme stresses in natural populations (Parsons 1987; Hoffmann and Parsons 1991).

Stress resistance limits will be sporadically tested at times of seasonal climatic stress so that a relatively small permanent climatic change could lead to species extinctions rather than adaptation within species. Studies on the diversity of *Drosophila* species in Australian rain forests now follow emphasizing distributional limits and edge effects, in order to enquire into biodiversity under climatic change.

RAIN FOREST *DROSOPHILA* DISTRIBUTIONS AND CLIMATIC STRESS

Species of the dipteran genus, *Drosophila*, are universally present in Australian rain forests ranging from the humid tropics to cool temperate southern regions (Parsons 1982). The high species diversity of the genus derives from a dependence upon a wide array of plant degradation products, including fungi, for resources. The widespread distribution of *Drosophila* indicates potential as indicator organisms of the status of ecosystems. Furthermore, comparative studies of rain forest species with widespread species occurring in or near human habitations can be carried out.

Drosophila species in rain forests are at risk to the extent that their habitats are disturbed, and conversely their conservation depends upon the maintenance of these habitats. The absence of rare and specialized species of *Drosophila* could therefore be an early sign of habitat deterioration. This could occur directly by man-made disturbances such as tree-felling, invasion by weedy plant species, or indirectly via global climatic change or more local change from habitat fragmentation. Such changes mean that the habitat, which is the templet for ecological strategies, is changed (Southwood 1988) with consequent alterations in the direction of evolutionary processes of indicator and of co-evolved species.

In a major biogeographic survey (Table 1) covering Australian rain forests of easy access 93 *Drosophila* species were obtained. These include 60 from the forests of the humid tropics of north Queensland where rare species, in particular of subgenus *Hirtodrosophila*, tend to occur in climatically non-stressful conditions around 18–20 °C under high humidities in the vicinity of forest fungi. As the temperature increases to a limit of about 25 °C, there is a tendency to find only commoner species that also occur at forest edges. This implies that the *Hirtodrosophila* species may be sensitive to a climatic shift comprising a temperature increase of 1–2 °C. Field collected flies of two patterned-wing *Hirtodrosophila* species have lethal tolerance LT_{50} values (time for 50 % of a sample of flies to die) at 25 °C and 0 % relative humidity less than one quarter of the 16 hrs obtained for the widespread, cosmopolitan species, *D. melanogaster* (Parsons 1985).

This combined laboratory – field approach suggests that a sequence of environment-sensitive and mainly rare *Drosophila* species will be threatened as high temperature/desiccation stress increases. This would reduce the species diversity of these forests towards that of more depauperate faunas characteristic of heat-stressed habitats. There would be a simultaneous change in the flora since there is a broad tracking by *Drosophila* species of the floristic characteristics of forests and their resources, with parallel gradients at the subgeneric and species group levels (Parsons 1982). In this regard, climate is an ecological variable of fundamental significance, although at a local level, variables such as soil type assume prominence (Table 1). Following an increase in high temperature/desiccation stress there would be more overlap between the fauna inside and outside rainforests as now happens following human disturbances. The time scale for the postulated greenhouse temperature change would pre-

Table 1 Number of north Queensland *Drosophila* species classified according to subgenus and rain forest type. Updated from Parsons (1985).

	Basalt			Granite/schist			
Rain forest type[a]	(1a)	(1b)	(5a)	(2a)	(8)	(9)	Total
Drosophila	6	4	2	4	2	–	7
Sophophora	11	7	3	6	4	3	13
Hirtodrosophila	15	4	2	2	4	3	16
Scaptodrosophila	15	7	4	4	6	3	24
Total	47	22	11	16	16	9	60

[a] (1a), (1b) complex mesophyll vine forests, (2a) mesophyll vine forests, (9) simple microphyll vine-fern forests as designated by Webb (1968).

clude reestablishment of equivalent forests at higher altitudes or latitudes, so that massive extinctions of *Drosophila* (and associated plant species) would appear likely for geographically localized and specialized species.

The small pockets of rain forest of the Northern Territory of Australia are climatically extreme for *Drosophila*. These are the wet – dry tropics where the climate is characterized by permanently high temperatures and markedly seasonal rainfall. Thermal optima are around 28 °C compared with optima in the 18–22 °C range in the floristically most diverse forests of the humid tropics of north Queensland (Nix 1981). Especially in low altitude regions of the Northern Territory, the upper limit of tolerance of *Drosophila* in terms of heat/desiccation stress is being approached, as shown by the collection of only two cosmopolitan species in Darwin, *D. ananassae* and *D. melanogaster*, compared with substantially more species in other Australian urban habitats (Parsons 1982). Nine additional species are known from the Northern Territory (Table 2); most of these have been found in north Queensland (and further afield) on the climatically stressed **edge** of rain forests or in lowland habitats.

The widespread nature of the Northern Territory species is presumably associated with their considerable resistances to heat stress. Experimental analyses of desiccation resistance of *D. melanogaster* from Darwin (Northern Territory) on the one hand, and Townsville and Brisbane (Queensland), on the other, show that Darwin flies are extremely desiccation resistant which is a reflection of the "heat stress status" of the Northern Territory (Table 3). For *Drosophila*, the Northern Territory is an outlier situation, with a fauna relatable to climatic harshness whereby widespread species predominate. In this way, a possible model of the biological effects of climatic change involving heat and desiccation stress emerges.

Table 2 Northern Territory *Drosophila* species. From data summarized in Parsons (1985).

Subgenus	Species	Comment[a]
Drosophila	*persicae*	Widespread rare
Sophophora	*ananassae*	Widespread cosmopolitan
	melanogaster	Widespread cosmopolitan
	flavohirta	Widespread rare
Scaptodrosophila	*anthemon*	Rare
	bryani	Widespread common
	metaxa	Widespread common
	hibisci	Widespread common
	fimbriata	Widespread rare
	glauca	Widespread rare
	ellenae	Related to southern colonists

[a] Any species listed as widespread has been found in north Queensland.

Table 3 Desiccation resistance of *Drosophila melanogaster* LT_{50} at 0 % RH, 25 °C. After Stanley and Parsons (1981).

	Climate	Males	Females
Queensland sites			
Brisbane	subtropical	16.94 ± 0.79	19.62 ± 1.62
Townsville	humid tropics	15.42 ± 0.86	19.95 ± 1.31
Northern Territory site			
Darwin	wet-dry tropics	17.76 ± 0.78	22.07 ± 0.96

EDGE EFFECTS

The combination of increasing habitat disturbance and climatic change focusses attention on species distributions and environmental changes at the edge of rain forests.

Habitat fragmentation is normally associated with decreased species diversity. The process of fragmentation in Amazonian forests was studied by Lovejoy et al. (1986) in many groups of organisms with the documentation of major effects. Abiotic changes included temperature, relative humidity, light, and exposure to wind. The biological responses to these changes were classified successively in three orders ranging from immediate effects upon plants and animals at the first order, to changes in insects and insectivorous species at the second and third orders. Differences in relative humidity between reserve edges and 100m inside ranged from 5 % in the early morning to 20 % at midday, with air temperature differences up to 5 °C. Substantial temperature/humidity gradients from the edge to the centre of a reserve may therefore occur, and can affect whole reserves ⩽ 10 ha in area.

In addition to "local" effects, general deforestation may cause climatic changes and so alter the level of heat/desiccation stress. Models of the global atmosphere and biosphere have been used to assess the effects of Amazonian deforestation on the regional and global climate, and predict a significant increase in surface temperature, a decrease in evapotranspiration and precipitation, and a longer dry season (Shukla et al. 1990). In this sense the Northern Territory *Drosophila* fauna is a case-study of what might be expected under such a climatic change.

North Queensland rain forests have climate-sensitive subgenus *Hirtodrosophila* species not found in the Northern Territory (Tables 1 and 2). Collection records obtained from sweeping about forest fungi (Ascomycetes, Basidiomycetes) at the edge of a north Queensland rain forest in Palmerston National Park indicate no *Hirtodrosophila* species compared with several such species 500 m inside the forest (Table 4a), suggesting that the forest edges are excessively stressful for these species.

Baiting with fermented fruits and rotted commercial mushrooms typically attracts an array of *Drosophila* species differing almost entirely from those obtained in the vicinity of forest fungi or by sweeping within forests. The eight species listed in Table 4b obtained by rotted commercial mushroom baiting do not show such edge effects; these species are mainly widespread within the forests and further afield including New Guinea. The generalist and widespread nature of these species is shown by simultaneous mushroom and fermented fruit collections in Mossman Gorge National Park in north Queensland giving similar species distributions for both baits including six of the eight species in Table 4b.

Table 4a *Drosophila* collection records, x, following sweeping rain forest fungi, Palmerston National Park, north Queensland.

		500 m inside forest	Forest edge
Microclimate 1 m from ground		16 °C RH 92 %	22 °C RH 80 %
Hirtodrosophila	*mixtura*	x	
	mycetophaga	x	
	baechli	x	
	hirtominuta	x	
	hirudo	x	
	laurelae	x	
	trifurca	x	
Scaptodrosophila	*brunnea*	x	x
	fungi	x	x

Table 4b *Drosophila* collection records, x, from baiting with rotted commercial mushrooms, Palmerston National Park, north Queensland. All species widespread with the possible exception of *D. progastor*.

		500 m inside forest	Forest edge	
Microclimate 1 m from ground		16 °C RH 92 %	22 °C RH 80 %	Recorded in New Guinea
Drosophila	*sulfurigaster*[a,c]	x	x	
	rubida[a,c]	x	x	x
	pseudotetrachaeta[a]	x	x	x
Sophophora	*serrata*[a,c]		x	x
	denticulata[a]		x	x
	progastor		x	
Scaptodrosophila	*metaxa*[b]	x	x	
	altera[a]		x	

[a] Also from Mossman Gorge National Park, north Queensland, where all species were collected from fermented fruit and mushroom baiting except *D. altera* which was only attracted to mushrooms.
[b] Also in the Northern Territory (Table 2).
[c] Also in urban collections, Townsville, north Queensland.

Species obtained by baiting at forest edges tend therefore not to be adapted to a particular habitat but are normally generalists adapted to a wide range of conditions including some overlap with urban habitats (Table 4b). In contrast, the subgenera (*Hirtodrosophila* and *Scaptodrosophila*) that have radiated most widely in Australia are little represented at forest edges which suggests that many of these species would be threatened by climatic change at an earlier stage than more generalist species.

For endemic species collectable by sweeping rather than by baiting, the actual microhabitat where flies are resting can often be assayed. *D. inornata*, the commonest of these species in south eastern Australia, can be readily swept in large numbers in tree fern habitats in tall open forests. These habitats are characterized by permanent water and a deep layer of permanent rotting leaf litter. Shading in the habitats is almost complete, and environmental extremes are uncommon. Collections of *D. inornata* over a range of temperatures in such habitats indicated that flies move closer to water in warm conditions to minimize temperature-/humidity stress when there are pronounced temperature gradients from the forest edge to permanent water (Parsons 1975). For example, in one series of collections, the fronds where flies were collected were in the temperature range 15 – 20 °C (mean 17.7 ± 2.0 °C) while ambient temperatures at the forest edges adjacent to the collection sites were 16 – 27 °C (mean 20 ± 3.4 °C). Therefore the temperature at which flies were collected is less variable than the general weather conditions at forest edges as would be expected assuming microhabitat selection based upon temperature/humidity relationships, and shows that flies migrate away from stressful microhabitats.

STRESS AND ENERGETIC LIMITS

Organisms unable to respond to an environmental factor enter a stressed state often measurable by an increase in respiration, or diversion of energy to maintenance from growth and reproduction. An organism may therefore increase its resistance to stress by a reduction in metabolic rate. Genetic associations between lowered metabolic rate and enhanced stress resistance have been found in *Drosophila*, mussels, mole rats and beef cattle (Nevo and Shkolnik 1974; Frisch 1981; Hoffmann and Parsons 1989a, b; Koehn and Bayne 1989). In *Drosophila melanogaster* a rapid selection response for desiccation resistance was associated with a lowered metabolic rate. Predictably, behavioural activity was reduced, since activity levels tend to be closely correlated with metabolic rate. In addition, fecundity was reduced which indicates that the cost of the lowered metabolic rate in the desiccation resistant lines is reduced fitness (Table 5). The underlying dependence upon metabolic rate implies that the desiccation resistant lines should be resistant to an array of generalized stresses. This has been verified for starvation, toxic concentrations of ethanol and acetic acid, heat shock, and gamma irradiation. These correlated resistances suggest that a substantial portion of the genetic variation in resistance to different environmental stresses has a common basis. This was confirmed by correlations between the desiccation, ethanol and starvation resistances of isofemale lines set up from the F_2's of a cross between one of the

Table 5 Activity, fecundity and metabolic rate in lines of *D. melanogaster* selected for desiccation resistance compared with unselected controls. After Hoffmann and Parsons (1989a).

	Activity[a] Low Humidity	Activity[a] High Humidity	Fecundity[b] (3 days)	Metabolic rate ($O_2 mg^{-1} hr^{-1}$)
Selected	10.4	6.0	131	2.1
Control	12.6	9.0	162	2.6
Significance level	<0.05	<0.05	<0.01	<0.01

[a] Number of flies (out of 15) that moved in a one-minute interval
[b] Number of eggs

selected and one of the control lines (Hoffmann and Parsons 1989b), as well as correlations between 26 isofemale strains set up from a natural population for these three resistance traits (Table 6).

Table 6 Correlations between 26 isofemale strains of *Drosophila melanogaster* for resistance to various environmental stresses.

	Ethanol	Desiccation	Starvation-1	Starvation-2
Ethanol	–			
Desiccation	0.65^{c}	–		
Starvation-1	0.41^{a}	0.53^{b}	–	
Starvation-2	0.44^{a}	0.53^{b}	0.98^{c}	–

^{a} $p < 0.02$, ^{b} $p < 0.01$, ^{c} $p < 0.001$, Hoffmann and Parsons (unpublished data) for traits defined in Hoffmann and Parsons (1989a, b).

A mammalian example providing a model of the situation likely towards the limits of the distribution of a species comes from cattle selected for growth rate in moderately stressful tropical conditions where growth is suppressed because of ecto- and endoparasites, high temperatures, poor nutrition and eye diseases (Frisch 1981). The selected line was more heat tolerant, had lower maintenance requirements and greater resistance to infection compared with the controls. As a consequence, this line had a higher growth rate in the presence of these stresses, which was achieved through increased resistance to environmental stresses and reduced metabolic rate. In other words, while increased resistance to stress appears readily achievable, reduced metabolic rate must be ultimately restrictive in limiting the range of a species. These results suggest that external environmental forces such as temperature are of primary importance in shaping biogeographic ranges. Root (1988) carried out a continent-wide analysis of avian (passerine) distributions and concluded that the locations of winter distribution and abundance patterns of several avian species are directly related to their physiological demands. An association of temperature extremes and physiological mechanisms was found so that birds are limited to regions where they do not raise their metabolic rates beyond about 2.5 times the basal level in order to remain warm. On a broad scale, therefore, the winter ranges of a large number of passerines are limited by the energy expenditure needed to compensate for colder ambient temperatures. Similar arguments for metabolic constraints upon distributional limits have been found in rodents (Bozinovic and Rosenmann 1989). Looked at in another way, species have a characteristic sustained metabolic rate covering normal activities, which falls between 1.5 – 7 times the resting metabolic rate, and when these limits are consistently exceeded, the existence of the species is threatened (Peterson et al. 1990). More generally, metabolic rates and costs at species boundaries need detailed investigation, especially as the increased cost of maintenance under stress can be substantial.

Energetic limitations at species borders would therefore restrict physiological and behavioural processes. Some of these processes are so energetically expensive that almost all of the energy intake of an animal is required; examples include vocalization by frogs associated with mating (Pough 1989), web-building in spiders (Tanaka 1989) and endothermy in actively flying insects (Chappell and Morgan 1987). Indeed, based upon reproductive success, arguments have been advanced that metabolic rate and fitness are associated (Feder 1987; Pough 1989). It follows that fitness at species borders would be less than in ecologically more benign regions because of the energetic costs from the substantial environmental perturbations and stresses at species borders. Therefore, while the allocation of energy to stress resistance would increase survival at species borders, less energy remains for reproduction, growth and other processes such as competitive relationships.

While greenhouse predictions are often expressed in terms of a temperature increase, this stressor cannot be considered in isolation. Reports of synergistic effects among atmospheric pollutants (Hutchinson 1984) indicate the importance of interacting combinations superimposed upon direct effects. For example, forest decline in Europe depends upon several primary and a multitude of secondary stress factors (Hinrichsen 1986). It is important to appreciate that interactions are most likely under conditions of severe stress, so that many negative and conflicting reports need to be assessed in terms of overall stress levels.

There is, however, a lack of information concerning the evolution of resistance to multiple environmental stresses which might be expected because of the underlying metabolic cost of many stresses. One suggested molecular mechanism concerns specific components of membrane lipids (Hochachka and Somero 1984). Experiments are required on genetic changes in response to multiple stresses, particularly on the interaction of high temperature stress with more specific stresses especially where hypotheses can be put forward based upon energetic considerations.

IMPLICATIONS FOR CONSERVATION

The greenhouse scenario emphasizes a need to consider phenotypic traits important in determining future survival under rapid climatic change. This means that the maximum potential of organisms to respond to directional selection needs to be preserved in populations for specified environmental stress factors. Populations from ecological margins are likely to have been selected for environmental stress resistance, and so are expected to contain genotypes useful in countering stress periods. The possibility of generalized stress resistance associated with metabolic rate means that genotypes selected under extremes of one type of environmental stress should exhibit increased resistance to other such stresses. However, the metabolic cost of increased stress resistance suggests that alleles increasing stress resistance will be at lower frequencies in populations from more favourable environments than at species margins. This suggests that conserving populations from more benign habitats may be a strategy of restricted use when the primary aim is the maintenance of high levels of stress resistance to ensure survival against future environmental changes (Hoffmann and Parsons 1991).

The maintenance of high levels of stress resistance in a population may therefore require natural selection under stressful conditions. This problem is particularly relevant to the release of captive-bred animals to the wild because individuals from these populations will not normally be exposed to extremes for abiotic factors. Indeed, there is evidence for a lowered capacity to adapt to extreme temperatures in captive animals, especially those housed in temperature-regulated facilities (Kohane and Parsons 1988).

Ideally, genetic variation in stress response traits should be monitored directly in populations; unfortunately measures of stress resistance often involve mortality following stress. Another difficulty is that heritabilities for stress resistance depend upon stress levels, and as for assessments of interactions between stresses, measurements of genetic parameters are best made under conditions of severe environmental stress. Useful alternative monitors include metabolic rate, heat shock proteins and fluctuating morphological asymmetry (Parsons 1990; Hoffmann and Parsons 1991).

Quantitative genetic studies are needed on such traits throughout the distribution of a species.

Such an emphasis upon the phenotypic level contrasts with approaches to conservation genetics based upon single loci with allozyme markers, which cannot be fully translated to the phenotypic level because of multiple loci, linkage, and interactions between loci. Furthermore, similar gene frequencies at a sample of allozyme loci do not indicate that populations will respond similarly to a given environmental stress. For example, populations of *D. melanogaster* and *D. pseudoobscura* show only minor allozyme differences associated with substantially divergent responses to selection (Cohan and Hoffmann 1989). This implies that approaches to conservation genetics that depend upon maintaining the overall level of genetic variation in populations as measured electrophoretically may be somewhat limited (Parsons 1989). Conservation genetics therefore needs to incorporate the analysis of stress response and related traits using populations from the varying habitats covering the range of a species.

Genetic variation for stress resistance traits should be maintained by incorporating heterogeneous conditions into management programs by maintaining habitat diversity. Relatively unfavourable habitats where species have low densities consequent upon physical stress should be included. In monitoring environmental stress traits among habitat types, a difficulty in animals is that most experiments have been carried out in the laboratory. However, the ecological significance of desiccation resistance has been demonstrated in organisms such as mosquitoes and *Drosophila* making this an appropriate trait for microhabitat studies.

Such studies should assist in the management question of whether species are best conserved in one large area or several small reserves (Diamond 1975; Lovejoy et al. 1986; Simberloff 1988). Increased microclimatic stress following fragmentation would argue against too much dependence upon small areas such as remnant patches of rain forest. In any case as shown above, flies may migrate away from stressful edges of rain forests. The potential role of spatial heterogeneity in maintaining genetic variation in stress resistance traits suggests that the best strategy for less restricted species is a reserve containing an array of habitats including marginal conditions. The same strategy applies to more specialist species especially those restricted to damp microhabitats deep inside rain forests because small reserves may not contain any such microhabitats. The high level of endemism of tropical Amazonian plant communities (Gentry 1986) and of *Drosophila* of Australian rain forests (Parsons 1982) provides a strong argument for such a strategy. Cohet et al. (1980) found three tropical endemic *Drosophila* species to have developmental ranges of $\leqslant$ 15 °C, while three *melanogaster* group species (two, *D. melanogaster* and *D. simulans*, also occur in temperate regions) have 18 – 20 °C ranges. Based mainly on resource utilization ranges, southern temperate zone Australian endemic and cosmopolitan species gave a range of around 14 °C (Parsons 1978). Temperature ranges for development are therefore wider in temperate than in tropical *Drosophila* species, but are less than those of terrestrial or amphibious animals which usually exceed 20 °C (Andrewartha and Birch 1954). This means that within a relatively narrow temperature range there is an enormous diversity of *Drosophila* species all involved with the exploitation of microbial degradation products, so that the species of the genus may be candidates for indicator species for assessing the level of disturbance in damper regions of forests. Permanent alterations in the *Drosophila* fauna would suggest substantial biotic changes at the level of plants, fungi, and hymenopterous wasps parasitic upon various life cycle stages of *Drosophila*.

CONCLUSION

Genetic responses to environmental change will be inadequate if the alteration is too rapid or too drastic, and failure to adapt to increases in environmental stress will result in species distribution changes, extinctions, and reduced biodiversity. This means that the predicted greenhouse changes, as well as the ongoing destruction of habitats by human activities, will lead to geographic range expansions by certain stress-resistant species into habitats such as rain forests, and simultaneously, stress-sensitive specialist species would face extinction so reducing biodiversity. Predicting the ability of a species to evolve in response to rapid environmental change is a major task for evolutionary biology. An increased understanding of the factors limiting species distributions, and of the characteristics of populations near the margins of forests, compared with more centrally-located populations, should therefore be given a high priority for future research on conservation strategies.

LITERATURE CITED

Andrewartha, H. G. and L. C. Birch. 1954. *The Distribution and Abundance of Animals.* University of Chicago Press, Chicago.

Bozinovic, F. and M. Rosenmann. 1989. Maximum metabolic rate of rodents: physiological and ecological consequences on distributional limits. Functional Ecol. 3: 173–181.

Chappell, M. A. and K. R. Morgan. 1987. Temperature regulation, endothermy, resting metabolism, and flight energetics of tachanid flies (*Nowickia* sp.). Physiol. Zool. 60: 550–559.

Cohan, F. M. and A. A. Hoffmann. 1989. Uniform selection as a diversifying force in evolution: evidence from *Drosophila.* Amer. Nat. 134: 613–637.

Cohet, Y., J. Vouidibio and J. R. David. 1980. Thermal tolerance and geographic distribution: a comparison of cosmopolitan and tropical endemic *Drosophila* species. J. Therm. Biol. 5: 69–74.

Darwin, C. 1859. *On the Origin of Species by Means of Natural Selection.* Murray, London.

Diamond, J. M. 1975. The island dilemma: lessons of modern biogeographic studies for the design of natural reserves. Biol. Conserv. 7: 129–146.

Feder, M. 1987. The analysis of physiological diversity: the prospects for pattern documentation and general questions in ecological physiology, pp. 38–75. In: M. E. Feder, A. F. Bennett, W. W. Burggren and R. B. Huey (eds) *New Directions in Ecological Physiology.* Cambridge University Press, Cambridge.

Frisch, J. E. 1981. Changes occurring in cattle as a consequence of selection for growth rate in a stressful environment. J. Agric. Sci. 96: 23–38.

Gentry, A. H. 1986. Endemism in tropical versus temperate plant communities, pp. 153–181. In: M. E. Soulé (ed) *Conservation Biology: The Science of Scarcity and Diversity.* Sinauer, Sunderland, MA.

Hinrichsen, D. 1986. Multiple pollutants and forest decline. Ambio 15: 258–265.

Hochachka, P. W. and G. N. Somero. 1984. *Biochemical Adaptation.* Princeton University Press, Princeton, NJ.

Hoffmann, A. A. and P. A. Parsons. 1989a. An integrated approach to environmental stress tolerance and life-history variation. Desiccation tolerance in *Drosophila*. Biol. J. Linn. Soc. 37: 117–136.

Hoffmann, A. A. and P. A. Parsons. 1989b. Selection for increased desiccation resistance in *Drosophila melanogaster*: additive genetic control and correlated responses for other stresses. Genetics 122: 837–845.

Hoffmann, A. A. and P. A. Parsons. 1991. *Evolutionary Genetics and Environmental Stress*. Clarendon Press, Oxford.

Hutchinson, T. C. 1984. Adaptation of plants to atmospheric pollutants, pp. 52–72. In: *Origins and Development of Adaptation*. CIBA Foundation Symposium 102, Pitman Books, London.

Koehn, R. K. and B. L. Bayne. 1989. Towards a physiological and genetical understanding of the energetics of the stress response. Biol. J. Linn. Soc. 37: 157–171.

Kohane, M. J. and P. A. Parsons. 1988. Domestication: Evolutionary change under stress. Evol. Biol. 23: 31–48.

Lovejoy, T. E., R. O. Bierregaard Jr., A. B. Rylands, J. R. Malcolm, C. E. Quintela, L. H. Harper, K. S. Brown Jr., A. H. Powell, G. V. N. Powell, H. O. R. Schubart and M. B. Hays. 1986. Edge and other effects of isolation on Amazonian forest fragments, pp. 257–285. In: M. E. Soulé (ed) *Conservation Biology: The Science of Scarcity and Diversity*. Sinauer, Sunderland, MA.

Nevo, E. and A. Shkolnik. 1974. Adaptive metabolic variation of chromosomal forms in mole rats *Spalax*. Experientia 30: 724–726.

Nix, H. A. 1981. The environment of Terra Australis, pp. 103–122. In: A. Keast (ed) *Ecological Biogeography of Australia*. W. Junk, The Hague.

Parsons, P. A. 1975. The effect of temperature and humidity on the distribution patterns of *Drosophila inornata* in Victoria, Australia. Env. Entomol. 4: 961–964.

Parsons, P. A. 1978. Boundary conditions for *Drosophila* resource utilization in temperate regions, especially at low temperatures. Amer. Nat. 112: 1063–1074.

Parsons, P. A. 1982. Evolutionary ecology of Australian *Drosophila*: a species analysis. Evol. Biol. 14: 297–350.

Parsons, P. A. 1985. Tropical *Drosophila*: Resistance to environmental stresses and species diversity. Proc. Ecol. Soc. Aust. 13: 43–49.

Parsons, P. A. 1987. Evolutionary rates under environmental stress. Evol. Biol. 21: 311–347.

Parsons, P. A. 1989. Environmental stresses and conservation of natural populations. Ann. Rev. Ecol. Syst. 30: 29–49.

Parsons, P. A. 1990. Fluctuating asymmetry: an epigenetic measure of stress. Biol. Rev. 65: 131–145.

Peterson, C. C., K. A. Nagy and J. Diamond. 1990. Sustained metabolic scope. Proc. Natl. Acad. Sci. USA 87: 2324–2328.

Pough, F. H. 1989. Organismal performance and Darwinian fitness: approaches and interpretations. Physiol. Zool. 62: 199–236.

Root, T. 1988. Energy constraints on avian ranges. Ecology 69: 330–339.

Schneider, S. M. 1989. The greenhouse effect: science and policy. Science 243: 772–781.

Shukla, J., C. Nobre and P. Sellers. 1990. Amazonian deforestation and climate change. Science 247: 1322–1325.

Simberloff, D. 1988. The contribution of population and community biology to conservation science. Ann. Rev. Ecol. Syst. 19: 473–511.

Southwood, T. R. E. 1988. Tactics, strategies and templets. Oikos 52: 3–18.

Stanley, S. M. and P. A. Parsons. 1981. The response of the cosmopolitan species, *Drosophila melanogaster*, to ecological gradients. Proc. Ecol. Soc. Aust. 11: 121–130.

Tanaka, K. 1989. Energetic cost of web construction and its effect on web relocation in the web-building spider, *Agelena limbata*. Oecologia 81: 459–464.

Webb, L. J. 1968. Environmental relationships of the structural types of Australian rainforest vegetation. Ecology 49: 296–311.

10 Conservation and sustainable use of Atlantic salmon

Kjetil Hindar
Norwegian Institute for Nature Research (NINA),
Tungasletta 2,
N-7005 Trondheim,
Norway

ABSTRACT

Wild Atlantic salmon (*Salmo salar*) have been an important resource throughout man's documented history in northern temperate areas and now form the basis for a rapidly increasing food production based on captive stocks. The current use of this biological resource is clearly not sustainable. Numerous populations have gone extinct during the last two centuries due to pollution, watercourse regulation, overfishing and disease introduction. A mixed-stock fishery in the North Atlantic Ocean accounts for the majority of the annual harvest of Atlantic salmon and continually depresses the sizes of natural spawning populations. Efforts to restore lost populations and enhance others through artificial reproduction and releases have been only partly successful. The practice of artificial reproduction now threatens all remaining anadromous populations of Atlantic salmon, because vast numbers of Atlantic salmon escape annually from fish farms in central parts of the distribution area. Escaped Atlantic salmon already outnumber wild fish in many Norwegian rivers. This situation calls for immediate action with respect to aquacultural practices and fishing regulations. Complete containment of farmed fish and protection of natural populations are suggested to be the two most important steps in a conservation strategy for maintaining the genetic and ecological integrity of extant Atlantic salmon populations.

INTRODUCTION

Conservation must be an integral part of any program for sustainable use of living natural resources, because future generations' use of the same resources may change. This paper presents some ecological, epidemiological and genetic problems for wild populations of a species which is in the process of being domesticated, and discusses how these problems can be dealt with from the perspectives of conservation and sustainable use of biological resources. I focus on Atlantic salmon (*Salmo salar*) in Norway, where the recent expansion of salmon aquaculture has been accompanied by an increasing number of escaped fish, now outnumbering wild ones in several

Conservation of Biodiversity for Sustainable Development.
Edited by O. T. Sandlund, K. Hindar and A. H. D. Brown. 1992. Scandinavian University Press, Oslo.

rivers (Gausen and Moen 1991; Lund et al. 1991). For a broader perspective, the reader is referred to a review by Hindar et al. (1991) of the genetic effects of cultured fish on natural fish populations, drawing on evidence from a number of salmonid species.

Artificial reproduction of Atlantic salmon was mastered around 1850 (it had been demonstrated by Jacobi as early as 1763) and rapidly led to high numbers of unfed fry being released in streams to enhance local populations for sport fishing (i.e. stock enhancement). Now juveniles are commonly fed through one winter in freshwater hatcheries to the smolt stage when they are released in rivers for recapture upon return as adults (i.e. sea ranching), or they may be placed in marine net pens for captive rearing throughout the life cycle (i.e. fish farming). Sea ranching of Atlantic salmon is based on releases of about 8 million smolts annually to the North Atlantic Ocean and the Baltic Sea. In the Baltic, released smolts currently outnumber the natural smolt production ten-fold (Isaksson 1988).

In the late 1960s, Norwegians began to explore the possibility of raising Atlantic salmon to marketable size in captivity, involving both the freshwater and the marine stages of the life cycle. The results were promising, and captive production of Atlantic salmon increased from less than 100 tons in 1970 to about 150,000 tons in 1990; about 100 times higher than the yield of wild Atlantic salmon in Norway.

The biological and management problems associated with the rapid increase in cultured production of Atlantic salmon have been growing even faster than production. Here I first present some background information on wild Atlantic salmon; second, the number of intentionally and accidentally released fish; third, the debate over the potential effects of these releases on wild populations; fourth, a strategy for long-term management of both captive and wild Atlantic salmon; and finally, some major problems for implementing this strategy. The lessons from this study may have relevance for the management of other species that contain wild populations alongside captive stocks at various stages of domestication.

NATURAL POPULATION STRUCTURE OF ATLANTIC SALMON

Geographical distribution and life cycle

Atlantic salmon are distributed in rivers and in the ocean along both sides of the North Atlantic Ocean (MacCrimmon and Gots 1979). There is no single overview of how many rivers that contain populations of reproducing Atlantic salmon, but it is likely more than 2,000 (Norway alone has about 500 rivers). During the last two centuries, salmon populations have been lost from many rivers in the urbanized areas in the southern parts of the distribution area (for example the Connecticut River, the Rhine and the Thames), the major reasons for local extinction being industrial and domestic pollution, and watercourse regulation. Local extinction continues to be a problem. In Norway, acid rain alone is responsible for Atlantic salmon being completely lost from 25 rivers since the 1930s (Hesthagen and Hansen 1991).

The life cycle of Atlantic salmon consists of the following stages (after Mills 1989): adult Atlantic salmon migrate upstream in rivers during spring and summer; eggs are laid in gravel in late autumn; alevins hatch in early spring and 3–4 weeks later start feeding as fry. Parr remain in fresh water for 2–3 years and migrate to the sea as smolt (12–18 cm and 20–60 g) in the spring of their 3rd or 4th year. Salmon travel long distances in the sea and feed there for 1–4 years before returning to fresh water to spawn in their parental stream at sizes ranging from 40 cm and 1 kg to 140 cm and 35 kg. Most (70–80 %) adult Atlantic salmon survive spawning, but only 3–6% return to spawn a second time (Mills 1989). Not all fish migrate; many males become sexually mature and spawn as parr, thereafter they may (or may not) smoltify and migrate to the sea (L'Abée-Lund 1989). Some locations not accessible from the sea have Atlantic salmon populations that lead their entire life in fresh water. Not all salmon home accurately; straying rates to other rivers are commonly less than 4 % in natural populations (Stabell 1984) but may be higher among small streams.

Population sizes

Altogether, Norwegian rivers contain in the order of 100,000 anadromous Atlantic salmon spawners per year (Ståhl and Hindar 1988). One river, the Tana, probably contains more than 20,000 spawners; in other rivers, the annual number of anadromous spawners appears to be from fewer than 100 to about 1,000.

Exploitation is and has been a major determinant to Atlantic salmon population sizes. Salmon have been exploited as a food source since palaeolithic times as evidenced by rock carvings from 3–6,000 B. C., and the earliest Norwegian legislation (from before 1274 A. D.) has commandments that regulated the river fishery for salmon along their migration route. In the 1820s, the Irish and English introduced rod and line fishing to Norwegians, together with a sense that the fishing was for sport rather than for food.

A new era of exploitation came with the high seas fisheries from the 1960s onwards, when salmon were not only exploited on their spawning migrations from the ocean to the rivers, but also directly in their oceanic feeding area. It has been estimated that exploitation rates in the sea may be as high as 80–90 % (Hansen 1986). In order to reduce exploitation rates, especially in the sea where stocks are mixed, drift-net fishing was prohibited along the Norwegian coast from 1989 onwards, and restrictions were imposed on the period of fishing for other coastal and river fishing methods. However, high-seas fishery off the Faroes and Greenland still accounts for a large proportion of the total catch of Atlantic salmon.

Genetic structure

Salmonid fishes are typically sub-structured into a series of local populations (Allendorf and Utter 1979; Ryman 1983). The amount of genetic differentiation among these populations as estimated from biochemical genetic data is quite large compared with other vertebrates. Atlantic salmon populations worldwide have a third of the total gene diversity resulting from genetic differences between populations (G_{ST} =0.36, Ståhl 1987; see Brown and Schoen, this volume, for explanation of terms); this differentiation among populations arises firstly from a major dichotomy between populations from either side of the North Altlantic Ocean, and secondly from genetic differences between European populations in Baltic and Atlantic drainages. On a more limited geographic range, say, the Norwegian coast or northern Sweden, genetic differences between local Atlantic salmon populations result in a G_{ST} of about 0.10 (Ståhl 1981; Ståhl and Hindar 1988), which is still as high as that between the three major races of man (Ryman et al. 1983) and more than ten times as high as the differentiation between populations of many marine fishes from a much wider geographic range (e.g. herring, Ryman et al. 1984). In absolute terms, Atlantic salmon are not highly variable genetically as estimated from protein coding loci; the total gene diversity in Ståhl's (1987) study was about 0.04.

Where looked for, it has been shown that a single river may contain several genetically different populations. Tagging studies suggest that homing is quite precise even within rivers, and early speculations of local populations within large rivers (e.g. the Tana River in Norway and the Miramichi River in Canada) have been substantiated by rearing studies (Riddell et al. 1981) and by biochemical genetic studies (Ståhl 1987; Ståhl and Hindar 1988). Whether or not this microgeographic genetic differentiation is stable over time remains to be seen; but preliminary data from the Tana River suggest that the local genetic structure is temporally stable (K. Hindar, unpublished data).

Adaptive differentiation

The distribution of genetic variation as estimated from biochemically detected loci needs not reflect that of the variation in morphological or ecological traits that are more subjected to natural selection (Crow 1986). The Norwegian priest Peder Claussøn Friis wrote 400 years ago that each river had its own characteristic stock of salmon. A number of morphological and ecological traits are now known to differ between Atlantic salmon populations. The following ones have a genetic basis in addition to being environmentally modified: body and fin morphology, egg number and size, survival to various life stages, individual growth rate, precocious male maturity, age at smolting, age and size at sexual maturity, migratory behavior at sea, timing of adult return from the sea, homing precision, and resistance to bacterial diseases and parasites (Riddell et al. 1981; Saunders 1981; Refstie 1987; Bakke et al. 1990; Hansen and Jonsson 1991).

Ricker (1972) concluded that there were both genetic and environmental components to variation in all morphological and ecological traits so far studied in Pacific salmon (*Oncorhynchus* spp.), and that this duality should be considered for all phenotypic traits that remain to be tested. Let us look more closely at the most conspicuous difference among anadromous Atlantic salmon populations, namely, the sea age (and thereby size) at which they mature sexually. Sea age at maturity varies considerably between populations and usually but not necessarily within populations (Figure 1). Many rivers contain salmon that mature after 1 year (1–3 kg) in the sea, and at least in some of these rivers there is no other sea age at maturity. In other rivers, the commonest sea age at maturity may be as high as 3 years (7–15 kg), but there are also fish (mostly or entirely males) in these rivers maturing after 1 year in the sea.

The adaptive significance of the variation in sea age at maturity within and between rivers is not well understood. The size of the river is often a good predictor of the sea age of its salmon. Small rivers have small, early-maturing fish and large rivers have large, late-maturing fish. One explanation for this is that the size of the fish increases with the difficulty of migration to the spawning sites, as estimated from the river length or other indices of river harshness (Schaffer and Elson 1975). Males are usually more variable in sea age at maturity than females. This has been suggested to reflect frequency-dependent opportunities for spawning among males (Gross 1985).

Breeding studies suggest that more of the genetic variation for sea age at maturity is distributed between rivers than between families within rivers, but the data for this conclusion are limited (Nævdal 1981). It is notable, however, that the variation in sea age at maturity observed among stocks in a common rearing environment reflected the differences in this trait between the wild source populations (Nævdal 1981).

We can summarize this overview by stating that there are genetic differences between Atlantic salmon populations both in

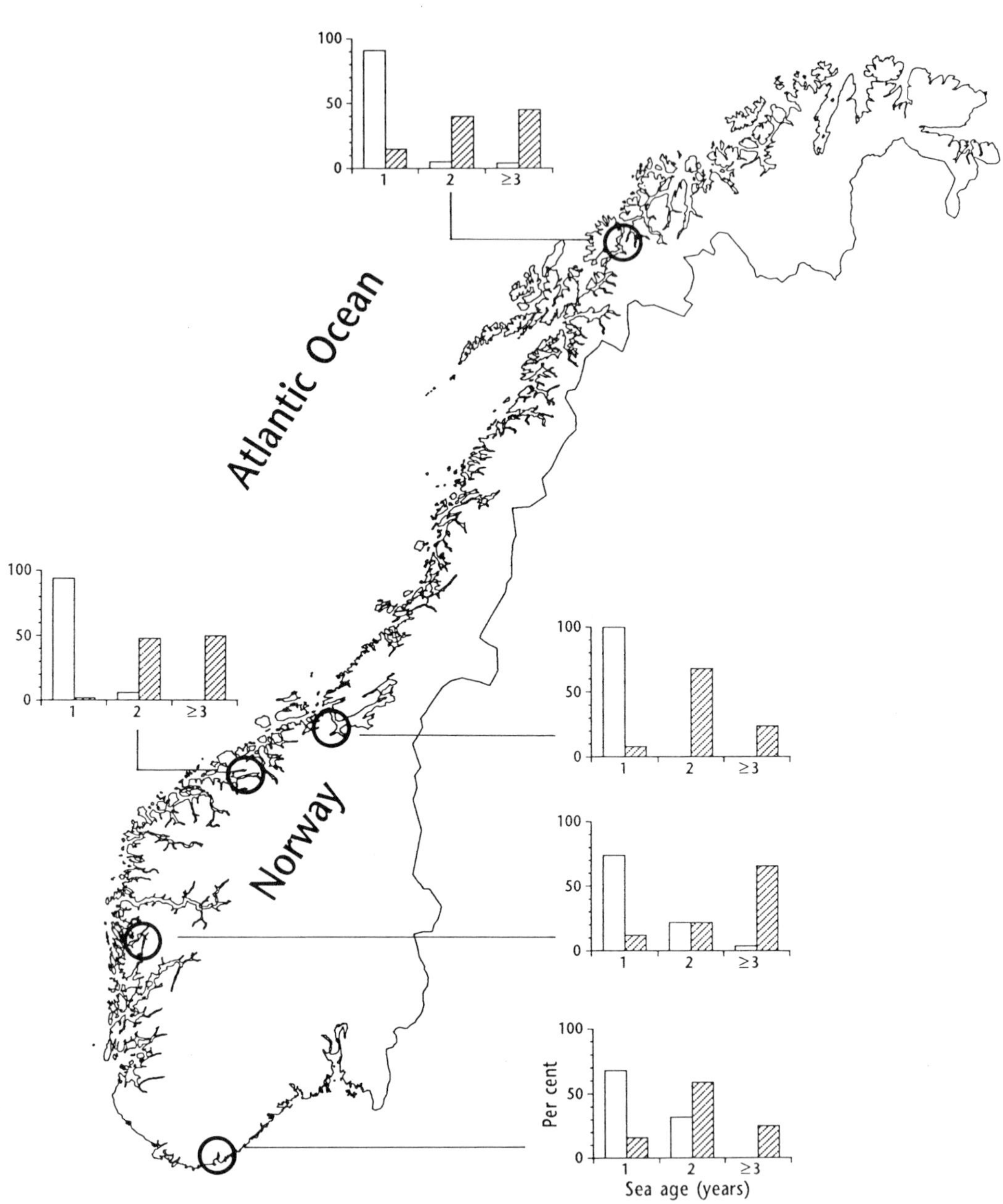

Figure 1. Sea-age at sexual maturity of Atlantic salmon from specific rivers in Norway. Contrasting sea-age distributions are found in populations from small (open columns) and large rivers (hatched columns) in close geographic proximity. The circles indicate localities where Atlantic salmon from one small and one larger river (i.e. having higher average waterflow) could be compared (mainly after Huitfeldt-Kaas 1946).

protein-coding loci where single genes and alleles can be identified and in ecological traits that reflect adaptation to different local environments. Precise homing appears to be one mechanism by which such differences can arise and be maintained. Therefore, one consequence of releases of non-native fish into salmon-producing rivers may be a breakdown of the genetic structure and of local adaptations. In the following, various forms of releases are presented and discussed in this context.

RELEASES OF ATLANTIC SALMON IN NORWAY

Intentional releases

Most intentional releases of Atlantic salmon in Norway are for local enhancement purposes. The largest releases are compensation measures taken to reduce the effects of watercourse regulation for hydroelectric power production. Sea ranching has hitherto been developed on an experimental scale only, but plans exist for upgrading of sea ranching to an industrial scale as it is for some of the Pacific salmon.

About 440,000 Atlantic salmon smolts resulted from intentional releases in Norway during 1986, as compared with an estimated natural production of about 6 million smolts (Ståhl and Hindar 1988). About 80 % of the releases in 1986 were derived from local populations; current legislation requires that all intentionally released fish be of local stock only.

Accidental releases

No hard data exist regarding the total number of accidentally released fish from contained culture, but it is evident that this number is considerable. For example, 1.2 million fish escaped from net pens along the Norwegian coast during storms in December 1988 and January 1989. The biomass of these fish alone (more than 1,200 tons) is equivalent to the total annual yield of wild Atlantic salmon in Norway in the sea and rivers combined (about 1,500 tons; Anonymous 1989). It has been estimated that 13, 28, and 35 %, respectively, of the Atlantic salmon spawners in 1987, 1988, and 1989 were escapees from marine net pens (Gausen 1988; Moen and Gausen 1989; Lund et al. 1991), based on differences in body and fin morphology and scale patterns between cultured and wild fish (Lund et al. 1989). In areas of intense fish culture, up to about 80 % of the spawning population consisted of escapees from net pens, and the 1989 estimates of proportions of escaped fish among the spawners suggest that most rivers in Norway received escaped fish (Figure 2), perhaps except in the farthest northeast where aquacultural production is still small scale.

The Atlantic salmon currently raised in fish farming (and presumably those escaping from it) are derived from wild Norwegian populations, selected for economically important traits in a national breeding program since the 1970s (Gjedrem et al. 1988). In the 1970s and 1980s, a shortage of smolts in the fish farming industry was compensated by imports from the Baltic countries, Iceland and Scotland, but due to increased Norwegian smolt production and the biological disasters following some of the imports (see below), this policy was abandoned and no import has occurred after 1988 (Egidius et al. 1991).

IMPACT OF RELEASES ON NATURAL POPULATIONS

The growing number of escaped Atlantic salmon in feeding areas and spawning populations has led to considerable conjecture about the genetic, epidemiological and eco-

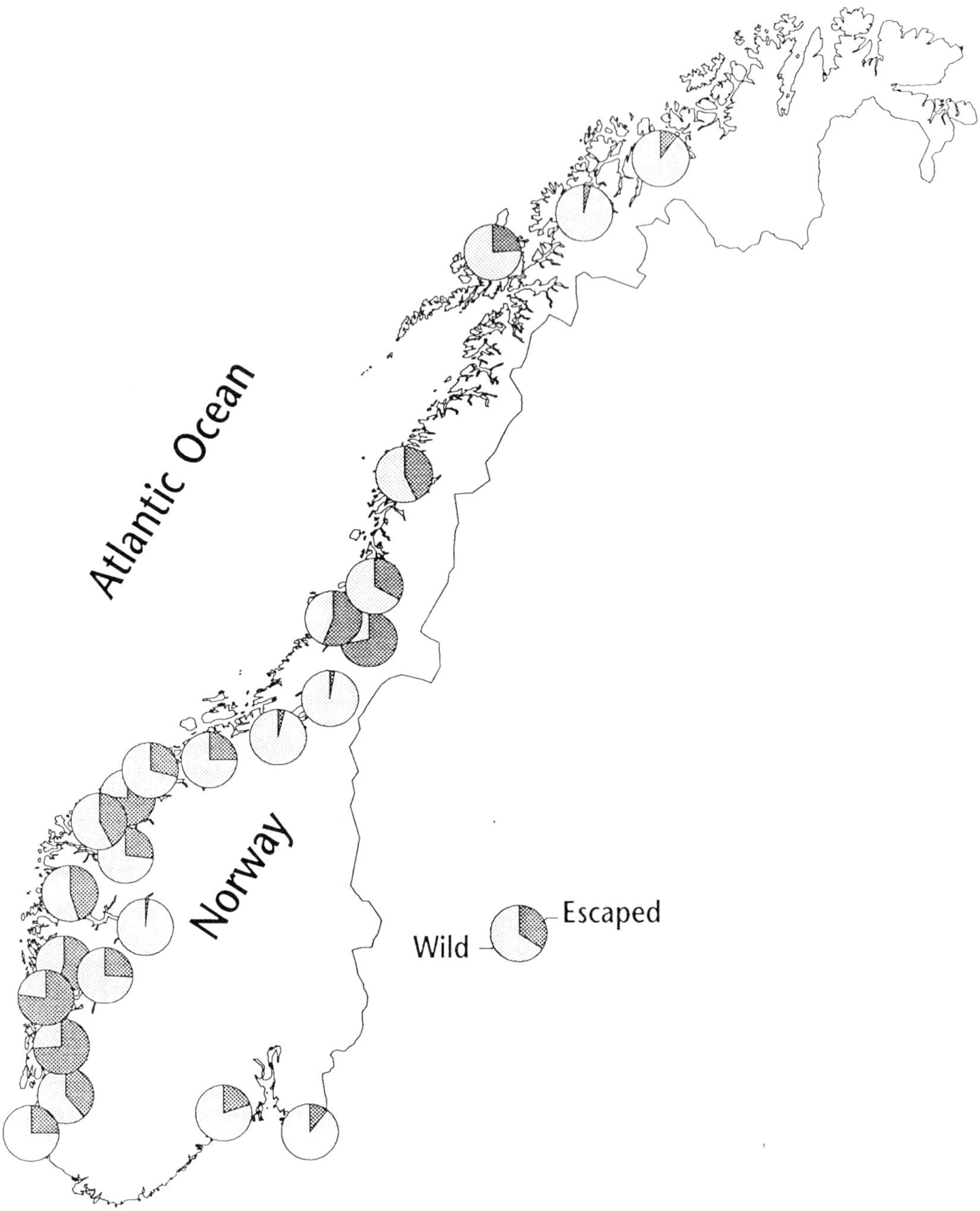

Figure 2. Proportions of net-pen escapees among spawning Atlantic salmon in Norwegian rivers during 1989 (after Lund et al. 1991).

logical effects of the cultured fish on native populations. The Norwegian debate has centered around two themes: (1) the reproductive success of escaped fish, and (2) the full spectrum of biological effects caused by escapees on native populations, whether or not the escapees spawn successfully.

Reproductive success

The argument of 'no documented successful spawning' has been used in favor of taking no action to reduce the number of escapees (cf. Sattaur 1989). This argument has now been shown to be false. Female escapees lay viable eggs that are fertilized by males and survive to the alevin stage in Norwegian rivers (Lura and Sægrov 1991a,b). This knowledge is based on comparisons of optic isomers of astaxanthin, which differ between the eggs of cultured and wild salmon so that each group can be identified until the start of exogenous feeding. Moreover, this study suggested that the reproductive success of escaped females was not very much lower than that of wild females (as long as could be followed by this method of investigation). Other experimental studies suggest that escaped male Atlantic salmon have a reduced spawning performance compared with wild males in competitive situations (T. Järvi and B. Jonsson, NINA, Trondheim, personal communication). Fish released at the smolt stage have been shown to have reduced spawning performance relative to wild fish (Jonsson et al. 1990).

Biological effects

Interbreeding

Referring to Atlantic salmon in Norway, Bentsen (1991) argued that the entire concept of genetic preservation of populations should probably be reconsidered, because the genetic diversification among populations has not reached a qualitative level (i.e. unique alleles). His argument was based on the concepts of quantitative genetics, which assume that most performance traits are the product of a number of (unidentified) genes, and that the same genetic performance may result from a large number of different allele combinations. According to Bentsen, a quantitative genetic approach might well suggest increased migration to threatened populations for which a qualitative approach would suggest protection by preventing migration.

This argument is not new (e.g. Kapuscinski and Lannan 1984; Wohlfarth 1986) and there are clear responses to it. First, the acceptable level of human-mediated gene flow must be quantified. Whereas it is true that small, isolated populations may benefit from a limited immigration (Allendorf 1983), the massive hybridization which is the probable outcome of current escape rates in Norway will result in a loss of genotypic variance among populations which is as irreversible as the loss of alleles (Nelson and Soulé 1987). Ryman et al. (in press) suggested to use the level of naturally occurring gene flow as a quantitative guideline for permissible introgression rates. Based on the relationship between observable genetic differentiation among populations (G_{ST}) and the number of effective migrants per generation in Wright's (1969) island model, the acceptable introgression rate for an average G_{ST} of 0.10 would be only 2 individuals per generation. This is a long way off the massive numbers of escaped cultured salmon now spawning in Norwegian rivers.

Second, in the small, isolated populations where increased gene flow is advisable to prevent inbreeding, I see no reason why the uncontrolled spread of escaped fish, being increasingly adapted to the captive environment, should be preferred to the controlled release of wild fish from adjacent localities. Third, as long as the escapes persist, there is no opportunity for the native populations to readapt to their local environment after the first influx of genes.

Finally, empirical observations from a number of salmonid fishes suggest that increased levels of gene flow following releases almost unerringly result in poorer performance of the recipient populations (Hindar et al. 1991). It was pointed out already by Haldane (1964) that theory alone cannot provide precise predictions as to the outcome of manipulating gene flow among populations. Predictions of the effects of large-scale releases must be based on empirical observations, a point which has been increasingly stressed in contexts such as species introductions (Drake et al. 1989), classical biological control (Howarth 1991), and the release of genetically engineered organisms (Levin, this volume).

The observed genetic effects of (intentionally or accidentally) released salmonids on natural populations are typically unpredictable; they vary from no detectable effects to complete introgression or displacement (Hindar et al. 1991). Where genetic effects on performance traits have been detected, they appear always to be negative in comparison with the unaffected native populations. Some empirical observations related to the actual or potential genetic effects of releasing Atlantic salmon are presented in Table 1. These observations suggest reduced performance following releases for performance traits such as territoriality, ocean survival, return rate to the river, and disease resistance. Ricker (1972) and Hindar et al. (1991) compiled evidence from other salmonid species indicating that a number of other traits had been negatively affected; survival, physical fitness and concealment behavior of juveniles, and straying rate and pre-spawning survival of adults. The typically negative effects of releases of non-native salmonids are not unexpected; if salmonid populations are adapted to their local environments, any introduced populations or crosses involving introduced populations should be expected to perform worse than the native ones (cf. Templeton 1986).

Disease introduction

Diseases are a growing problem for Atlantic salmon populations in Norway. Salmon populations in 32 Norwegian rivers have been greatly reduced since the mid 1970s following introductions of fish from hatcheries infected by the monogenean parasite *Gyrodactylus salaris* (Johnsen and Jensen 1986). This parasite appears not to occur naturally in Norway, and was probably imported here with salmonid eggs or juveniles from infected localities in the Baltic. Its occurrence in the wild may be traced back to a few infected hatcheries from where fish were transported to other hatcheries or released directly into rivers. Laboratory studies suggest that there is a genetic basis for the different effect of the parasite in Norway and in the Baltic; in the latter area populations resistant to the parasite have been found whereas the Norwegian populations tested are highly susceptible to the parasite (Bakke et al. 1990).

Furunculosis, caused by the bacterium *Aeromonas salmonicida*, may lead to similar dramatic effects on natural populations. In 1985, this pathogen was introduced to Norwegian fish farms with infected smolts from Scotland, and spread rapidly from the first few infected farms to reach about 350 fish farms (45 % of the total) and 50 hatcheries by October 1990. Of the 1.2 million smolts that escaped from fish farms during the winter of 1988–89, more than 250,000 fish were from farms infected with furunculosis. Not unexpectedly, the disease was found among spawning salmon in the autumn of 1989, firstly among farmed escapees and later also among wild fish. By October 1990, furunculosis had been registered in 30 rivers in Norway, including wild fish in nine rivers. In one river, the Aursunda, the disease reached epidemic proportions; the majority of the anadromous spawners (more

Table 1. Observations bearing on the genetic effects that cultured Atlantic salmon (*Salmo salar*) may have on indigenous populations of the same or closely related species.

Trait studied	Observation (region)	Reference
Genetic structure	Landlocked Atlantic salmon not detectably affected by releases of offspring of anadromous Atlantic salmon. (Norway)	1.
	Some interbreeding between introduced hatchery stock of Scottish origin and local stocks. (Spain)	2.
Hybridization	Increased rate of hybridization between Atlantic salmon and brown trout (*Salmo trutta*) in areas where brown trout have been introduced. (Canada)	3.
	High hybridization rates with brown trout where exogenous Atlantic salmon have been planted. (Spain)	4.
Stock size	Eradication of local stocks following introduction of fish from hatcheries infected by *Gyrodactylus salaris*. (Norway)	5.
Territorial behavior	Weaker territorial behavior in offspring from hatchery stock than from wild stock. (Sweden)	6.
Ocean survival	Hatchery-reared Atlantic salmon have lowered survival when released in rivers other than their own. (Canada)	7.
Ocean and river recapture rate	Hatchery stock from the nearest river similar to the local one; recapture rate of hatchery stock from distant river 50 % lower. (Norway)	8.
River recapture rate	Recapture rate of transplanted hatchery stock about one fifth or less of that of local stocks. (Spain)	2.
Return rate	Lower return rates following crossbreeding of stocks. (Canada)	9.
Disease resistance	Susceptibility to vibriosis much higher in one stock from the Baltic than in 14 stocks from Norway. (Norway)	10.
	Susceptibility to *G. salaris* much higher in two Norwegian stocks than in one stock from the Baltic. (Norway)	11.

References: 1. Vuorinen and Berg (1989). 2. Garcia de Leániz et al. (1989). 3. Verspoor (1988). 4. Garcia de Leániz and Verspoor (1989). 5. Johnsen and Jensen (1986). 6. Norman (1987). 7. Ritter (1975). 8. Hansen et al. (1989). 9. Bailey (1987). 10. Gjedrem and Aulstad (1974). 11. Bakke et al. (1990).

than 700 fish) were found dead and there is evidence that the disease was transmitted to young Atlantic salmon in that river (Rikstad 1991).

Some diseases believed to be "imported" may well be due to local pathogens occurring at low abundance in the wild while causing noticeable disease in the high-density and well-monitored captive environment. But it must not be forgotten that the high-density captive environments in themselves may cause epidemics in local wild populations. Because of high pathogen densities in the culture situation, the resulting transmission rates may become extremely high relative to the natural situation. For example, in the 1960s furunculosis was registered in one wild population of Atlantic salmon in southeastern Norway, where it did not cause an epidemic or any apparent long-term reduction in population size.

The present situation on the Norwegian west coast is dramatically different. The farms are now all along the coast. Fish are often transported among hatcheries and net pens, and raised in densities high enough to lead to epidemics in captivity. The high numbers of escapees can introduce high numbers of the pathogen to wild populations. Simple epidemiological theory predicts epidemics to follow in the wild under these circumstances, even though (the same theory predicts that) an epidemic is usually not the outcome under natural conditions (e.g. Anderson and May 1986).

Displacement

A decline of the local wild population has been observed to accompany releases without any detectable introgression or disease introduction (Hindar et al. 1991). The reasons for this decline remain obscure, but it is possible that both competition between released and wild fish, and increased predation rates contribute to the decline. It has for example been observed that escaped female Atlantic salmon may spawn later in the autumn than wild ones and destroy the spawning redds of the wild fish in the process (Lura and Sægrov 1991b). If the survival of offspring of escaped fish is poor, the outcome will be a decline in population size.

LONG-TERM MANAGEMENT OF WILD AND CAPTIVE STOCKS

In order not to compromise the natural genetic diversity of Atlantic salmon while developing closed culture and future sea ranching, a number of measures must be taken. The following ones are particularly relevant to the Norwegian situation; other measures to minimize the genetic impact of cultured fish on wild populations are discussed by Hindar et al. (1991).

Contained rearing

The most obvious way to reduce gene flow and spread of pathogens from a closed environment is to keep escape rates at a minimum. Completely contained rearing, costly as it may seem, may be the least expensive way of reducing the impact of cultured fish on native populations. Complete containment also has some obvious positive effects on fish farming itself. Among these are (1) the economic benefit of not raising fish that escape, (2) the possibility to control incoming and outflowing water, thereby avoiding to a large extent the import and spread of pathogens and reduce organic contamination of the locality, (3) better quality control of the flesh of the cultured fish because of a more controlled environment with respect to temperature and salinity, and (4) reduced feed and labour costs. The increased initial cost of complete containment results partly from containment itself, and partly from securing a water supply to a contained environment. It is quite expensive to pump water to a land-based operation, but recent experiments suggest that the extra costs are outweighed by the increased profitability of production (Arve Berg, Norwegian Hydrotechnic Laboratory, Trondheim, personal communication).

Sterilization of salmonid fishes is possible on a large scale (Chourrout 1987) and constitutes a complete "containment" with respect to direct genetic effects. However, sterilization is not preferred to better physical containment because it does not deal with the potential spread of pathogens. Also, there are unresolved questions about the ecology and behavior of sterile fish: the females and some males stay in the sea much longer than their normal life span and their ecological interaction with wild fish must be critically evaluated before large-scale releases of sterile fish can be encouraged (Hindar et al. 1991).

Area planning, including protected areas

Escaped Atlantic salmon are rare in northeastern Norway where aquacultural production is small. Even in areas of intense fish culture rivers located more than 20 km from the nearest fish farm receive fewer escapees than rivers closer to a fish farm (Gausen and Moen 1991). Hence, localization of culture operations may help to protect natural populations. A nationwide plan for the use of the Norwegian coast has recently been issued (Pedersen et al. 1988). Accompanying this plan was a list of salmon rivers worthy of protection and a preliminary establishment of protection zones around these rivers in order to stop new fish farms or hatcheries from being established there during a 5-year period (1989–1994).

Do these protection zones work? In Møre and Romsdal county where production is large and escapees numbered more than 1 million fish during 1988–89, escaped fish made up an average of 40 % of the spawners in eight streams sampled during the autumn of 1989, all of which are situated within a protection zone (Lund et al. 1991). These data show that protection zones are virtually worthless in the proximity of large numbers of escaping fish, and that only very large protection zones (e.g. all of northeastern Norway, as suggested by Ståhl and Hindar 1988) may offer protection from escapees. The evident conclusions are that the number of unintentionally released fish must be reduced, and that natural populations in areas of intense fish culture must be protected by moving culture operations away from these rivers and preferably inland.

Sea ranching should preferably be developed in the large, acidified rivers of southern Norway which have lost their original salmon populations, and which could support a sea ranching program by liming of the water. In rivers having native populations, local stocks only should be used in releases, and the release of juvenile Atlantic salmon should always be compared with habitat restoration – including building fishways to increase the productive area – with respect to increasing production with minimal genetic consequences.

Reduced or selective fishing

Many spawning populations of Atlantic salmon are so small that it seems inevitable that a reduction of the total fishing pressure on these populations is needed. It is possible that increased knowledge about behavioral differences between escaped and wild fish could be used to increase the fishing pressure on the former while reducing it on the latter. For example, it is known that escaped salmon (and also intentionally released ones if released in estuaries or directly into the sea) ascend rivers later in the season than wild fish and that they are less prone to migrate to the upper reaches of a stream (Hansen et al. 1987, 1989; Moen and Gausen 1989; Jonsson et al. 1990). Thereby, a coastal fishery in the period from September to November would probably be more efficient for capturing escaped fish and less so for wild fish than the present fishing season from June to August.

Gene banks

Extinction of local populations can be counteracted by the establishment of gene banks, i.e. natural or artificial facilities established for maintaining population gene pools (Allendorf and Ryman 1987) or gametes (Stoss 1983). In Norway, Atlantic salmon sperm has been cryopreserved from 2,800 males from 128 localities (Gausen in press). There is not yet any method available for freezing salmonid eggs or embryos, and diploid genomes will have to be restored from cryopreserved sperm by androgenesis (Parsons and Thorgaard 1985). One gene bank based on captive rearing throughout the life cycle has been established in Norway to secure 14 populations that are close to extinction from *Gyrodactylus* infection (Gausen in press).

Viable, wild populations are usually the safest, most certain method of maintaining genetic diversity and protecting small or endangered stocks. Increased gene flow is a threat to the preservation of genetic variation both *in situ* and *ex situ* because of the problems associated with its timely detection (Allendorf and Phelps 1981), and because all wild populations are potentially at risk. Therefore, it is imperative that gene flow from cultured to wild Atlantic salmon is minimized to prevent the loss of genetic diversity arising from genetic differences between natural populations.

MANAGEMENT FOR SUSTAINABILITY

"The future evolution of salmonid species now lies chiefly in the hands of the people who are beginning to breed them on a large scale. The task embodies challenges. One challenge is to maintain genetic diversity by providing appropriate sanctuaries to ensure the survival of wild stocks. Another is to utilize the enormous adaptability of these remarkable animals by selective breeding ...." (Donaldson and Joyner 1983). This statement by two of the pioneers in salmon breeding is remarkably relevant to the current situation in Norway. The latter of the two challenges is adequately taken care of in breeding programs run by research stations and the fish farming industry (Gjedrem et al. 1988). The former challenge, however, appears not to be appreciated as a task which has to be addressed and dealt with by the industry (cf. Sattaur 1989).

So long as the fish survive, Atlantic salmon culture is here to stay. Only sound management of both the aquaculture industry and the natural populations can secure sustainable use of the biological resources they represent. Several steps must be taken to reach sustainability. First, the long-term management goals must be defined. Second, the action required to reach these goals must be identified and prioritized. Third, the authorities, agencies or individuals responsible for their implementation must be identified. Finally, the results reached must be checked against the management goals and the required action adjusted accordingly. Norway may serve as an example of how this process is hindered by inefficient management.

Atlantic salmon in Norway are managed by three government departments that have different responsibilities and different management goals. Wild salmon are managed by the Ministry of Environment, cultured salmon by the Ministry of Fisheries, and salmon possibly infected with diseases by the Ministry of Agriculture, whether these fish are wild or cultured. As this paper shows, wild Atlantic salmon are threatened by massive accidental releases of cultured fish. Thus, the most important step for successful long-term management of wild salmon: **stop unintentional releases**, has to be taken by authorities other than the one responsible for managing wild fish. It is unlikely that this step will be taken until action is required for more immediate purposes, for example to stop unintentional releases for economic reasons in the industry itself, or in order to reduce the spread of diseases among captive fish. I conclude by suggesting that sustainable use of Atlantic salmon is not possible until the responsibility for reaching the combined goals of protection of wild stocks and domestication of the species in contained culture is localized to one authority which assumes this dual responsibility.

ACKNOWLEDGMENTS

I am indebted to Nils Ryman and Fred Utter for discussion, Richard L. Saunders for comments on the manuscript, Hans B. Bentsen, Dagfinn Gausen, Lars Petter Hansen, Torbjörn Järvi, Bror Jonsson, Roar A. Lund, Harald Lura, Vidar Moen, Harald Sægrov and Finn Økland for letting me cite unpublished work, and Lars Kvenild, Eva M. Schjetne and Finn Økland for making the figures.

LITERATURE CITED

Allendorf, F. W. 1983. Isolation, gene flow, and genetic differentiation among populations, pp. 51–65. In: C. M. Schonewald-Cox, S. M. Chambers, B. MacBryde and W. L. Thomas (eds) *Genetics and Conservation: A Reference for Managing Wild Animal and Plant Populations.* Benjamin/Cummings, Menlo Park, CA.

Allendorf, F. W. and S. R. Phelps. 1981. Isozymes and the preservation of genetic variation in salmonid fishes, pp. 37–52. In: N. Ryman (ed) *Fish Gene Pools. Preservation of Genetic Resources in Relation to Wild Fish Stocks.* Ecological Bulletins, Stockholm.

Allendorf, F. W. and N. Ryman. 1987. Genetic management of hatchery stocks, pp. 141–159. In: N. Ryman and F. Utter (eds) *Population Genetics and Fishery Management.* University of Washington Press, Seattle, WA.

Allendorf, F. W. and F. M. Utter. 1979. Population genetics, pp. 407–454. In: W. S. Hoar, D. J. Randall and J. R. Brett (eds) *Fish Physiology*, Vol. VIII. Academic Press, New York.

Anderson, R. M. and R. M. May. 1986. The invasion, persistence and spread of infectious diseases within animal and plant communities. Phil. Trans. R. Soc. Lond. B 314: 533–568.

Anonymous. 1989. Fishing and rearing of salmon, etc. 1987. Central Bureau of Statistics of Norway, Oslo. (In Norwegian, English summary)

Bailey, J. K. 1987. Canadian sea ranching program (east coast), pp. 443–448. In: K. Tiews (ed) *Proceedings of the World Symposium on Selection, Hybridization and Genetic Engineering in Aquaculture*, Vol. II. Heeneman, Berlin.

Bakke, T. A., P. A. Jansen and L. P. Hansen. 1990. Differences in host resistance of Atlantic salmon, *Salmo salar* L., stocks to the monogenean *Gyrodactylus salaris* Malmberg, 1957. J. Fish Biol. 37: 577–587.

Bentsen, H. B. 1991. Quantitative genetics and management of wild populations. Aquaculture 98: 263–266.

Brown, A. H. D. and D. J. Schoen. 1992. Plant population genetic structure and biological conservation. This volume.

Chourrout, D. 1987. Genetic manipulations in fish: a review of methods, pp. 111–126. In: K. Tiews (ed) *Proceedings of the World Symposium on Selection, Hybridization and Genetic Engineering in Aquaculture*, Vol. II. Heeneman, Berlin.

Crow, J. F. 1986. *Basic Concepts in Population, Quantitative, and Evolutionary Genetics*. Freeman, New York.

Donaldson, L. R. and T. Joyner. 1983. The salmonid fishes as a natural livestock. Scient. Amer. 249(1): 48–56.

Drake, J. A., H. A. Mooney, F. di Castri, R. H. Groves, F. J. Kruger, M. Rejmánek and M. Williamson (eds) 1989. *Biological Invasions: A Global Perspective*. Wiley, Chichester, UK.

Egidius, E., L. P. Hansen, B. Jonsson and G. Nævdal. 1991. Mutual impact of wild and cultured Atlantic salmon in Norway. J. Cons. Int. Explor. Mer 47: 404–410.

Garcia de Leániz, C. and E. Verspoor. 1989. Natural hybridization between Atlantic salmon, *Salmo salar*, and brown trout, *Salmo trutta*, in northern Spain. J. Fish Biol. 34: 41–46.

Garcia de Leániz, C., E. Verspoor and A. D. Hawkins. 1989. Genetic determination of the contribution of stocked and wild Atlantic salmon, *Salmo salar* L., to the angling fisheries in two Spanish rivers. J. Fish Biol. 35 (Suppl. A): 261–270.

Gausen, D. 1988. Registrering av oppdrettslaks i vassdrag, pp. 58–69. In: B. Lindgren (ed) *Fagmøte om Sikringssoner for Laksefisk*. LENKA, Miljøverndepartementet, Oslo, Norway. (In Norwegian)

Gausen, D. (in press). The Norwegian gene bank programme for Atlantic salmon (*Salmo salar*). In: J. G. Cloud and G. H. Thorgaard (eds) *Genetic Conservation of Salmonid Fishes*, NATO Advanced Study Institute, Series A, Life Sciences. Plenum, New York.

Gausen, D. and V. Moen. 1991. Large-scale escapes of farmed Atlantic salmon (*Salmo salar*) into Norwegian rivers threaten natural populations. Can. J. Fish. Aquat. Sci. 48: 426–428.

Gjedrem, T. and D. Aulstad. 1974. Selection experiments with salmon. I. Differences in resistance to vibrio disease of salmon parr (*Salmo salar*). Aquaculture 3: 51–59.

Gjedrem, T., B. Gjerde and T. Refstie. 1988. A review of quantitative genetic research in salmonids at AKVAFORSK, pp. 527–535. In: B. S. Weir, M. M. Goodman, E. J. Eisen and G. Namkoong (eds) *Proceedings of the Second International Conference on Quantitative Genetics*. Sinauer, Sunderland, MA.

Gross, M.R. 1985. Disruptive selection for alternative life histories in salmon. Nature 313: 47–48.

Haldane, J. B. S. 1964. A defense of beanbag genetics. Persp. Biol. Med. 7: 343–359.

Hansen, L. P. 1986. The data on salmon catches available for analysis in Norway, pp. 73–78. In: D. Jenkins and W. M. Shearer (eds) *The Status of the Atlantic Salmon in Scotland.* ITE Symposium No. 15, Huntingdon, UK.

Hansen, L. P., K. B. Døving and B. Jonsson. 1987. Migration of farmed adult Atlantic salmon with and without olfactory sense, released on the Norwegian coast. J. Fish Biol. 30: 713–721.

Hansen, L. P. and B. Jonsson. 1991. Evidence of a genetic component in the seasonal return pattern of Atlantic salmon, *Salmo salar* L. J. Fish Biol. 38: 251–258.

Hansen, L. P., B. Jonsson and R. Andersen. 1989. Salmon ranching experiments in the River Imsa: is homing dependent on sequential imprinting of the smolts?, pp. 19–29. In: E. Brannon and B. Jonsson (eds) *Proceedings of the Salmonid Migration and Distribution Symposium*. School of Fisheries, University of Washington, Seattle, WA.

Hesthagen, T. and L. P. Hansen. 1991. Estimates of the annual loss of Atlantic salmon, *Salmo salar* L., in Norway due to acidification. Aquacult. Fish. Manage. 22: 85–91.

Hindar, K., N. Ryman and F. Utter. 1991. Genetic effects of cultured fish on natural fish populations. Can. J. Fish. Aquat. Sci. 48: 945–957.

Howarth, F. G. 1991. Environmental impacts of classical biological control. Ann. Rev. Entomol. 36: 485–509.

Huitfeldt-Kaas, H. 1946. Tribes of salmon in Norway. Nytt Mag. Naturvidensk. B 85: 115–159.

Isaksson, A. 1988. Salmon ranching: a world review. Aquaculture 75: 1–33.

Johnsen, B. O. and A. J. Jensen. 1986. Infestations of Atlantic salmon, *Salmo salar*, by *Gyrodactylus salaris* in Norwegian rivers. J. Fish Biol. 29: 233–241.

Jonsson, B., N. Jonsson and L. P. Hansen. 1990. Does juvenile experience affect migration and spawning of adult Atlantic salmon? Behav. Ecol. Sociobiol. 26: 225–230.

Kapuscinski, A. R. D. and J. E. Lannan. 1984. Application of a conceptual fitness model for managing Pacific salmon fisheries. Aquaculture 43: 135–146.

L'Abée-Lund, J. H. 1989. Significance of mature male parr in a small population of Atlantic salmon (*Salmo salar*). Can. J. Fish. Aquat. Sci. 46: 928–931.

Levin, B. R. 1992. DNA technology and the release of genetically engineered organisms: some implications for the conservation of genetic resources. This volume.

Lund, R. A., L. P. Hansen and T. Järvi. 1989. Identification of reared and wild salmon by external morphology, size of fins and scale characteristics. NINA Research Rep. 1: 54 pp. (In Norwegian, English summary)

Lund, R. A., F. Økland and L. P. Hansen. 1991. Farmed Atlantic salmon (*Salmo salar*) in fisheries and rivers in Norway. Aquaculture 98: 143–150.

Lura, H. and H. Sægrov. 1991a. A method of separating offspring from farmed and wild Atlantic salmon (*Salmo salar*) based on different ratios of optical isomers of astaxanthin. Can. J. Fish. Aquat. Sci. 48: 429–433.

Lura, H. and H. Sægrov. 1991b. Documentation of successful spawning of escaped farmed female Atlantic salmon, *Salmo salar*, in Norwegian rivers. Aquaculture 98: 151–159.

MacCrimmon, H. R. and B. L. Gots. 1979. World distribution of Atlantic salmon, *Salmo salar*. J. Fish. Res. Bd Can. 36: 422–457.

Mills, D. 1989. *Ecology and Management of Atlantic Salmon.* Chapman & Hall, London.

Moen, V. and D. Gausen. 1989. Rømt oppdrettsfisk i vassdrag 1988. Direktoratet for naturforvaltning, Trondheim, Norway, Rapport 3–1989. 26 pp. (In Norwegian)

Nævdal, G. 1981. Fish rearing in Norway with special reference to genetic problems, pp. 85–93. In: N. Ryman (ed) *Fish Gene Pools. Preservation of Genetic Resources in Relation to Wild Fish Stocks.* Ecological Bulletins, Stockholm.

Nelson, K. and M. Soulé. 1987. Genetical conservation of exploited fishes, pp. 345–368. In: N. Ryman and F. Utter (eds) *Population Genetics and Fishery Management.* University of Washington Press, Seattle, WA.

Parsons, J. E. and G. H. Thorgaard. 1985. Production of androgenetic diploid rainbow trout. J. Hered. 76: 177–181.

Pedersen, T. N., J. Aure, B. Berthelsen, S. Elvestad, A. S. Ervik and H. Kryvi. 1988. A nation-wide analysis of the suitability of the Norwegian coast and watercourses for aquaculture. A coastal zone management programme. ICES C. M. 1988/F: 11, Copenhagen, Denmark.

Refstie, T. 1987. Selective breeding and intraspecific hybridization of cold water finfish, pp. 293–302. In: K. Tiews (ed) *Proceedings of the World Symposium on Selection, Hybridization and Genetic Engineering in Aquaculture*, Vol. I. Heeneman, Berlin.

Ricker, W. E. 1972. Hereditary and environmental factors affecting certain salmonid populations, pp. 19–160. In: R. C. Simon and P. A. Larkin (eds) *The Stock Concept of Pacific Salmon.* H. R. MacMillan Lectures in Fisheries, University of British Columbia, Vancouver, B. C.

Riddell, B. E., W. C. Leggett and R. L. Saunders. 1981. Evidence of adaptive polygenic variation between two populations of Atlantic salmon (*Salmo salar*) native to tributaries of the S. W. Miramichi River, N. B. Can. J. Fish. Aquat. Sci. 38: 321–333.

Rikstad, A. 1991. Furunkulose i laksevassdrag. Norsk Fiskeoppdrett 1–91: 62. (In Norwegian)

Ritter, J. A. 1975. Lower ocean survival rates for hatchery-reared Atlantic salmon (*Salmo salar*) stocks released in rivers other than their native stream. ICES Ana. Cat. Comm. C. M. 1975/M: 26, Copenhagen, Denmark.

Ryman, N. 1983. Patterns of distribution of biochemical genetic variation in salmonids: differences between species. Aquaculture 33: 1–21.

Ryman, N., R. Chakraborty and M. Nei. 1983. Differences in the relative distribution of human gene diversity between electrophoretic and red and white cell antigen loci. Hum. Hered. 33: 93–102.

Ryman, N., U. Lagercrantz, L. Andersson, R. Chakraborty and R. Rosenberg. 1984. Lack of correspondence between genetic and morphologic variability patterns in Atlantic herring (*Clupea harengus*). Heredity 53: 687–704.

Ryman, N., F. Utter and K. Hindar (in press). Introgression, supportive breeding, and genetic conservation. In: J. D. Ballou, T. Foose and M. Gilpin (eds) *Population Management for Survival and Recovery.* University of Chicago Press, Chicago.

Sattaur, O. 1989. The threat of the well-bred salmon. New Scient. 122 (1662): 54–58.

Saunders, R. L. 1981. Atlantic salmon (*Salmo salar*) stocks and management implications in the Canadian Atlantic provinces and New England, USA. Can. J. Fish. Aquat. Sci. 38: 1612–1625.

Schaffer, W. M. and P. F. Elson. 1975. The adaptive significance of variations in life history among local populations of Atlantic salmon in North America. Ecology 56: 577–590.

Ståhl, G. 1981. Genetic differentiation among natural populations of Atlantic salmon (*Salmo salar*) in northern Sweden, pp. 95–105. In: N. Ryman (ed) *Fish Gene Pools. Preservation of Genetic Resources in Relation to Wild Fish Stocks.* Ecological Bulletins, Stockholm.

Ståhl, G. 1987. Genetic population structure of Atlantic salmon, pp. 121–140. In: N. Ryman and F. Utter (eds) *Population Genetics and Fishery Management.* University of Washington Press, Seattle, WA.

Ståhl, G. and K. Hindar. 1988. Genetisk struktur hos norsk laks: status og perspektiver. Rep. 1988–1, Fiskeforskningen, Direktoratet for naturforvaltning, Trondheim. 57 pp. (In Norwegian)

Stabell, O. B. 1984. Homing and olfaction in salmonids: a critical review with special reference to the Atlantic salmon. Biol. Rev. 59: 333–388.

Stoss, J. 1983. Fish gamete preservation and spermatozoan physiology, pp. 305–350. In: W. S. Hoar, D. J. Randall and E. M. Donaldson (eds) *Fish Physiology*, Vol. IX B. Academic Press, New York.

Templeton, A. R. 1986. Coadaptation and outbreeding depression, pp. 105–116. In: M. E. Soulé (ed) *Conservation Biology: The Science of Scarcity and Diversity.* Sinauer, Sunderland, MA.

Vuorinen, J. and O. K. Berg. 1989. Genetic divergence of anadromous and nonanadromous Atlantic salmon (*Salmo salar*) in the River Namsen, Norway. Can. J. Fish. Aquat. Sci. 46: 406–409.

Wohlfarth, G. W. 1986. Decline in natural fisheries – a genetic analysis and suggestion for recovery. Can. J. Fish. Aquat. Sci. 43: 1298–1306.

Wright, S. 1969. *Evolution and the Genetics of Populations*, Vol. 2. *The Theory of Gene Frequencies.* University of Chicago Press, Chicago.

SECTION 4

Strategies for *ex situ* conservation

So far we have considered biodiversity in the wild – its present vulnerability to continuing depredation and the need to conserve it *in situ* in reserves. We now turn to a group of adjunct approaches to the conservation of biodiversity that are *ex situ* procedures. Such procedures remove samples of species and their genes from their present precarious situation at their site of origin for conservation elsewhere under institutional protection. The conserved organisms can be maintained in the actively growing state in botanical or zoological gardens, or stored as dormant samples in gene banks. The former are more important for wild taxa, whereas the latter are employed for the genetic resources of domesticated plants and animals.

For wild plants, Heywood (Chapter 11) stresses the need to conserve germplasm of the many species of any actual or potential use to humans. This amounts to tens of thousands of species, very few of which currently have even a modest sampling in gene banks. The world's botanic gardens have a key role to play in this task. They contain an enormous array of plant species, the listing of which is itself a major job, let alone obtaining data on provenance and sampling. The Botanic Gardens Conservation Secretariat holds data on about 10,000 species, mainly rare and endangered species, in cultivation in 350 botanic gardens around the world.

Gene banks for plants have a history of over six decades since Vavilov pioneered the vision of large collections that include as much as possible of the genetic diversity of economic plants and their related wild species. Blixt (Chapter 12) outlines the work of the Nordic Gene Bank as an institution with conservation responsibilities in agricultural species and wild relatives, particularly the Nordic endemics. He concludes by calling for closer cooperation between gene banks of domesticated species and those working in other areas of plant conservation.

In Chapter 13, de Boer deals with propagation of populations of endangered animal species in zoos, with the ultimate goal of sustaining wild populations in nature. He stresses that maintenance and mating procedures should aim to preserve the demographic and population genetic structure, its allelic frequencies, and heterozygosity as much as possible. In the future, zoos could administer breeding programs for about one thousand species of higher vertebrates worldwide. This is a substantial challenge, but worthwhile as a concerted political commitment to conservation and ideally to saving the habitats of these species as well, and thereby thousands more. He concludes with the thought that the escalating fragmentation of natural populations *in situ* makes them increasingly resemble the small populations in zoos. The experience of managing the zoo populations should improve our management of their natural counterparts.

Economic and biological reasons combine with cultural and historical ones prominently for the conserving of the genetic resources of domesticated animals, as Maijala and Kolstad explain in Chapter 14. Success will depend on a composite of methods employing live animals and frozen semen and embryos. The maintenance of the genetic integrity of each conserved breed is a fundamental objective, despite the cost entailed.

Thus four chapters cover four kinds of resources, wild or domesticated, plants or animals. The institutions involved differ in emphases and techniques. Yet they are united by the underlying principles of sampling strategies, of maintenance without genetic change (static conservation), of interactive coordinated activities, of adequate documentation and of future use of these resources back in the wild or in productive systems.

11 Conservation of germplasm of wild plant species

Vernon H. Heywood
Plant Conservation Office,
IUCN – The World Conservation Union,
Descanso House, 199 Kew Road,
Richmond, Surrey, TW9 3BW,
UK

ABSTRACT

The conservation of genetic resources of wild species, including crop relatives, both *in situ* and *ex situ*, is an important aspect of biological diversity. In the conservation of plant genetic resources, attention has been focussed mainly on the small number of species that provide our basic food needs, plus plantation or industrial crops, forages, green manures and timber trees. While there are adequate collections in gene banks for most of these major crops and some of the lesser ones, as defined in agricultural terms, the number of accessions of wild relatives is small. Greater efforts need to be devoted to the conservation of genetic resources of the tens of thousands of wild or semi-domesticated species that are used in local indigenous agriculture in many parts of the world, especially in lesser developed countries. Suggestions are made for a programme of action, including developing the role of botanical gardens and arboreta as centres for the collection, storage, and conservation of wild plant resources.

INTRODUCTION

The conservation of germplasm of wild species, including crop relatives, both *in situ* and *ex situ*, is an important if neglected aspect of biological diversity. In the conservation of plant genetic resources, attention has been focussed mainly on the small number of species, in the range of 200 – 250 that provide our basic food needs (such as cereals, food legumes, rootcrops, fruits, vegetables, oils, spices), plus plantation or industrial crops (such as cacao, oil palm, rubber, jute, sugar cane, beet, cotton, tobacco) and forages, green manures and timber trees. While there are adequate collections in gene banks for most of these major crops and some of the lesser ones as defined in agricultural terms, the number of accessions of wild relatives of crop plants included is small. As Plucknett et al. (1987) note "In spite of the importance of wild species in crop breeding, they have been generally neglected in germplasm collecting". The reasons for this neglect include the fact that widecrossing is time-consuming and often difficult, and the lack of

Conservation of Biodiversity for Sustainable Development.
Edited by O. T. Sandlund, K. Hindar and A. H. D. Brown. 1992. Scandinavian University Press, Oslo.

available information about their potentially useful features such as disease resistance, drought tolerance, etc. The following factors are also important: seed yield is often low, especially during regeneration, which makes maintenance in seed banks difficult; the seed cases shatter causing loss, and many wild species are photoperiod sensitive. Other difficulties are the prolonged period of seed and the variation in flowering periods.

As well as these technical reasons for the neglect of wild germplasm, there are historical and political factors involved as will be discussed later. But whatever the reasons, the fact is that samples of wild species typically account for less than 2 % of the accessions in crop germplasm collections (Plucknett et al. 1987, and see Table 1). In 1988 IBPGR, IUCN and WWF launched a call to governments and other agencies to intensify their efforts to conserve the wild relatives of crops both *in situ* and *ex situ* (Hoyt 1988). As much wild germplasm as possible should be collected before it becomes extinct.

The main agencies concerned with the conservation of crop germplasm are FAO, IBPGR and other CGIAR Institutes, and other institutions such as the Nordic Gene Bank, CENARGEN (Brazil), the SADCC Gene Bank (Zambia), the National Bureau of Plant Genetic Resources (India), United States National Germplasm System and some 150 others (see Plucknett et al. 1987 Appendix 2). Their operation and functio-

Table 1. Only the wild relatives of a few crops such as wheat, potato and tomato have been widely collected and preserved in seed banks. In most cases, wild germplasm represents less than 2 percent of the seed bank holdings for each crop and most wild relatives of crops still thrive only in the wild. IBPGR is helping to focus world-wide efforts to collect and conserve them. Source: IBPGR estimates, 1988.

Crop	Wild species held in all seed banks as % of total holdings	Estimated % of wild species still to be collected
CEREALS		
Rice	2	70
Wheat	10	20–25
Sorghum	0.5	90
Pearl millet	10	50
Barley	5	0–10
Corn (maize)	5	50
Minor millets	0.5	90
ROOT CROPS		
Potato	40	30
Cassava	2	80
Sweet potato	10	40
LEGUMES		
Beans	1.2	50
Chickpea	0.1	50
Cowpea	0.5	70
Groundnut	6	30
Pigeonpea	0.5	40

ning is covered by Blixt (this volume). Most of the accessions in these gene banks are of cultivars. Many of them, however, have a limited number of accessions of wild species. Thus the US National Plant Germplasm System contains samples of several thousand species: according to figures obtained from the GRIN (Genetic Resources Inventory) database, there are 9,045 species, 1,789 genera in 207 families represented in the 387,864 accessions in the system (A. Storer in litt.). Of these species 437 are truly domesticated (representing 5 % of the total) and these account for 311,513, i.e. 80 % of the total accessions. Thus wild species in this case account for 20 % of the total number of accessions. The Southern Region Plant Introduction Station (SRPIS), for example, currently maintains a collection in excess of 60,000 accessions representing 270 genera and 1,320 species (Jarret et al. 1990).

The germplasm collection of the N. I. Vavilov All-Union Scientific Research Institute of Plant Industry (VIR), in the Soviet Union, is reported to have the world's largest collection of germplasm in terms of genera, species and cultivars (Zaitzev 1990). According to its Director, Krivchenko (1988) it "holds more than 375,000 samples relating to 155 botanical families, 304 genera and 2,539 species."

It is clear that germplasm of thousands of wild species exists in leading gene banks but there is no complete inventory, a subject considered in more detail below. Some of these are crop relatives while others are undefined. The Germplasm Acquisition Policy for the US National Plant Germplasm System which regards the latter as "species not even considered part of a tertiary gene pool of present crops..." and with "the potential to become new crops, contribute to genetics research or, as technology advances, become part of the tertiary gene pool of current crops" (See Diversity, Vol. 5, no.4, p. 31, 1990).

We should not, however, be misled by narrow definitions of crops and underestimate the problems of conserving wild resources. My concern in this paper is with plants useful to humankind, other than the 500 or so crops covered by the agricultural germplasm systems: some of them cultivated although not the subject of a breeding programme, some semi-domesticated, some wild-harvested, and some of them of potential use (see later).

PLANTS USED BY HUMANKIND

Introduction

Of the 250,000 known species of higher plants, up to 100,000 of them have been used at some time by humankind. The number in current use is still enormous. In California, more than 250 crops are grown commercially. In addition to the major crop plants and their wild relatives, efforts need to be devoted to the exploration, collection and conservation of germplasm of wild or semi-domesticated species and local crops in use in indigenous agriculture, especially in the lesser developed countries. Estimates as to the numbers of these species vary considerably, according to how widely the net is cast, and no inventory has ever been attempted. Even if we exclude ornamentals, but include medicinals we may be dealing with tens of thousands.

The PROSEA (Plant Resources of South-East Asia) project, for example estimates that there are some 5–10,000 species useful for mankind in Southeast Asia. By extrapolation this would give a figure of at least 25–50,000 for the tropical regions of the world. Of course, the majority of these species have little economic or nutritional value. Yet many of these plants form an important element in the diet of local communities without, in most cases, entering into trade statistics.

A good example of the use of local plant (and animal) resources is given by Calestous Juma in his important study *Biological Diversity and Innovation, Conserving and Utilizing Genetic Resources in Kenya* (Juma 1989a). Kenya is one of the few African countries with an advanced agricultural system, and one which owes a great deal to its colonial past. In addition to well-known indigenous crops such as coffee, sesame, castor oil, sorghum, teff, millet and okra, Kenya is rich in other crops which form part of the traditional diet including numerous pulses such as *Vigna*, amaranths, and members of the Solanaceae and Brassicaceae. However, during the period of modern agriculture in Kenya, the country increasingly relied on exotic crops and tree species to earn revenue. Over the period 1964–85 it imported about 64 % of the genetic resource accessions needed for breeding purposes.

Very little attention has been paid to research on local fruits and vegetables in Kenya, despite their importance in the nutrition and lifestyle of many local communities. In a discussion of the agricultural situation in the Bugoma area on the slopes of Mount Elgon, Juma (1989a) notes that the residents use a wide range of plants, including 100 different species of fruits and vegetables, drawn from 70 genera and 35 families. What is more, the Bukusu people have some understanding of genetic resource conservation and utilization as part of their local knowledge system that links natural, social and metaphysical factors.

The emphasis in breeding programmes on increasing food production has led to a narrower genetic base in the major commercial food crops. This, says Juma (1989b), does not reflect the preferences of the consumers who opt for greater variability in their food sources when able to exercize their choice.

Underexploited species

Examples of new or underexploited plants range from well known groups such as palms (which in terms of general usefulness to humankind are the most important plant family in the tropics; Johnson 1989) and rattans, medicinal plants and fruit trees, to weeds such as *Vernonia galamensis* (Perdue 1988; Perdue et al. 1989), a promising new crop for third world farmers and one that may reduce air pollution.

A good review of the attention focussed on new plant sources for food or industry is the symposium volume *New Crops for Food and Industry* (Wickens et al. 1989). Reviews of underexploited or promising new species for development are being compiled for many countries or regions. One example is a series of volumes on promising plant species of the Andean subregion (Correa and Bernal 1990). The editors note that many of the species reported on are practically unknown in other parts of the world but are vital to native populations of the region. However, genetic erosion is already serious in this area of the Andes.

In India, the Hindustani Gene Centre houses some 160 cultivated species, 56 species of lesser-known cultivated food plants and about 320 species of wild and weedy economic forms (Arora and Nayar 1984). Paroda and Mal (1989) list a large number of useful plant species from this region which have yet to be domesticated and used more effectively for the benefit of humankind. This includes 81 important little known cultivated food plants awaiting exploitation in India, plants for extreme environmental/emergency situations, and industrial plants. Research on the collection, evaluation, utilization and conservation of such underexploited plants is being undertaken by the All India Coordinated Programme and by the National Bureau for Plant Genetic Resources, New Delhi. The latter maintains active collections of such species at its headquarters as well as in its regional stations in different agro-climatic zones.

Brazil is an example of a tropical country with enormous potential for new food and industrial crops, especially from the Amazon. Arkcoll and Clement (1989) review Amazon plants that can provide crops for carbohydrate, for oil, for protein, for fruit and nuts and for diverse uses. They draw attention to the apparent lack of little-known vegetables in the Amazon and the very low consumption of leafy spinaches and suggest that the abundant fruit available in the forest is eaten in sufficient quantity to satisfy vitamin requirements. Several species of palm hearts are eaten by local communities in Brazil. Most of the accessible *Euterpe edulis* which is used for commercial canning in the south of Brazil in the past has been eliminated. Canners have now moved up the Amazon estuary where they are exploiting wild *Euterpe oleracea* at the rate of almost 100,000 tonnes per annum. Other species are used similarly on a smaller scale and experiments are being made with commercial plantations of *Bactris gasipaes* which produces good quality palm hearts on a sustainable basis. In addition to the domestic market palm hearts are considered to have good export potential.

In South and Southeast Asia wild and semidomesticated fruit trees, such as mango and durian, have considerable economic importance and export potential. There are several initiatives to study the genetic resource conservation and utilization of such species. Genetic diversity of the wild relatives of citrus and mango species has been studied by IBPGR, IUCN and WWF in Southeast Asia (IBPGR 1985a). A project involving IBPGR, USAID and the Central Research Institute for Horticulture, Indonesia, is underway to identify and assess areas of high genetic diversity of such species in the forests of Sumatra, Indonesia.

Forestry species

Similar considerations apply to forest species where imported species dominate the reafforestation programmes, despite the studies by local botanists of local tree species. Sometimes there are good reasons for this; sometimes it is done by default as when the only seed stocks available are of exotic species. In Kenya, as in many other countries, however, the growing awareness of the value of native plant diversity has stressed the importance of local species and their conservation. FAO has produced a list of important forestry species (FAO 1981), and 2,000 woody species have been identified and listed as useful "multipurpose trees" by the International Council for Research in Agroforestry (ICRAF) (Huxley and Westley 1989).

Edible trees and other woody plants play an important role in the rural economy of the Nigerian forest zone (Okafor 1978, 1980). In addition to their direct dietary benefits, their value includes cash revenue, significance in traditional farming practices, timber supplies, firewood, chewing sticks, fibres, drugs, dyes, socio-cultural roles and religious purposes. Studies have been published on shrubs and trees important for energy production (NAS 1980) and on tropical leguminous trees and shrubs as potential future resources (NAS 1979).

Shrubs

Shrubs as a group deserve further attention as regards germplasm conservation. The Shrub Research Consortium (SRC) was founded by five North American Institutions in 1983 to develop shrub plant materials, methods of cultivation, seeding and management in natural settings. It now consists of 20 member institutions and organizes wildland shrub symposia which have considered the ecology, management and propagation of species such as *Artemi-*

sia tridentata and allies (sagebrush), *Atriplex confertifolia* (shadscale), *Purshia* spp. (bitterbrush), *Kochia prostrata* (forage kochia).

Medicinal plants

Another important group of plants lacking in germplasm provision are the medicinal plants. The WHO estimates that 80 % of peoples in developing countries still rely on plant-based medicines as their principal source of health care. An increasing number of the species utilized are becoming endangered as habitats are lost and demand increases. The conservation needs of medicinal plants were discussed at an interdisciplinary International Consultation on Medicinal Plant Conservation, organized by WHO, IUCN and WWF (Akerele et al. 1991). A set of Guidelines on the conservation of medicinal plants was agreed at the meeting.

Some 25,000 plant species have been used in herbal medicines so that their conservation is a problem of major dimensions. The conservation of many samples of all these species is neither possible nor desirable. As in other classes of plants used by humankind, priority must be given to those that are most at risk of extinction, totally or locally.

Genetic erosion is widespread and can be as significant in medicinal plant species as in food crop species. This is because particular variants or alleles within populations are needed for selection and breeding. This applies especially at the level of chemical variation in medicinal plants whose use is often site-specific and substitutions may be undesirable or even dangerous.

A general problem affecting most medicinal plant species which are used locally is that information on supply and demand is often lacking. We need to know which plants are used, how they are harvested, by whom, how they are traded, whether the supply is sufficient to meet the demand, whether the plants are in or susceptible to cultivation, whether they are endangered in the wild and if so what conservation measures are appropriate. Similar considerations apply to aromatic or oil plants (Atal and Kapur 1982a,b). Some detailed investigations of the local use of medicinal and aromatic plants in traditional systems are underway and should be undertaken for every country that relies on such herbal medicines.

Other species

Wild or semidomesticated species also provide plants of economic value for fibres, oils, tannins, waxes, beverages, intoxicants and spices. With certain conspicuous exceptions such as oil palm, sunflower (*Helianthus* spp.), oilseed rape, jute, and coffee, little attention has been paid to the exploration for and conservation of germplasm of such species.

Ornamental species

Another very important group of plants is that used for ornament in parks, private gardens and public places. While some have been extensively bred and modified, the majority are wild species. The number of species involved is very large and difficult to estimate precisely. Many species are cultivated only in botanic gardens. The widespread misidentification of the cultivated material complicates the problem. The European Garden Flora plans to include all species likely to be found in general collections (i.e. excluding specialist collections and botanic gardens), whether grown outside or under glass. It expects to cover some 12,000 species. While many of them are also cultivated outside Europe, probably as many again are in cultivation in other parts of the world. There is no germplasm system as such for these plants, many of which are of direct wild origin.

In addition, thousands of species are grown in specialist nurseries and in private collections. These cover plants from virtually all families and regions and contain a bewildering diversity. As no inventory of these species or their exact provenance exists, the size and value of these collections as a source of germplasm cannot be judged.

Keystone species

Another important category of plants that tends to be overlooked in considerations of genetic conservation are keystone or target species. These are species whose disappearance would significantly decrease the biodiversity of an area or on whose presence the stability or very existence of an ecosystem depends. With the expected effects of global change, both in climate and population size, it is clear that we ought to consider now what the demands on germplasm will be that will be needed to counter or tolerate the stressful environmental conditions expected.

BOTANIC GARDEN COLLECTIONS

Introduction

The world's botanic gardens collectively contain the largest assemblage of biodiversity outside nature. There are approximately 1,500 botanic gardens and arboreta in the world, ranging from major research institutions with large collections, and extensive facilities such as herbaria, laboratories, libraries, as at the Berlin Botanic Garden, Royal Botanic Gardens, Kew, Missouri Botanical Garden, New York Botanical Garden and Rio de Janeiro Botanic Garden, to small municipal botanic gardens with small collections and few facilities (Heywood 1987, 1990; WWF/IUCN/BGCS 1989) (Table 2). As Raven (1981) comments, these collections of plants constitute an important resource for pure and applied research in many fields of plant science that depend on growing material.

Number and range of accessions

The number of species cultivated in botanic gardens is difficult to assess, partly because the gardens themselves often do not have accurate statistics. Raven (1981) suggests that about 35,000 species are presently cultivated in botanic gardens but I believe the figure is substantially higher. Many botanic gardens have published catalogues of their holdings, such as Bogor (about 3,500 species) and the National Botanic Garden of Belgium at Meise which recently issued a catalogue of its greenhouse holdings (17,000 accessions, many of which are of known wild origin, belonging to 10,235 taxa). Such catalogues publicized the wealth of new accessions from overseas territories during the 18th and 19th centuries when gardens competed in the range of material they cultivated. A difficulty with many of these catalogues is that many of the plants listed have not been taxonomically verified. Moreover the lists soon become out of date.

Other sources of information on the acccessions held by botanic gardens are their databases. Already many of them have computerized their holdings (Cullen et al. 1987).

About half of the world's botanic gardens and arboreta regularly produce a Seed List (Index Seminum), listing those taxa for which seed (or in some cases vegetative propagules) are available for free exchange. This system has operated for over two centuries. It indicates the important role the gardens have long played in the exchange and movement of germplasm around the world.

An increasing number of botanic gardens now include seed of wild origin in their Seed Lists and recommendations have been published as to the nature and quality of the information that they should contain. An analysis published in 1983 indicated that 432 out of 600 Seed Lists reviewed from 25 countries included seed collected from wild habitats. Unfortunately few tropical botanic gardens were included in this number.

A new source of information on the numbers and nature of the accessions held by botanic gardens is the new edition of the International Directory of Botanical Gardens (Heywood et al. 1990). This Directory includes, as far as possible, for each garden, entries for the number of taxa in cultivation and a summary of special collections, including conservation collections.

Often overlooked but extremely important kinds of collection are those built up by experimental taxonomists or cytologists in university or other laboratories associated with botanic gardens. These collections contain a wide range of wild collected material of the genera or species groups concerned. Unfortunately they are often not included in the botanic garden's accessions database or record system and are omitted from global surveys like those of the Botanic Gardens Conservation Secretariat (BGCS). They probably cover several thousand species. As Poppendieck (1976) has pointed out, these collections often become dismantled or neglected after the person who assembled them leaves the institution or changes projects. University botanic gardens in particular are full of the debris of former theses. Raven (1981) suggests that the possibility of storing samples in seed banks should be considered if facilities are available, as a means of conserving these valuable collections for future use.

It is not yet possible to summarize usefully all this information about botanic garden collections. The data in the Directory database are still being analyzed but it is clear that botanic gardens contain enormous resources of germplasm of a very diverse nature. These resources stand in urgent need of reappraisal in terms of its coverage and value as part of the national and international effort to collect and conserve genetic material of wild species.

A sample of the size of accessions of different botanic gardens is given in Table 2. The size ranges from over 30,000 to a few hundred with the "average" collection of 1,000–5,000 taxa. Yet it is impossible to arrive at a cumulative figure because of the very high degree of overlap between many of the accessions in different gardens.

Specialized collections

Many gardens contain special collections of plants of their own region such as the Rancho Santa Ana Botanic garden which has 2,000 taxa, all of which are native to California; the Jardín Botánico Viera y Clavijo, Gran Canaria, has conservation collections mainly from the Macronesian region; the National Botanic Gardens of South Africa between them cultivate about 7,000 species representing much of the very rich flora of the region; the University of British Columbia Botanic Gardens includes about half of the native species of the province amongst the 14,000 taxa it cultivates; the Conservatoires of Brest, Nancy, Porquerolles and Mascarin in France maintain conservation collections of native species of French metropolitan and overseas territories as part of their policy; for Australia, a recent estimate suggests that between a third and a half of the native vascular flora of 20,000 species is in cultivation in the country's botanic gardens.

Many gardens have built up special conservation collections of particular groups. For example, the Parque Exotique de Prafrance, Generargues, France grows more than 100 species of bamboo; the Marie Selby Tropical Garden has a collection of 4,658 epiphytic plants; several very extensive orchid collections exist such as the Wheeler Orchid Collection and Species

Table 2 Accession figures for selected botanic gardens and arboreta.

Country/ Botanic garden	No. of accessions	Country/ Botanic garden	No. of accessions
Australia		**Singapore**	
Adelaide	4,500	Singapore	3,000
Canberra	6,000		
Sydney	7,000	**South Africa**	
Toowong	14,000	Kirstenbosch	6,200
Brazil		**Sri Lanka**	
Rio de Janeiro	7,000	Peradeniya	4,000
São Paulo	1,000		
		Taiwan	
China		Heng-chun	1,100
Beijing	3,500		
Nanjing	3,000	**United Kingdom**	
Guangzhou	4,500	Edinburgh	17,500
Xiamen	4,000	Glasgow	10,000
Xian	2,000	Kew	72,000
Hungary		**United States**	
Budapest	3,000	Brooklyn, New York	14,000
		Chicago	4,500
India		Denver	2,500
Calcutta	5,000	Fort Worth	2,500
Ootacamund	1,100	Honolulu	5,000
Lucknow	6,000	Longwood	12,000
Trivandrum	2,200	Missouri	30,000
		Santa Barbara	900
Indonesia		US Nat.Arboretum	7,600
Bogor	53,000	Waimea	7,000+
Japan		**USSR**	
Hiroshima	9,800	Moscow	21,000
Kyoto	5,500	Tblisi	300
Tokyo	3,000		
Tokyo University	4,000		

Bank at Bell State University with 3,000 orchid species in cultivation of which 150 are rare or endangered; also in the USA are the Orchid Jungle, Florida, with conservation collections of 2,000 species as well as 15–20,000 hybrids, and the Smithsonian National Orchid Collection which is part of the Smithsonian Institution's Office of Horticulture in Washington DC. This latter has some 300 genera and 2,000 orchid species in cultivation. With so many other sizeable orchid collections in botanic gardens, both in temperate and tropical countries the total number of orchid species in cultivation could well exceed half the total known in the wild. In addition several botanic gardens maintain *in situ* orchid sanctuaries as in the case of the Botanic Garden, Orchid Research and Development Centre, Tippi in Arunachal Pradesh, India which also grows 400 species, representing 93 genera.

Germplasm collections in botanic gardens

An increasing number of botanic gardens hold germplasm collections of particular groups of economically important species, including crop relatives. Examples include the IBPGR-designated germplasm collection of *Citrus* and relatives at the Rimba Ilmu, University of Malaya, Kuala Lumpur, the taro (*Colocasia*) germplasm collections at the Lyon Arboretum, Honolulu, and the extensive tropical fruit collections at Lancetilla Botanic Garden and Experimental Station, Honduras. The Experimental Botanic Gardens of the Botanical Survey of India have been established to build up germplasm collections of wild relatives of crop plants as well as rare and endangered species. The US National Arboretum acts as a germplasm repository for ornamental trees and shrubs and is currently developing a two hectare site, known as China Valley, as a depository for wild-collected Chinese germplasm to be used for plant improvement.

Thus botanic gardens cultivate a very sizeable proportion of the valuable and interesting plant species, including many rare and endangered species. The cultivation database of the Botanic Gardens Conservation Secretariat holds data on about 30,000 accessions, representing 10,000 species, mainly of rare and endangered species, in cultivation in some 350 botanic gardens.

Of course most of the accessions in botanic gardens are too small to be germplasm collections. Often the accession, as in the case of trees, consists of one or a few specimens only. But even these may be of great importance in situations where the species concerned is reduced to very small populations in the wild. In such instances it is essential to assemble all the available material, whether in cultivation or in the wild, and devise the most effective means of enhancing the genetic stock for conservation and possible reinforcement of the field populations. An example of this is on the island of Mauritius where WWF and IUCN are supporting the Government's Wildlife and Conservation Programme through a plant conservation project that has been in operation since 1983. The flora of Mauritius is among the most endangered in the world. The only hope for the continued survival of most of over 300 endemic species is intensive management of the pockets of remaining natural vegetation as mini-reserves supplemented by *ex situ* cultivation and propagation (Jackson et al. 1988).

W. Strahm (quoted in FAO/IUCN/UNESCO/UNEP 1989: p. 8) estimates that at least 50 species of Mauritian plants do not reproduce adequately in the wild and therefore need urgently to be propagated in nurseries *ex situ*. Some of these are attractive plants with potential as ornamentals such as *Gaertnera longifolia* and *Hibiscus genevii*, the latter known from only one locality. Other species are wild relatives of crops such as an endemic wild coffee, *Coffea macrocarpa*, and *Solanum torvum*, wild eggplant. Similarly on the adjacent island of Rodrigues the flora is highly endangered. Of the 49 endemic species, 22 are endangered and 8 extinct and altogether 96 % of the endemic flora is extinct or threatened (Strahm 1989). To manage the most endangered species a programme of fencing, gardened nature reserves and propagation has been carried out by Wendy Strahm and

Karen Thingsgaard under an IUCN-WWF project in tandem with the work on Mauritius. Several botanic gardens have been closely involved, including the Conservatoire Botanique de Brest, Trinity College Dublin Botanic Garden, and the Sir Seewoosagur Ramgoolam Botanic Gardens, Mauritius. Altogether 40 species of threatened Rodriguan plants are in cultivation between 29 botanic gardens. The rescue of the Rodriguan endemic "café marron" (*Ramosmania heterophylla*), through the cultivation and propagation at the Royal Botanic Gardens Kew, of a cutting taken from the single known wild individual and flown to England in April 1986, has entered conservation lore. A rooted cutting of the rescued plant was returned to Mauritius in 1989.

CONSERVATION STRATEGIES

It is clear that the above range of species and situations departs quite widely from the traditional approach to genetic resource conservation. For any realistic strategy to be proposed there has to be a great deal of selectivity and clear statement of objectives and priorities. In accordance with the objectives of the World Conservation Strategy (IUCN/WWF/UNEP 1980), conservation is "The management of human use of the biosphere so that it may yield the greatest sustainable benefit to present generations while maintaining its potential to meet the needs and aspirations of future generations". Yet conservation bodies such as IUCN – The World Conservation Union and WWF, and agricultural or genetic resource organizations such as FAO, IBPGR and the IARC institutes differ greatly in their emphasis. In the view of Grossmann (1988) this has not led to cooperation and complementarity but to a division of labour and struggle over resources. Recently encouraging signs of collaboration have emerged under the auspices of the Ecosystems Conservation Group and other structures.

In situ conservation

It is generally agreed that for the majority of situations, *in situ* conservation is the best method of conserving wild plant genetic resources (FAO/IUCN/UNESCO/UNEP 1989). It has many advantages, the most notable of which is that it is dynamic and permits the populations of the species concerned to continue to evolve. But we have to be clear that while ecosystem conservation is by definition, *in situ*, it does not necessarily conserve the genetic resources of the species contained therein adequately. Yet such a romantic or unfocussed approach to genetic resource conservation is widely adopted by many conservationists. On the other hand the crop breeding and forestry sectors do not in practice rely much on *in situ* conservation but focus on *ex situ* methods.

The IBPGR concluded in 1983 that the existing system of protected areas was inadequate for the survival of numerous valuable species and their intraspecific diversity. They set up a taskforce to consider the proper conservation of genetic resources of species *in situ*. It decided that adequate geographical and ecological data need to be collected for the better collection of genetic resources, so as to sample adequately the allellic variation within the species' populations. They recommended ecogeographical surveys (IBPGR 1985b) to determine (1) the distributions of particular species in particular regions and ecosytems; (2) patterns of infra-specific diversity; (3) relationships between survival and frequency of variants and associated ecological conditions. Roche and Dourojeanni (1984) reached similar conclusions for the conservation of genetic resources of tropical woody species (see also Palmberg 1987; FAO/IUCN/UNESCO/UNEP 1989).

For most protected areas, inventories have not been made of their floristic composition, let alone detailed population studies within the species. Their specific value for genetic resource conservation cannot be assessed. They must be regarded, therefore, as of potential value until their composition can be determined and action taken to ensure the continued survival of targetted species and their populations. Genetic resource conservation differs from ecosystem conservation in that for the former easy and rapid accessibility is important. Plant breeders need to have access to their breeding materials, not somewhere in nature. The availability of seed in the bank or plants in the field gene bank is important.

To carry out ecogeographic surveys and *in situ* conservation requires cooperation between botanists, ecologists, conservation biologists and planners as well as crop genetics professionals. To achieve greater cooperation, understanding of the different aims of the several different practitioners is needed, especially between the genetics resource sector and the conservation sector.

Frequently *in situ* genetic resource conservation of individual species will require management intervention in the ecosystem. It may not be enough to fence the areas, expecting that the populations will continue to evolve normally. Burning, for example, may be needed but this could be incompatible with the management of the ecosystem as a whole. Much more research is needed on the management techniques for genetic resource conservation *in situ*. Assessments need to be made of the costs of this approach, both economic and social, which can be quite considerable and need well-trained personnel and outside support.

Unfortunately the number of reserves set up specifically for genetic conservation of plant species is small. We desperately need more experience before fully assessing the overall value of this approach.

Examples of *in situ* reserves set up specially for plants are the gene sanctuaries for wild relatives of *Citrus* in the Garo hills of India and similar reserves for fruit trees in other parts of India, as well as the orchid sanctuaries already mentioned, maintained by the Botanical Survey of India and other bodies. There are similar reserves in China, the USSR, Germany and Poland, for example, often for native fruit trees. In Brazil conservation of germplasm *in situ* was started in 1982 and more than ten genetic reserves have been set up, covering a range of species including *Hevea* spp., *Bertholettia excelsa*, numerous other forest species, native fruit trees, palms, forage species, *Manihot* and *Arachis* spp. Ethiopia has special conservation areas set aside for coffee and other wild relatives of crops. In addition there are many small scale reserves in various countries, such as the mini-reserves for endangered species in Mauritius and Rodrigues mentioned above, often managed by botanic gardens, and in the USA the plant reserves established by The Nature Conservancy and other bodies to preserve endangered native species of the North American flora.

This kind of special purpose, often quite small scale reserve is likely to become more frequent as interest increases in genetic conservation *in situ*. There is no overall inventory of such plant reserves to assess their full extent at this stage. It is a priority for action.

Ex situ conservation

This approach has been applied most extensively and successfully to crop species by the crop genetic resource sector, through the establishment of gene banks for seed, tissue or pollen or as growing collections in plantations, field gene banks and botanic gardens. Botanic gardens, as we have seen are *ex situ* centres par excellence and hold enormously diverse, if numerically inadequate, collections. *Ex situ* is appropriate for the conservation of germplasm of wild species in determined circumstances, for example, when wild populations are reduced to small numbers or where habitats with all their component species are in imminent danger of destruction. Its disadvantages are well documented and it is never an adequate substitute for *in situ* conservation if the latter is properly carried out. All too often *in situ* conservation is not possible, feasible or practicable. In practice both approaches are complementary: neither is ideal and an integrated approach to conservation and management is strongly to be recommended. The aversion of many conservationists to *ex situ* conservation is somewhat irrational. If we are to face the challenges of preserving a reasonable sample of plant diversity for future generations, all available methods must need to be employed.

CONCLUSIONS

The importance of germplasm of wild species is now more widely appreciated both by conservationists and crop genetics experts. Many more species are currently used as food and in human activities than is generally acknowledged. They do not benefit from the attentions of plant breeders because of their low total monetary value. Many of them are not traded except locally and do not appear in official statistics. No official organization is responsible for conserving germplasm of the majority of these species. It is only the Botanic Gardens Conservation Secretariat that is making a concerted effort to bring many of them into cultivation and into conservation, both *in situ* and *ex situ*, through its network of member gardens. The undertaking is a vast one and will need a great deal of research and financial support. It will also require a considerable change in attitude. The conservation sector will need to learn to appreciate the importance of genetic conservation and the genetic resource sector will need to learn to appreciate the importance and value of conservation in the wild. Conservation with development as advocated in the World Conservation Strategy requires more than words if it is to be fully implemented. Now is the time for action.

LITERATURE CITED

Akerele, O., V. H. Heywood and H. Synge (eds) 1991. *Conservation of Medicinal Plants*. Cambridge University Press, Cambridge, UK.

Arkcoll, D. B. and C. R. Clement. 1989. Potential new food crops from the Amazon, pp. 150–165. In: G. E. Wickens, N. Haq, and P. Day (eds) *New Crops for Food and Industry*. Chapman and Hall, London.

Arora, R. K. and E. R. Nayar. 1984. *Wild Relatives of Crop Plants in India*. NBPGR Science Monograph No 7, New Delhi.

Atal, C. K. and B. M. Kapur (eds) 1982a. *Cultivation and Utilization of Aromatic Plants*. Regional Research Laboratory, CSIR, Jammu-Tawi, India.

Atal, C. K. and B. M. Kapur (eds) 1982b. *Cultivation and Utilization of Medicinal Plants*. Regional Research Laboratory, CSIR, Jammu-Tawi, India.

Blixt, S. 1992. Gene banks for plant conservation. This volume.

Correa Q, J. E. and H. Y. Bernal. 1990. *Especies Vegetales Promisiarias de los Paises del Convenio Andrés Bello, tomo III*. SECAB, Bogotá, Colombia.

Cullen, J., M. Lear, D. C. Mackinder and H. Synge. 1987. Principles and standards for the computerization of garden record schemes, as applied to conservation, with proposals for an International Transfer Format, pp. 301–337. In: D. Bramwell, O. Hamann, V. Heywood and H. Synge (eds) *Botanic Gardens and the World Conservation Strategy*. Academic Press, London.

Food and Agricultural Organization. 1981. *Report of the Fifth Session of the FAO Panel of Experts of Forest Gene Resources*. FAO, Rome.

FAO/IUCN/UNESCO/UNEP. 1989. *Plant Genetic Resources. Their Conservation in situ for Human Use*. Food and Agriculture Organization of the United Nations, Rome.

Grossman, R. 1988. Equalizing the flow: institutional restructuring of germplasm exchange, pp. 255–273. In: J. R. Kloppenburg (ed) *Seeds and Sovereignty*. Duke University Press, Durham and London.

Heywood, V. H. 1987. The changing role of the botanic garden, pp. 2–18. In: D. Bramwell, O. Hamann, V. Heywood and H. Synge (eds) *Botanic Gardens and the World Conservation Strategy*. Academic Press, London.

Heywood, V. H. 1990. Botanic gardens and the conservation of plant resources. Impact of Science on Society 158: 121–132.

Heywood, C., V. Heywood and P. Wyse Jackson. 1990. *International Directory of Botanical Gardens V*. Koeltz Scientific Books, Konigstein.

Hoyt, E. 1988. *Conserving the Wild Relatives of Crops*. IBPGR/IUCN/WWF, Rome and Gland.

Huxley, P. A. and S. B. Westley 1989 (eds) *Multipurpose Trees: Selection and Testing for Agroforestry*. ICRAF, Nairobi.

International Board for Plant Genetic Resources. 1985a. *Systematic and Ecogeographic Studies on Crop Genepools: 1* Mangifera *L*. IBPGR, Rome.

International Board for Plant Genetic Resources. 1985b. *Ecogeographical Surveying and in situ Conservation of Crop Relatives*. IBPGR, Rome.

IUCN/WWF/UNEP. 1980. *The World Conservation Strategy*. International Union for the Conservation of Nature, Gland.

Jackson, P. Wyse, W. Strahm, Q. C. B. Cronk and J. A. N. Parnell. 1988. The propagation of endangered plants in Mauritius. Journal Irish Garden Plants Society, Moorea: 35–45.

Jarret, R. L., M. Spinks, G. Lovell and A. G. Gillaspie. 1990. Profile: The S-9 Plant Germplasm Collection at Griffin, Georgia. Diversity 6: 23–25.

Juma, C. 1989a. *Biological Diversity and Innovation, Conserving and Utilizing Genetic Resources in Kenya*. African Centre for Technology Studies, Nairobi.

Juma, C. 1989b. *The Gene-hunters. Biotechnology and the Scramble for Seeds*. Princeton University Press, Princeton, NJ.

Krivchenko, V. I. 1988. The role of Vavilov in creating the national Soviet program for plant genetic resources. Diversity 16: 5–7.

National Academy of Sciences. 1979. *Tropical Legumes: Resources for the Future*. NAS, Washington, DC.

National Academy of Sciences. 1980. *Firewood Crops. Shrub and Tree Species for*

Energy Production. NAS, Washington, DC.

Okafor, J. C. 1978. Development of forest tree crops for food supplies in Nigeria. For. Ecol. Mgmt. 1: 225–247.

Okafor, J. C. 1980. Edible indigenous woody plants in the rural economy of the Nigerian forest zones. For. Ecol. Mgmt. 3: 45–55.

Palmberg, C. 1987. Conservation of genetic resources of woody species. In: *Simposio Sobre Silvicultura y Mejoramiento Genetico*. CIEF, Buenos Aires.

Paroda, R. S. and B. Mal. 1989. New plant sources for food and industry in India, pp. 135–149. In: G. E.Wickens, N. Haq and P. Day (eds) *New Crops for Food and Industry*. Chapman and Hall, London.

Perdue, R. E. 1988. Systematic botany in the development of *Veronica galamensis* as a new industrial oil seed crop for the semi-arid tropics, pp. 125 – 135. In: I. Hedberg (ed) *Systematic Botany – a Key Science for Tropical Research and Documentation*. Symbolae Botanicae Upsalienses 28(3).

Perdue, R. E., E. Jones and C. T. Nyati. 1989. *Veronica galamensis*: a promising new industrial crop for the semi-arid tropics and sub-tropics, pp. 197–207. In: G. Wickens, N. Haq and P. Day (eds) *New Crops for Food and Industry*. Chapman and Hall, London.

Plucknett, D. L., N. J. H. Smith, J. T. Williams and M. N. Anishetty. 1987. *Gene Banks and the World's Food*. Princeton University Press, Princeton, NJ.

Poppendieck, H.-H. 1976. Mesembryanthemums and the problems of their cultivation, pp. 55–60. In: J. B. Simmons, R. I. Beyer, P. E. Brandham, G. Ll. Lucas and V. T. H. Parry (eds) *Conservation of Threatened Plants*. Plenum Press, NY.

Raven, P. H. 1981. Research in botanical gardens. Bot. Jahrb. 102: 53–72.

Roche, L. and M. Dourojeanni. 1984. *A Guide to in situ Conservation of Genetic Resources of Tropical Woody Species*. FORGEN/MISC/84/2. FAO, Rome.

Strahm, W. 1989. *Plant Red Data Book for Rodrigues*. IUCN/Koeltz Scientific Books, Konigstein.

Wickens, G. E., N. Haq and P. Day (eds) 1989. *New Crops for Food and Industry*. Chapman and Hall, London.

WWF/IUCN/BGCS. 1989. *The Botanic Gardens Conservation Strategy*. WWF, IUCN, BGCS, Gland and Richmond.

Zaitzev, V. A. 1990. Long-term seed storage in the USSR: modern genebank at the Kuban Seed Testing Station. Diversity 6: 14–17.

12 Gene banks for plant conservation

Stig Blixt
Nordic Gene Bank for Agricultural and Horticultural Plants,
P.O. Box 41,
S-230 53 Alnarp,
Sweden

ABSTRACT

The mandate of the Nordic Gene Bank for Agricultural and Horticultural Plants (NGB), an institution of the Nordic Council of Ministers and started in 1979, is to preserve the genetic variation of agricultural and horticultural plants and their wild relatives. The gene pools of interest to NGB thus include the important endemic and indigenous wild relatives of cultivated plants. The NGB policy with regard to preserving genetic resources is to preserve, whenever possible, these endemic and indigenous wild plants *in situ* and the standard cultivars and genetic stocks *ex situ*. This creates an overlap between "conventional" gene banks for agricultural and horticultural plants and plant conservation in general. For the benefit of further discussions, the handling of the material and the organization of the NGB collections is briefly presented. It is concluded that due to a common interest and a certain overlap in terms of material to conserve, the time is ripe for gene banks and institutions for general plant conservation to cooperate on aims and mechanisms.

INTRODUCTION

The Nordic Gene Bank for Agricultural and Horticultural Plants (NGB) is an inter-Nordic institution with the mandate to preserve genetic resources of agricultural and horticultural plants and wild relatives of those plants. The institution formally began its activities January 1, 1979, and was initially located at Lund, Sweden, but moved in 1984 to its present location at Alnarp, Sweden. It should therefore be emphasized that NGB does not carry out plant conservation *per se*; the mandate of NGB covers a limited sector of plant genetic diversity of the Nordic region. The aim of this paper is to present the organization and concept of NGB in fulfilling its present mandate and to introduce considerations on flora conservation in this context.

Conservation of Biodiversity for Sustainable Development.
Edited by O. T. Sandlund, K. Hindar and A. H. D. Brown. 1992. Scandinavian University Press, Oslo.

ORGANIZATION

NGB is an institution under the Nordic Council of Ministers, which may be said to be an executive body under the Nordic countries. NGB is governed by a Board with one full and one deputy member from each country. The Board has a Technical Advisory Board, with national sections. Much of the actual work in the different countries is carried out by or through the Crop Working Groups, of which NGB at the moment has nine (Figure 1).

NGB has presently 11 staff members at headquarters, divided into three departments dealing with the actual plant material, the information relating to the material and international projects, respectively, as well as some administrative functions.

NGB has one international project: it acts as management consultant to the Nordic development agencies in establishing a regional gene bank in Southern Africa, where the countries of the Southern African Development Coordination Conference (SADCC; Angola, Botswana, Lesotho, Malawi, Mozambique, Swaziland, Tanzania, Zambia, Zimbabwe and Namibia) cooperate in ways similar to the Nordic countries (Figure 2). Particularly in this context NGB has to consider the problem of plant conservation. The emphasis is on the efficient use of the limited resources both in funds and competence available. The aim is to

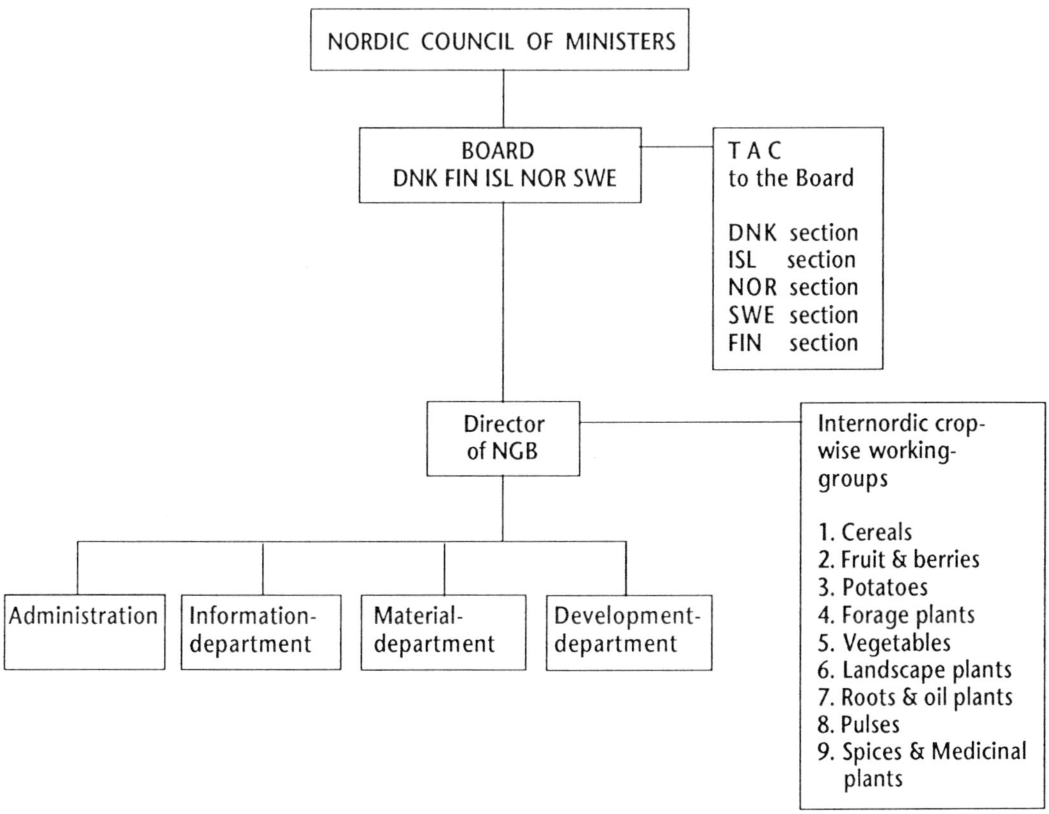

Figure 1. Nordic Gene Bank (NGB) organisational scheme (after NMR 1988). DNK, Denmark; FIN, Finland; ISL, Iceland; NOR, Norway; SWE, Sweden.

establish, over a 20-year period, an infrastructure with national centres coordinated by a regional gene bank. The role of NGB is to monitor the administration of funds provided by the five Nordic countries, coordinated by the Swedish International Development Authority (SIDA), and to transfer experience and know-how. The regional gene bank as well as the national centers are run by their own boards and own staff, trained by the project, determining their own priorities. The NGB is not involved in the actual work, such as explorations, collections, etc., in any other way than as advisor. The only Nordic staff in SADCC is a Project Manager, as required by the Development Authorities.

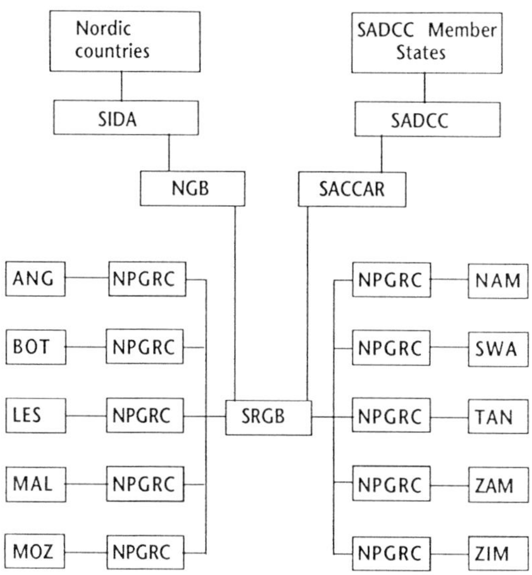

Figure 2. Organization model of the project SADCC Regional Gene Bank (SRGB). SACCAR, Southern African Center for Co-operation in Agricultural Research; NPGRC, National Plant Genetic Resources Center; ANG, Angola; NAM, Namibia; BOT, Botswana; SWA, Swaziland; LES, Lesotho; TAN, Tanzania; MAL, Malawi; ZAM, Zambia; ZIM, Zimbabwe; MOS, Mosambique.

GENE POOLS

The conservation of genetic resources of agricultural and horticultural plants as well as plant conservation as a whole, requires clarity that what is to be preserved is genes, the hereditary material. The total hereditary material of a species is generally referred to

Table 1. Gene pools: populations in which the full range of genetic variation is maintained; genetic gene pools (after Harlan and Wet 1971).

Gene pool	Definition
Primary	Corresponding to the biological species concept; crossing within the gene pool easy; hybrids generally fertile; chromosome pairing and gene segregation approximately normal; gene transfer generally easy.
Secondary	Approximate an experimentally defined coenospecies; includes all biological species crossing with the crop; gene transfer possible after overcoming barriers separating biological species.
Tertiary	Crosses can be made with the crop but: hybrids tend to be anomalous, lethal or completely sterile; gene transfer is either not possible with traditional techniques; or rather extreme or radical measures are required, e.g. embryo culture, grafting, doubling of chromosome number, using bridging species.

as its gene pool. The term gene pool can have different specific meanings. The primary, secondary and tertiary gene pools, according to Harlan (Table 1), reflect the genetic affinity of species and the ease with which genes can be transferred from species to species. This is important for gene banks dealing with domesticated plants and perhaps even more so for plant conservation as soon as *ex situ* flora conservation enters the picture.

For agricultural and horticultural plants within the mandate of NGB, two other gene pools are of major interest, namely, the indigenous and the exogenous gene pools (Table 2). The endemic gene pool on the species level is of little consequence to NGB, since the number of endemics in the Nordic countries is small and none of these is within the mandate of NGB; examples are *Papaver laestadianum* and *Helianthemum oelandicum* (Sjörs 1967). However, at the subspecific and genotypic level, the endemism in the Nordic countries is probably extensive, particularly among the apomicts, such as *Ranunculus* and *Hieracium*, but also among the important fodder-grasses, such as *Festuca* and *Poa*.

Table 2. Gene pools: other categories of gene pools.

Gene pool	Definition
Domesticated	The genetic variation within the domesticated forms of a species.
Wild	The genetic variation within the wild forms of a species.
Endemic	The genetic variation within the populations endemic to a certain area, country or region.
Indigenous	The genetic variation within the populations indigenous to a certain country or region, i.e. wild growing populations as well as cultivated ones.
Exogenous	The genetic variation within the populations of species introduced to and adapted to a certain country or region, i.e. no wild growing populations exist.

This is of particular importance also when considering plant conservation. Many plant species are at the fringe of their distribution in the Nordic countries. It might be tempting to reason that extinction of these small populations is of no consequence; and this might be so. On the other hand, it is often just in such border areas that evolution takes new directions. The concerns of general plant conservation and of the agricultural gene banks have here a common interest and currently the greatest experience on how to handle genetic diversity *ex situ* may rest with the gene banks.

The exogenous gene pool comprises material introduced to the region which, after a long cultivation period, has been adapted to the conditions of the region; the species may thus have formed a unique infra-specific gene pool in that region. Such is the case for most crops important for Nordic agriculture over thousands of years, e.g. barley, wheat and rye, but also for some recent additions, like soybean (Holmberg 1973). In addition, NGB needs to preserve other materials of great practical and scientific value, for instance genetic stock collections of *Pisum* and barley. These gene pools are highly specialized and of less interest for plant conservation.

A quite important part of the NGB concept of genetic resources preservation is that each gene bank should be responsible for preserving the endemic and indigenous plants of their area and that all material should be freely available. This is by no means a generally accepted view, but in accordance NGB preserves only material of

Table 3. Distribution of requests from the NGB.

Year	Distributed to:						
	Nordic	Europe	America	Africa	Asia	Australia	Total
1979	2	–	–	–	–	–	2
1980	6	1	–	–	–	–	7
1981	12	–	–	–	–	–	12
1982	9	2	–	–	–	–	11
1983	15	–	3	–	–	–	18
1984	22	27	6	1	5	1	62
1985	23	22	4	2	10	3	64
1986	17	33	7	1	2	7	67
1987	29	22	4	2	3	5	65
1988	22	29	6	1	2	5	65
1989	25	24	4	1	1	5	60
Total	182	160	34	38	23	26	433
Total number of samples							8999

Nordic origin and, in the case of genetic stocks, material on which specific information of Nordic origin has been produced. All material is distributed free of charge and without restrictions (Table 3). When considering plant conservation, this view seems rather self-evident.

With respect to the indigenous Nordic material (Figure 3), the NGB has been commissioned to preserve the genetic resources of agricultural and horticultural plants and wild relatives in the Nordic countries. The indigenous part of this gene pool comprises many different plant species which migrated into the area some 15,000 years ago after the end of the latest ice period, and now occur as part of the wild flora. Examples are most of the important forage plants, e.g. *Festuca, Dactylis, Phleum, Poa, Agrostis, Trifolium* spp., as well as *Prunus* spp. (i.e. *P. avium;* cherry), *Ribes* spp., *Rosa* spp., *Rubus* spp., *Vaccinium* spp., and *Brassica campestris.*

Looking on this from a global point of view, no fewer than 530 out of the 4,000 indigenous Nordic plant species are cultivated to some extent somewhere. In reality, NGB preserves *ex situ* less than 100 species currently. Considering international undertakings, such as the FAO undertaking for the conservation of genetic resources and the European Cooperative Programme on Genetic Resources, it is arguable that NGB should deal with more species than at present, but current resources limit such activities. This shows, however, that the area of overlap between gene bank preservation and plant conservation in the Nordic region is considerable.

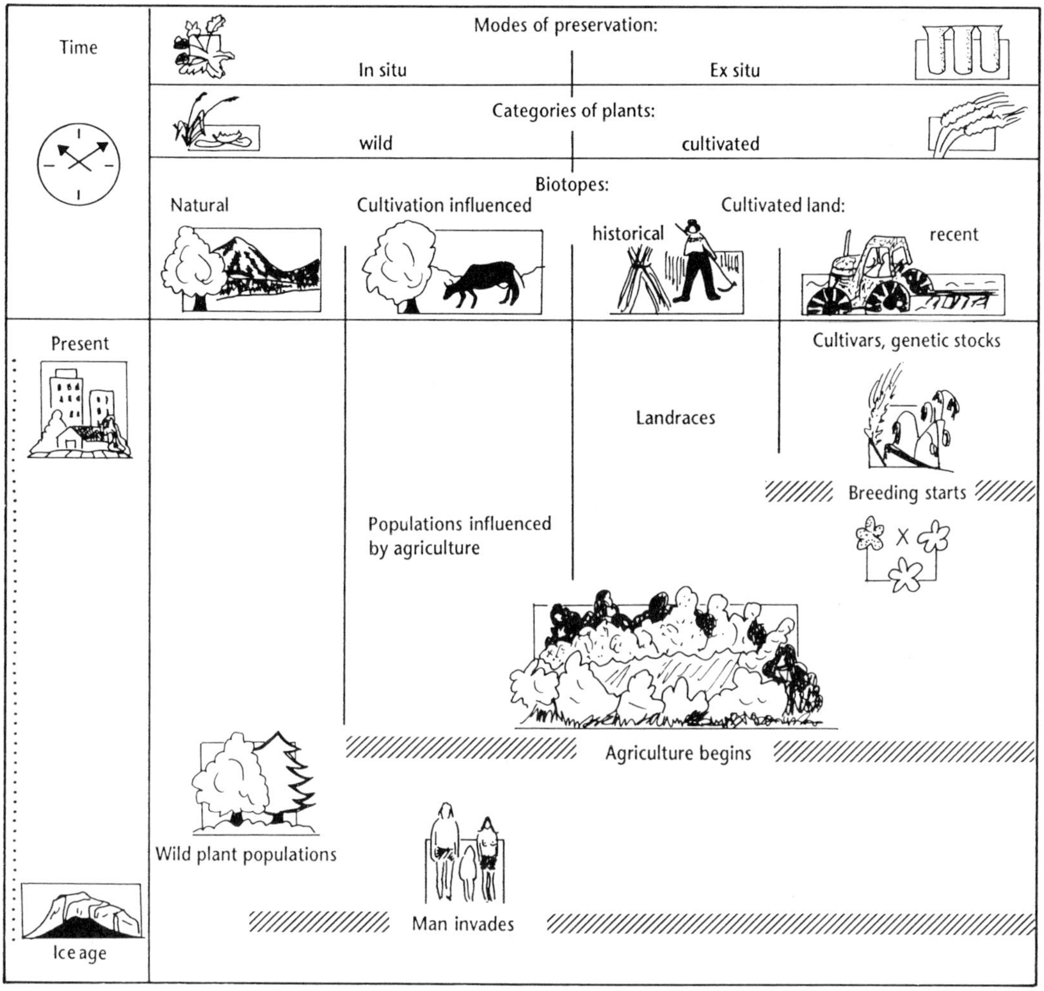

Figure 3. The Nordic gene pool; biotopes, material and modes of preservation.

MODE OF PRESERVATION

Different categories of plant material require different handling in order to preserve their genetic integrity and identity. The material may be preserved (1) *in situ*, i.e. in the original habitat, or (2) *ex situ*, as propagules (mainly seeds) stored in the gene bank or as plants growing in a nursery or clonal archive, or (3) *inter situ*, as landraces cultivated under historical agricultural management patterns.

For those species having an indigenous gene pool in the Nordic countries, information on the genetic diversity and its distribution within and between populations is largely lacking. Since the wild forms of many of these species have such a wide distribution in the Nordic area, NGB can manage a representative preservation of the genetic diversity, for practical as well as scientific reasons, only by *in situ* conservation, i.e. conservation in the original habitat.

Cultivars developed to be grown over the largest possible area in the Nordic countries and which can be multiplied or rejuvenated in any farm field, can preferably be preserved *ex situ*, i.e. as seed samples stored at the gene bank.

Landraces of outbreeding crops, forages, fruits and vegetables indigenous to the Nordic countries constitute a particular problem since natural selection is so important for maintaining the genetic identity of these forms. These landraces ideally require multiplication in a farming system characteristic of its time and at or near the area of origin, i.e. a form of *ex situ* preservation in a historical form of farming or gardening, here called *inter situ*.

MAINTAINING COLLECTIONS

In the case of *ex situ* preservation, the organism or parts of it such as seeds, tubers, etc., are removed from the original habitat and transferred to a gene bank. The collected material constitutes an accession which, in the case of whole organisms, may be planted in clonal archives or, in the case of seeds, stored under suitable conditions. The quantity of material which can be preserved in this way is limited.

In maintaining the genetic diversity of its commissioned area, the gene bank has certain obligations. The material should, for instance, be kept safe; it should maintain its genetic integrity; and it should be available for study and for use. In order to fulfill such obligations efficiently, the material within the gene bank is divided into different Collections. With reference to seed collections, IBPGR recognizes Base, Active and Working Collections. In addition to the IBPGR definitions, NGB wants to include the *in situ* preserved populations of wild forms of crop species which are expected to form a major part of the material within the framework of its collections. NGB has also a different definition of Base collection and Working collection, as specified below.

NGB is presently considering four categories or Collections (Figure 4) in its organisation, viz.:

- the Safety Collection
- the Base Collection
- the Active Collection and
- Working Collections

A central issue for a major portion of the accessions to be maintained by NGB, is the maintenance of the genetic integrity of the samples. Maintaining genetic integrity means maintaining the original genotype of an apomictic, cleistogamic or clonally propagated accession, or the total allelic composition collected in the original samples of outbreeders (of any degree). It may not always be practical to maintain the genetic integrity of plants which usually or always outbreed in order to reproduce, but one should at least aim for it. Commercial seed producers have developed a system over many years to handle the well-known difficulties in maintaining the genetic integrity of a cultivar which requires constant maintenance breeding, as well as changes in the composition of the cultivar which can occur during the process of seed multiplication. There seems every reason to apply available knowledge on this to gene bank management. In this context the Safety Collection is analogous to material deposited by breeders with official authorities; the Base Collection is analogous to Breeders' seed quality; and the Active Collection corresponds to different commercial seed qualities. The Working collections is analogous to seed produced by farmers for their own use.

The Safety Collection at NGB is a safeguard against any accidental loss of material. Consequently, limited samples of most of the original seed collections are stored separately from the Base and Active Collec-

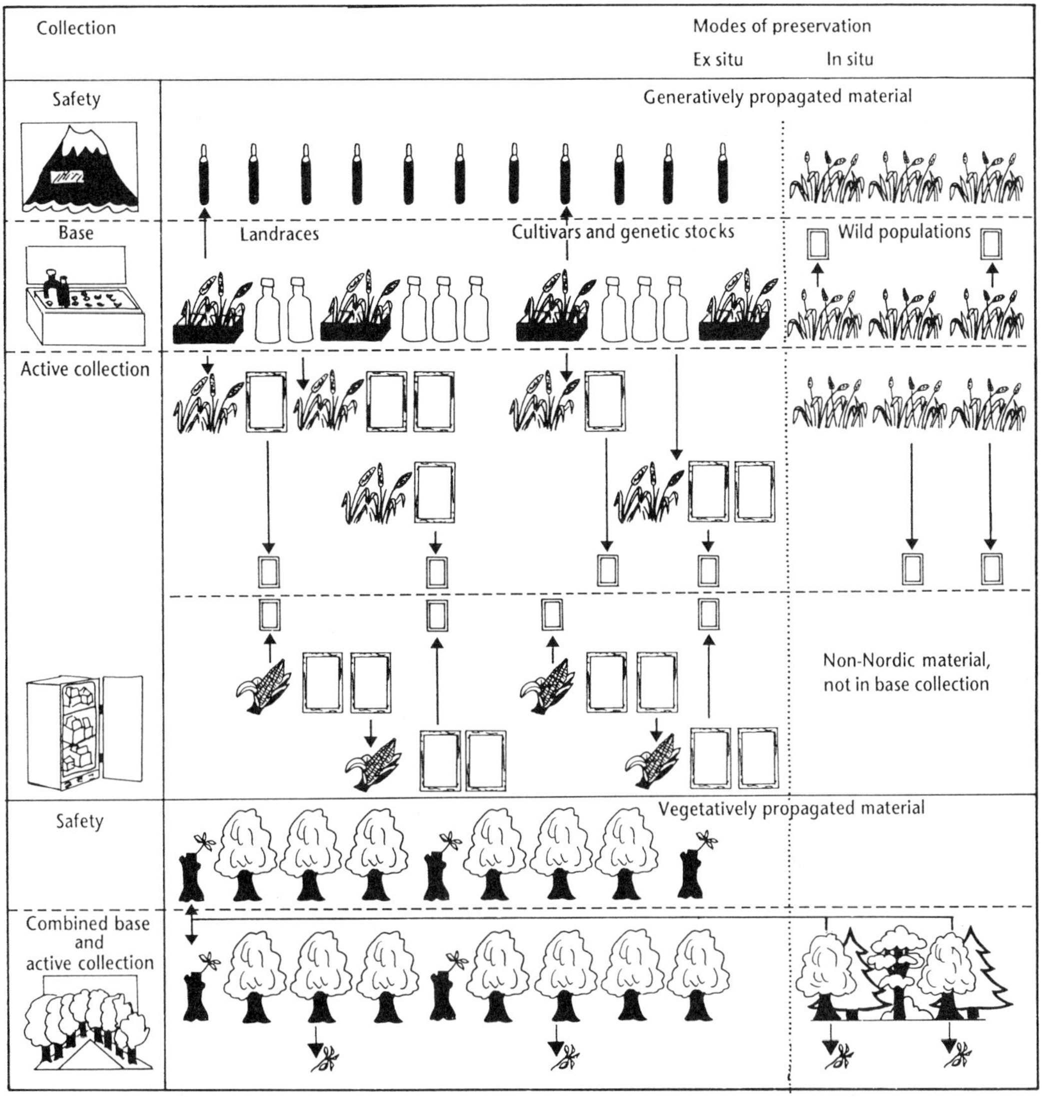

Figure 4. Organization of the Nordic Gene Bank (NGB) material.

tions, in a non-accident-prone location requiring little, if any, artificial energy input. These conditions are met at the NGB permafrost store at Svalbard (Spitsbergen) and allow NGB to fulfill many of the objectives of a traditional Base Collection.

The main objective of the Base Collection of NGB is to maintain maximum genetic quality in the material. The number of rejuvenations are therefore kept to a minimum and the handling of the material occurs under strict quality demands, corresponding to guidelines for the production of Breeders' Seed Quality material. The Base Collection is meant to be the source of seed for the Active Collection. The Base Collection is also maintained under long-term storage conditions.

The purpose of the Active Collection is to provide seed for distribution and for characterisation; in short, seed for all work relevant to gene bank operation. Consequently, multiplication of the material is a major undertaking. This, again, corresponds to the production of Original Seed Quality from Breeders' Seed and the subsequent production of lower grades from Original Seed.

The Working Collection is material held by breeders and assumed to be represented in gene banks. There is a clear distinction between a working collection and breeding material.

With regard to the vegetatively propagated crops, some species are space-demanding trees and bushes. It is therefore, as a rule, not practical to keep a collection specifically for the purpose of safeguarding against accidental losses. NGB has therefore chosen to maintain each accession in at least two locations so that the duplicated accessions can act as safety collections for each other. We still lack practical experience from long-term running of such collections, but feel it is important to design one as a Base and the other as an Active Collection, so that in the case of limited financial resources, a correct allocation of funds will be possible.

With respect to *in situ* conservations, it should be noted that the reserve set aside for preserving the population(s) will be designated the Base Collection for that (those) accession(s) representing that (those) population(s). Seeds or vegetative parts may be collected for *ex situ* preservation in the Safety Base Collection, or for study and use in the Active Collection. In case of imminent threat to the reserve, material may be removed and placed in the *ex situ* Base Collection as well as in the Safety Base Collection. Collecting seeds or vegetative parts in the reserves for direct distribution is also feasible, but has as yet only been implemented to a limited extent.

Plant conservation is presently dealt with mainly by organizations concerned with the conservation of natural resources, working with mechanisms such as nature reserves of different categories. Information is, at least in the Nordic countries, normally available on these nature reserves regarding location, size, care and purpose. In many cases comprehensive and detailed information on plant taxa growing in the reserve will already be available. This information is expected to be mainly at the species level, but it is assumed to yield a picture of the geographical distribution of populations of species of interest to NGB. Together with geological and meteorological data, and in collaboration with universities and other institutions, it might be possible to asses the genetic diversity covered by the existing reserves. The present discussion within NGB on how to handle the *in situ* conservation, therefore, includes the proposition that NGB does not expect to set up any "gene bank reserves". In case of major gaps being discovered, NGB wants rather to offer information on the specific needs relating to wild forms of domesticated plants and to try to influence the local and national authorities for nature reserve to consider these needs when establishing new reserves.

CONCLUSION

The handling of *ex situ* collections has been dealt with in some detail since it is relevant to any further discussion on the possible needs of *ex situ* preservation in conserving the wild flora. The title of this volume, Conservation of Biodiversity for Sustainable Development, implies, it seems, both conservation of ecosystems in order to maintain the environment and the use of these ecosystems for development. The title seems further to indicate that the time may be ripe for discussing a closer cooperation between gene banks for domesticated plants and plant conservation in general, as well as the possibility and utility, if any, of an entire or partial integration. After all, with respect to plant genetic resources, the domesticated plants are but specialized species and/or forms of the flora, and thus, the preservation of domesticated plants and their wild relatives but a sector of the conservation of the entire flora.

LITERATURE CITED

Nordiska Ministerrådet 1988. Organisationsschema för Nordiska Rådet och Nordiska Ministerrådet. Nordiska Samarbetsorgan 1988: 267. NORD 1988: 57, NMR Copenhagen.

Harlan, J. R. and J. M. J. de Wet. 1971. Toward a rational classification of cultivated plants. Taxon 20: 509–517.

Holmberg, S. A. 1973. Soybeans for cool temperate climates. Agri Hortique Genetica 31: 1–20.

Sjörs, H. 1967. *Nordisk växtgeografi*. Scandinavian University Books, Stockholm.

13 *Ex situ* propagation programmes as a contribution to the conservation of biodiversity

Leobert E. M. de Boer
National Foundation for Research in Zoological Gardens/EEP Executive Office,
c/o Amsterdam Zoo (Artis),
PO Box 20164,
NL-1000 HD Amsterdam,
The Netherlands

ABSTRACT

Ex situ propagation plays an increasingly important role in the conservation of endangered animal species. The ultimate goal of *ex situ* propagation programmes is to provide support for survival of populations in their natural environments. The foundations for achievement of this goal are the preservation of the species' natural genetic structure and variability, and the prevention of demographic imbalance. Both of these require population management measures, which in their turn require a sound theoretic basis, the availability of management tools and the availability of organizational structures. The current status of these three prerequisites is briefly reviewed. The development of theoretic considerations in the preservation of small and fragmented *ex situ* and *in situ* populations has been rapid during the past decade. Management tools have been developed abundantly in the past years, especially in the form of various types of computer software packages for population management. *Ex situ* propagation programmes in the zoo world are currently organised along geographic regions, and are co-ordinated per species within these regions. Global co-ordination of regional breeding programmes is provided by the Captive Breeding Specialist Group of IUCN. This Specialist Group also aids in proper choice of taxa requiring *ex situ* propagation, and stimulates the interaction between *ex situ* and *in situ* conservation. The role of biotechnology in *ex situ* conservation and current achievements in reintroduction of captive bred animals are briefly discussed. It is concluded that: (1) *Ex situ* propagation of endangered species allows highly intensive population management and exclusion of various stochastic influences; (2) Research conducted in the *ex situ* situation may be highly beneficial to conservation of *in situ* animal populations; (3) Application of biotechnological methods may facilitate survival of both captive and wild populations; and (4) Reintroduction of captive-bred animals into the wild may support further conservation of a species,

Conservation of Biodiversity for Sustainable Development.
Edited by O. T. Sandlund, K. Hindar and A. H. D. Brown. 1992. Scandinavian University Press, Oslo.

provided that a number of factors are taken into consideration.

INTRODUCTION

There are some early examples of the potential of captive breeding in preventing extinction of highly endangered animal species in the zoo world. Species such as the Przewalski's horse, Pere David's deer and Hawaiian goose, would no longer exist but for the fact that their populations were maintained and propagated for many decades in captivity.

However, the concept that zoos could serve the goal of preserving biological species through propagation of captive stock at a more than incidental scale is much more recent. The zoo world, roused by the rapidly accelerating destruction of the world's habitats and increasing endangerment of biological diversity, started to explore this idea and its consequences only some 15 years ago. This resulted in the organization of the first co-operative and co-ordinated breeding programmes in the early 1980s. Since then considerable progress has been made with respect to theoretical considerations, practical implications in captive breeding programmes and organizational structures required for their proper co-ordination. Simultaneously, the number of captive breeding programmes have increased exponentially. It can now be stated beyond doubt that captive propagation of endangered species constitutes one important tool in conservation of biological and genetic diversity (Foose 1983; Foose et al. 1987; de Boer 1989a; see also the recent collection of papers in Endangered Species Update 1990).

GOALS AND PREREQUISITES

The ultimate goal of captive, or *ex situ* breeding programmes is to provide support for the survival of species, subspecies or defined populations in natural environments (Foose et al. 1986; Seal 1986). To achieve this goal, the primary objectives of *ex situ* propagation are:

1. To propagate and manage *ex situ* populations of highly endangered taxa with prescribed levels of genetic diversity and demographic stability for defined periods of time to prevent extinction of such taxa.
2. To employ these *ex situ* programmes as parts of conservation strategies that interactively manage captive and wild populations to ensure ultimate survival of these taxa in natural environments. This means that an important objective will be to use animals from *ex situ* populations to re-establish or re-inforce natural populations.
3. To develop self-sustaining captive populations of rare or endangered taxa for education programmes which may be very beneficial to survival of conspecifics in the wild.

The foundations for the achievement of such objectives for every captive breeding programme species can be defined on three levels:

1. At the level of individuals, the *ex situ* situation should ensure sufficient longevity and physical, physiological and psychological well-being: animals should live long lives of acceptable quality. This is obviously contingent upon species-specific zootechnical, medical and biological knowledge and research.
2. At the level of breeding pairs and groups, sufficient reproduction should

occur in order to guarantee continuity over generations. This, of course, requires species-specific knowledge and research on reproductive biology, ethology, husbandry etc.

3. At the level of populations, the preservation of a population which resembles the natural genetic structure as closely as possible should be assured, while a demographically stable population structure should also be established. Whereas the first two levels specifically deal with the size of the population (the number of animals maintained in captivity and thus its quantity), this third level refers to its quality.

It is axiomatic that preservation of a population's genetic structure is crucial to its survival (de Boer 1989a, b). The retention of genetic variability and of the approximate original allelic frequencies and levels of heterozygosity are particularly important. The preservation of such genetic structure determines the probability of success when the *ex situ* population is used to re-establish or re-inforce wild populations that are exposed to natural selection at some stage in the future.

Similarly, it is self evident that the achievement of a demographically stable population structure is crucial to a population's survival (Foose 1981; Foose and Ballou 1988; Albers 1989). Demographic instability not only affects genetic structure, but also hampers optimal harvesting from the population for reintroduction. Demographic imbalance may even cause a population's extinction.

Neither the preservation of genetic structure in *ex situ* populations, nor the achievement of demographic stability of such populations can be attained without population management procedures (Foose and Ballou 1988). The obvious reasons for this are:

1. That such populations are much smaller than original, unthreatened wild populations, rendering them subject to severe risks of genetic and demographic stochasticism.
2. That they are nearly always extremely fragmented, scattered among the animal collections of many dozens of zoos all over the world, which considerably increases risks of inbreeding in the subunits.
3. That in the *ex situ* situation they are subject to risks of unnatural (i.e. artificial) selection.

The effects of these risks can be summarized as degeneration, domestication and vulnerability to extinction. Population management procedures for *ex situ* propagation programmes, intending to minimize such risks, greatly rely on three conditions:

1. Development of a sound theoretical basis with respect to genetics and demography of small and fragmented populations.
2. Development of what may be called "tools" to analyse individual breeding programme populations in view of theoretical knowledge.
3. Development of organizational structures to implement the management proposals resulting from analyses of the genetic and demographic status in individual populations.

Major progress has been made in these three areas since the start of co-operative breeding of endangered species in zoos and related institutions on a larger scale some 10 years ago. Let us therefore briefly review these developments and their current status.

THEORY

At least a dozen population geneticists and demographers attached to zoos and universities have been intensively working on the theory of preservation of small and fragmented animal populations in zoos as well as in the wild during the past decade. This has resulted in much clearer insight into ongoing processes of such populations, especially processes of loss in genetic variability, of inbreeding and of demographic stochasticism. Simultaneously, this work has resulted in formulation of guidelines for population management and breeding strategies that enable prevention, or at least minimisation of the negative effects of these processes. In general terms, it is now known how many founder animals are needed to start a captive breeding programme, what the influences of sex ratio and family sizes in the population are, what the minimum size of the final population in captivity should be, how rapidly the original founder population should be expanded to reach this final size, how generation time influences several of these parameters for specific species, and so on (for reviews and specific references, see Hedrick and Miller this volume; Frankel and Soulé 1981; Schonewald-Cox et al. 1983; Soulé 1986; Foose and Ballou 1988; de Boer 1989c). Schemes have been developed for prevention of inbreeding as well as for retention of a maximum percentage of the original, wild genetic variability over prolonged periods of time. Basically genetic variability and original allelic frequencies are retained by preventing any unnatural (artificial) selection during captive propagation and by reducing the influence of genetic drift as much as possible. It has been more or less arbitrarily agreed that a minimum of 90 % of the original genetic variability should be maintained during 200 years of *ex situ* propagation (AAZPA 1983; Foose et al. 1986).

On the basis of this general theoretical knowledge, currently "refinements" are being worked out, particularly regarding multiple approaches in population management that are tailored to species-specific social and reproductive systems (Princée in press). A great deal of work remains to be done in this area, but regular meetings, symposia and workshops are organised which enable the theoreticists to exchange thoughts and to further develop their ideas (e.g. Ralls and Ballou 1986; Ballou et al. in press).

TOOLS

Much excellent work has also been done in recent years concerning development of tools, enabling the translation of theory into practically usable expedients. Of primary importance in this respect are the systems that have been developed to collect and record individual animal data that are basic to co-ordinated population management. The majority of the large zoos of the world currently use a uniform computerized system for this purpose, the ARKS, or Animal Records Keeping System (Scobie and Flessness 1987; Flesness and Mace 1988). This system not only allows standardized detailed record keeping within each of the individual zoos, but also the compilation of the individual zoo records in a global database, the ISIS or International Species Information System (Seal et al. 1977; Jones 1990). Currently, this database comprises data on hundreds of thousands of zoo animals in more than 300 of the world's zoos. Several countries or regions, such as Great Britain, The Netherlands and Australia, have developed national or regional databases of zoo animal records that serve as intermediates between the databases of the individual zoos and the global ISIS database (Flesness and Mace 1988; F. P. G. Princée, personal communication).

As a second step in data collection regional and international studbooks for endangered species are compiled on the basis of the databases of the individual zoos and the national and global databases (Glatston 1986). Regional/international studbooks for a particular species include all relevant data on all animals of that species that are living, or have been living, in any of the zoos of the world. These studbooks form the basis of genetic and demographic analyses. Studbook management software packages have been developed in order facilitate compilation of studbooks and performance of relevant analyses (Scobie and Flesness 1989; Flesness and Mace 1988; Princée 1989). Complex calculations with respect to inbreeding, representation of the original founder animals in descendent populations, demographic structures and trends in populations, etc., that formerly took many weeks of work by an expert, can now be completed within one day. In addition, computer simulation software for so called gene drop or gene flow simulations can analyse the genetic erosion in complex pedigrees, and can also determine the effects of possible future mating combinations (MacCluer et al. 1986; Thompson 1986; Princée 1988; Flesness and Mace 1988; Lacy 1989). All these software packages, which are still expanding and developing towards a world standard, are powerful tools for the formulation of population management proposals for captive breeding programme species.

REGIONAL ORGANIZATIONS

Developments in the establishment of organizational structures to implement population management proposals, have been as rapid as progress in theory and software tools. For purely logistic reasons breeding programmes are organised on a regional basis. Co-operation and exchange of animals and knowledge within regions of limited size is much less difficult than on a global scale, involving distant continents and countries with different languages and economic, legislative and political backgrounds. The first region that started formal co-operative breeding programmes was the North American Association of Zoological Parks and Aquaria, the AAZPA, by implementing its so called SSP, or Species Survival Plans, in the early 1980s (AAZPA 1983). Shortly afterwards, Great Britain and Ireland followed with their Joint Management of Species Programs, as well as Australia with its Species Management Programs (Baker and George 1988). More recently, the Australian organization evolved into an Australasian regional structure including New Zealand and New Guinea as well. In 1985 continental Europe established a regional breeding programme organization, the EEP organization, from the German term Europäisches Erhaltungszucht Programm (European breeding programmes for the conservation of endangered species) (Nogge 1988). Since then the Japanese Association of Zoos has initiated regional breeding programmes (JAZGA 1990), while India is preparing to follow the previous examples. Quite recently it was announced that in 1991 similar programmes will begin on the African continent, and the Southeast Asian region is also planning to start such programmes.

Apart from some minor organizational differences between the regions, regional breeding programmes for endangered animal species in captivity generally are of a similar structure. The regional programme for a particular species is always co-ordinated by a species co-ordinator, who has a complex set of tasks. First of all he or she should see to the compilation of a regional studbook for the species concerned. Then the co-ordinator should approach all the zoos in the region that keep the species concerned to ask for their participation in the co-operative programme. Together with a species committee, elected from the participating zoos and supplemented with specific experts, the co-ordinator analyses the regional studbook data genetically and demographically. The results form the basis of proposals for the year-to-year breeding strategy and the compilation of a master-plan that details recommendations, such as which individual animals should be exchanged between which institutions, which individuals should breed more or less than the average, how fast the population should grow, and how many new animal spaces are required (Foose and Ballou 1988). Regional breeding programmes and their co-ordinators, however, not only deal with genetics and demography. They also compile knowledge and experience of their species, especially with respect to reproduction and husbandry. By drawing up husbandry and reproduction recommendations, and by carrying out or stimulating appropriate research into these subjects, they try to create optimal circumstances for the survival and wellbeing of their *ex situ* populations.

Regional species coordinators generally are zoo collaborators (though there are also co-ordinators from outside the zoo world) with a special interest and knowledge regarding the species whose population they co-ordinate. In order to equip them with the relevant non-species specific knowledge required, most of the regional breeding programme organizations organise species co-ordinators training courses. These courses provide training in genetic and demographic theory, in the use of computer software tools, the production of husbandry guidelines and masterplans, and so forth. Additionally, species coordinators' manuals have been published (e.g. de Boer 1989c).

As indicated previously, the first regional breeding programme organizations were initiated in the early 1980s. Within each of the individual regions only a few species were initially chosen for programmes. Recently, however, the number of programmes has been increasing rapidly. Currently North America has regional breeding programmes for 55 species, Europe for 68, Great Britain for 41, Australia for 73 and Japan for 32 species (Table 1). In India, regional studbooks are being compiled for some 60 species in anticipation of future breeding programmes. When we look at the total number of different animal species for which there are breeding programmes in at least one of the regions, mammals (with 105 species) and birds (with 58 species) are far ahead of reptiles, amphibians, fish and invertebrates, for which currently there are only a handful of regional breeding programmes. Altogether there is at least one

Table 1. Regional *ex situ* propagation programmes; numbers of taxa within each of the regions and world wide (data as of August 1990 after T. J. Foose in Seal 1990).

Region[a]	Mammals	Birds	Reptiles	Fish	Invertebrates	Total
North America	35	13	5	1[b]	1	55
Europe	55	11	2	0	0	68
United Kingdom	28	13	0	0	0	41
Australia-New Zealand	44	27	1	1	0	73
Japan	22	9	0	1	0	32
Total number of different taxa	105	58	8	3	1	175

[a] India is initiating regional *ex situ* propagation programmes as well. In anticipation of this, studbooks are being established for 44 mammals, 15 birds and several reptiles.

[b] This programme in fact includes a number of Lake Victoria cichlid species.

regional breeding programme for 175 endangered species. Looking at the distribution of programmes over the various regions, it appears that for the majority of species there is only one regional programme; 56 species have programmes in two of the regions, 28 species have programmes in three of the regions, while only three species have breeding programmes in all of the five currently existing regional organizations.

GLOBAL CO-ORDINATION

Given the increasing number of regional breeding programmes, and the increasing number of endangered species for which there are programmes in two, three or more regions, the need is growing for co-ordination at the inter-regional and global level. Such co-ordination is partly performed by the international studbook keepers. In 1990 the number of species for which there were international studbooks or registers – including the data of all animals in captivity at a global level – increased to 101 (Olney 1990). The standardisation and regular updating of these international studbooks are supervised by Dr. Peter Olney of the Zoological Society of London in his function of International Studbook Co-ordinator. The majority of the 100 international studbook keepers, however, so far are not actively involved in inter-regional co-ordination of breeding programmes. Their task is mainly to compile the studbook data on the basis of which such co-ordination can be carried out. Only a relatively small number of them include in their studbooks guidelines and directives for breeding within the various regions on the basis of analyses of the world population of their studbook species.

Two organizations are of crucial importance to a development towards more international and global co-ordination of breeding programmes. Firstly, the IUDZG – the International Union of Directors of Zoological Gardens – is very supportive to co-operative propagation of endangered animal species in zoos. Its role is mainly stimulative as it tries to activate the zoo world to become involved in breeding programmes and their co-ordination. Secondly, there is the CBSG or Captive Breeding Specialist Group, one of the many Specialist Groups of the Species Survival Commission of the IUCN, the great international nature conservation organization (Jones 1990). The CBSG currently consists of some 175 members, all of which are in some way involved in captive breeding, and the majority – though not all – of which are attached to zoos or related institutions. The CBSG is one of the most active specialist groups of IUCN/SSC, and in 1990 it even established a permanent professional executive office. Its role in global co-ordination of regional breeding programmes is a more practical one. For each species for which there are breeding programmes in more than one region, CBSG aims to establish a global co-ordination committee. Such a committee should consist of the regional species co-ordinators of the species concerned, the international studbook keeper for that species and an international species co-ordinator supervising the work of its regional counterparts. This committee, supplemented with specific experts, should formulate a global masterplan with respect to the breeding strategy for the species concerned. Such a masterplan should include proposals for the total world population size, the population sizes of the various regions required, the rate of exchange of genetic material between the regions and so forth. So far, global masterplans of this type are prepared for no more than a handful of species, such as the Przewalski's horse, the okapi, the golden lion tamarin and the red or lesser panda (e.g. Seal et al. 1990; van Puijenbroeck 1990). It is expected, however, that their number will rapidly increase in the next few years.

SPECIES SELECTION AND THE SPACE PROBLEM

Global co-ordination of regional breeding programmes is not the only task of the Captive Breeding Specialist Group. It plays a very active part in various other aspects of captive breeding as well. Firstly it stimulates the further development of theoretical models with respect to the conservation of small and fragmented populations, by inviting researchers to work on this problem and by organising workshops and symposia (Ralls and Ballou 1986; Ballou et al. in press). Secondly, the CBSG is much involved in developing a system that enables a proper choice of taxa which require captive breeding programmes for their survival. This involves assessment of the chance of survival of remnant wild and captive populations of endangered species using so called population viability analyses (e.g. Soulé 1987; Lacy et al. 1989; Foose et al. 1987). Action plans are formulated for the broader taxa such as birds, primates and sea mammals (see Seal 1990). These plans roughly indicate which species would benefit most from the support of captive breeding in light of the available space, knowledge and experience. Space is a particularly limiting factor, thus taxa for captive breeding should be selected very carefully. The broad taxon action plans supply guidelines for this selection.

It is estimated that in the future zoos and related institutions could deal with breeding programmes for one thousand species of higher vertebrates (Conway 1986; Foose et al. 1986; Soulé et al. 1986). Achievement of this goal is an enormous challenge which will take at least several decades of very hard work by many people. If, however, developments in the next years are as rapid as they have been in the past decade, there is a realistic chance of success.

Lending support to the survival of one thousand endangered species by captive breeding seems impressive, especially when one takes into account how much work and organization are involved in every single breeding programme. Nevertheless, 1,000 species are but a few when we consider that currently many tens of thousands of species are seriously threatened. Some conservationists even fear that within the next century a million species are bound to disappear. Does it make sense then to try to save one thousand of them through the efforts of captive breeding? The answer to this question is yes. Firstly, captive breeding programmes create public awareness of the necessity to save biological species by all means. The public relations value of captive breeding projects can be enormous as has already been shown on several occasions (e.g. with respect to the programs for the golden lion tamarin and the black-footed ferret; Kleiman et al. 1985, in press; Thorne et al. 1988). Secondly, captive breeding mostly involves taxa that are representative, or "key" species of their habitats; when such species are saved, this is highly beneficial to the preservation of their entire habitats, often including many other threatened species.

INTERACTIVE MANAGEMENT OF *EX SITU* AND WILD POPULATIONS

The last mentioned function of captive breeding depends of course on a close interaction between the propagation of endangered species in captivity and the conservation of wild populations and habitats. Assuring this achievement is a primary goal of the Captive Breeding Specialist Group.

That is why, in choosing species for captive breeding, the group works in close contact with the other SSC Specialist groups, most of which are involved in the conservation of wild populations of certain animal and plant groups. That also is why CBSG aims at an interactive management of captive and wild populations in the future. In fact, the captive and wild remnant populations of many endangered species should be considered as parts of "metapopulations". The wild elements of such metapopulations could considerably benefit from genetic and demographic re-inforcement from the captive counterparts in the various breeding programme regions. In this way, support via captive breeding is much more effective: one should not wait until the last wild population becomes completely extinct before using the genetic and demographic reserve in captivity. Rather, this reserve should be used much earlier in order to prevent extinction in the wild.

BIOTECHNOLOGY

An aspect of particular interest is the possible application of advanced biotechnological methods in *ex situ* propagation of endangered animal species. Biotechnology has been developing very rapidly in recent years. Three categories of such technologies are of importance to *ex situ* propagation programmes: (1) technologies for artificial reproduction; (2) cryopreservation of sperm, ova and embryos; and (3) genetic technologies.

Artificial reproduction techniques such as artificial insemination, and embryo transfer may be of crucial importance in furthering reproduction in species for which not enough experience so far exists to breed them sufficiently in captivity. In addition, such techniques enable management of captive populations according to the guidelines formulated on the basis of genetic and demographic theories. They enable very rapid increase of populations with critically low numbers of founder animals. They enable the exchange of genetic material between breeding program regions without risks of loosing genetically important individuals during transport, and enable influencing of sex ratio and family size. With the exception of artificial insemination, which has been applied in several dozens of zoo species (e.g. in giant panda, gorilla, birds of prey and cranes), to date the use of such techniques has been explored in no more than a handful of exotic animal species. Their number, however, is gradually increasing. Intra-specific embryo transfer has been carried out, for instance, in Siberian tigers (Johnston et al. 1990), lion-tailed macaques (Cranefield et al. 1988) and bongo antelope (Dresser 1986). Inter-specific embryo transfer was accomplished for instance, between bongo antelope and eland antelope (Dresser 1986), the Przewalski's horse and domestic horse and the Grant's zebra and domestic horse (Kydd et al. 1985), and between gaur and domestic cattle (Strover et al. 1981). Several institutions (i.e. Wildlife Research Federation of Cincinnati Zoo, and the Institute of Zoology of the Zoological Society of London) are actively carrying out research in this field. In the future, artificial reproduction techniques may also be of importance to the interactive management of *ex situ* and *in situ* populations. They enable transfer of genetic material and vice versa without necessitating movement of animals from one continent to the other.

Apart from the combined usage with artificial reproduction techniques, cryopreservation may be of importance in preserving important genetic material in the *ex situ* population (e.g. founder lineages which are crucial to a population's genetic make up). Cryopreservation methods for sperm and embryos have been explored in many species. Several institutions experimentally store germplasms of large numbers of exotic species (e.g. Cincinnati, San Diego, London). Nevertheless true "gene banks" for endangered animal species do not yet exist, and cryopreservation techniques still warrant improvement (Summers 1986; Whittingham in press; Holt in press).

Two categories of genetic technology are of major importance to *ex situ* propagation of endangered animal species (de Boer 1982a; Ryder 1986). The first category includes techniques for proper assessment of the taxonomic units (species, subspecies, intra-subspecific geographic forms) that should be preserved as "closed" *ex situ* populations. In many cases, genetic techniques such as studies of chromosomes, blood proteins and mitochondrial DNA offer more accurate assessment of intra-specific divergence than traditional taxonomic methods (de Boer 1982b; de Boer and Seuánez 1982; Ryder et al. 1988; Avise et al. 1987). By applying such techniques in cases of uncertain phylogenetic distance between intra-specific forms, conclusions can be drawn concerning the need of their combined or separate preservation in captivity. (This, by the way, not only applies to the *ex situ* situation, but also to the preservation of wild populations.) Secondly, genetic techniques that enable precise assessment of genealogical relationships are of importance. Knowledge of exact relationships is crucial to intensive population management. Nevertheless, even under the controlled conditions of captivity, parentages are not always known, especially in species that are kept in multi-male/multi-female groups. Various genetic techniques are available to trace genealogies, e.g. studies of chromosomal variants, blood protein variants and DNA fingerprints (Lacy and Foster 1988; Ryder 1986, in press; Ely and Ferrell 1990; Amos in press). Application of these techniques enables intensive management of "black box" *ex situ* populations. In addition, they also may give some insight in genealogical relationships in wild remnant populations, allowing a certain degree of management even in the wild.

REINTRODUCTION

As indicated in the beginning of this paper, the ultimate goal of *ex situ* propagation programmes is to support the survival of populations of endangered animal species in their natural environments. This implies that at some stage of each programme genetic material of the *ex situ* population should be transferred to the wild in order to either reinforce remnant populations, if these still exist, or to re-establish wild populations. Even if biotechnology – particularly artificial reproduction techniques – is considerably improved in the future, the usefulness of these techniques for the transfer of genetic material to reinforce wild populations will be rather limited. In the majority of cases reinforcement will consist of transfer of live animals that have been bred and raised in captivity to the natural habitat of their species. Populations that have already become extinct in the wild can only be reestablished by such reintroductions. Thus, the ultimate goal of *ex situ* propagation programmes largely depends on the possibility of reintroduction.

Reintroductions of captive-bred animals were already undertaken in the past century; the Alpine ibex serves as one of the most well known examples (see Stüwe and Nievergelt 1989). Since then, many dozens of species have been reintroduced with various degrees of success. Recently Arzdorf (1990) presented an extensive review, summarizing reintroduction attempts of captive-bred animals of 129 species (51 mammals, 71 birds and 7 reptiles), 11 of which were completely extinct in the wild at the time of reintroduction. Thirty of these attempts resulted in the establishment of self-sustaining populations (10 mammals, 16 birds, and 4 reptiles), a rate of success that probably will increase considerably as many of the 129 reintroduction projects were still in their initial stages at the time of Arzdorf's review. Two other recent collections of examples of success and failure of reintroductions can be found in Endangered Species Update (1990) and in Gipps (in press).

The literature on reintroduction projects currently includes many hundreds of references (see Arzdorf 1990) and is increasing rapidly with the number of new projects undertaken. Some authors have discussed reintroduction in more general terms (e.g. Brambell 1977; Scott and Carpenter 1987; Ounsted in press; Lyles and May 1987), particularly emphasizing prerequisites to success. These may be summarized as follows:

1. The *ex situ* population, which is the source of animals for reintroduction, should be managed according to genetic and demographic guidelines as summarized in this paper. Preservation of original genetic variability is of particular importance, providing material for natural selection to act on. Animals for reintroduction should be carefully selected from the *ex situ* population, especially with respect to their genetic contribution.
2. There should be sufficient suitable habitat available. Several reintroductions failed because animals were released in suboptimal habitat which did not meet the species' requirements.
3. The factors that originally caused the species' extinction or decimation in the wild should be known and should be under control before reintroduction is attempted. Various reintroduced populations have also been destroyed because of threatening factors that were not eliminated.
4. Great care should be taken to avoid the introduction of "foreign" diseases via reintroduced animals, especially if there is still a resident wild population of the species in the reintroduction area or an adjacent area.
5. Care should be taken to reintroduce animals of the same geographic form, if there still is a resident population of the species in the reintroduction area, and the resulting population density should not be beyond that what the area can support.
6. Special attention should be given to non-genetically controlled behaviours crucial to the survival of reintroduced animals that might have been lost during the period (often many generations) of propagation in the *ex situ* situation. Animals may have to be trained in order to regain these behavioural capabilities.
7. Individual animals to be reintroduced should be gathered into groups that reflect the social structure of the species prior to their release.
8. Intensive study and monitoring preceding the reintroduction, during the actual release and during a long period afterwards is crucial to success.
9. Support is required from all relevant national and local authorities and nature conservation organizations, as well as that of the local populations.

Several recent examples, such as those of the Arabian oryx (Stanley Price 1989) and the golden lion tamarin (Kleiman et al. 1985, in press), have shown that consideration of all the above conditions, together with great effort, may lead to successful reintroductions. These, and all other experiences with animal reintroductions are currently collected and analysed by the IUCN/SSC Reintroduction Specialist Group (Stanley Price in press). This will doubtlessly lead to a much better insight into the possibilities and impossibilities of reintroduction and to a higher rate of success in future projects. It will also substantiate the conclusion that reintroduction is not simply a matter of releasing animals in the wild, but rather is a long lasting, intensive process. It should be born in mind that extinction of species often took several, and sometimes many generations. Why then should reintroduction succeed within a single year? More logically, successful reintroduction should be allowed two, three or even ten generations' time!

CONCLUSIONS

To conclude, we note that captive, or *ex situ* propagation programmes, only play a relatively minor role in the preservation of genetic and biological diversity. However, this role should not be underestimated, since such programmes do have some clear advantages:

1. For endangered species, *ex situ* propagation allows highly intensive population management, excluding all kinds of stochastic influences that can never be excluded in the wild.
2. Similarly, the *ex situ* situation enables the conduction of intensive research into the biology of the species concerned.
3. *Ex situ* propagation allows the development and application of biotechnological methods which may be supportive to the species' survival.
4. *Ex situ* propagation projects should aim at supporting the survival of populations in their natural habitats, involving interactive management of captive and wild popualtions, and carefully using advanced techniques of reintroducing captive-bred animals into the wild.

The knowledge and experience gained in these highly intensively managed *ex situ* populations is not only valuable for prolonging the existence of the species concerned, but also for the survival of many other threatened species in the wild. Because of the increasing fragmentation of habitats, more and more wild populations are going to resemble zoo populations, scattered around in their original distribution areas as many tiny subpopulations. Such highly fragmented populations can only survive if they are managed, and for such management the lessons that can be learned from the *ex situ* projects are highly beneficial.

LITERATURE CITED

American Association of Zoological Parks and Aquariums. 1983. *Species Survival Plan.* AAZPA, Wheeling, WV.

Albers, K. B. M. 1989. Demography and breeding programs, pp. 1–27. In: L. E. M. de Boer (ed) *EEP Co-ordinators' Manual.* Stichting Nationaal Onderzoek Dierentuinen, Amsterdam.

Amos, W. (in press). Deducing population structure and breeding system in natural populations using hypervariable sequences. In: D. M. Moore, W. V. Holt, G. M. Mace and O. A. Ryder (eds) *Biotechnology and the Conservation of Biodiversity.* Zoological Society of London, Symposium, September 1990.

Arzdorf, W. 1990. *Stand, Möglichkeiten und Grenzen von Zucht und Auswilderung bedrohter Tierarten als Beitrag zum Artenschutz.* Zoological Garden of Cologne, Cologne.

Avise, J. C., J. Arnold, R. M. Ball, E. Bermingham, J. E. Neigel, C. A. Reeb and N. C. Saunders. 1987. Intraspecific phylogeography: the mitochondrial DNA bridge between population genetics and systematics. Ann. Rev. Ecol. Syst. 18: 498–522.

Baker, R. M. and G. G. George. 1988. Species management programmes in Australia and New Zealand. Int. Zoo Yearb. 27: 19–26.

Ballou, J. D., T. Foose and M. Gilpin (eds) (in press). *Population Management for Survival and Recovery*. University of Chicago Press, Chicago.

Boer, L. E. M. de. 1982a. Genetics and the conservation of the orang utan, pp. 39–60. In L. E. M. de Boer (ed) *The Orang Utan: its Biology and Conservation*. Junk, The Hague.

Boer, L. E. M. de. 1982b. Karyological problems in breeding owl monkeys. Int. Zoo Yearb. 22: 119–124.

Boer, L. E. M. de. 1989a. Preservation of species in zoological and botanical gardens, pp 1–23. In: L. E. M. de Boer (ed) *EEP Co-ordinators' Manual*. Stichting Nationaal Onderzoek Dierentuinen, Amsterdam.

Boer, L. E. M. de. 1989b. Genetics and breeding programs; genetic guidelines and their background for EEP-coordinators, pp. 1–86. In: L. E. M. de Boer (ed) *EEP Co-ordinators' Manual*. Stichting Nationaal Onderzoek Dierentuinen, Amsterdam.

Boer, L. E. M. de (ed) 1989c. *EEP Coordinator's Manual*. Stichting Nationaal Onderzoek Dierentuinen, Amsterdam.

Boer, L. E. M. de and H. N. Seuánez. 1982. The chromosomes of the orang utan and their relevance to the conservation of the species, pp. 135–170. In: L. E. M. de Boer (ed) *The Orang Utan: its Biology and Conservation*. Junk, The Hague.

Brambell, M. R. 1977. Reintroduction. Int. Zoo Yearb. 17: 112–117.

Conway, W. G. 1986. The practical difficulties and financial implications of endangered species breeding programmes. Int. Zoo Yearb. 24/25: 210–219.

Cranefield, M. R., S. E. Kempske and N. Schaffer. 1988. The use of in vitro fertilisation and embryo transfer techniques for the enhancement of genetic diversity in the captive population of the lion-tailed macaque, *Macaca silenus*. Int. Zoo Yearb. 27: 149–159.

Dresser, B. L. 1986. Embryo transfer in exotic bovids. Int. Zoo Yearb. 24/25: 138–143.

Ely, J. and R. E. Ferrell. 1990. DNA "fingerprints" and paternity ascertainment in chimpanzees (*Pan troglodytes*). Zoo Biol. 9: 91–98.

Endangered Species Update. 1990. Special Issue: Captive propagation and reintroduction: A strategy for preserving endangered species? Endangered Species Update 8(1): 1–88.

Flesness, N. R. and G. M. Mace. 1988. Population databases and zoological conservation. Int. Zoo Yearb. 27: 42–49.

Foose, T. J. 1981. The relevance of captive populations to the conservation of biotic diversity, pp. 374–401. In: C. M. Schonewald-Cox, S. M. Chambers, B. MacBryde and W. L. Thomas (eds) *Genetics and Conservation: A Reference for Managing Wild Animal and Plant Populations*. Benjamin/Cummings, Menlo Park, CA.

Foose, T. J. 1983. Demographic problems and management in captive populations. Proc. 1980 Ann. AAZPA Conf., American Accociation of Zoological Parks and Aquariums, Wheeling, WV.

Foose, T. J. and J. Ballou. 1988. Population management: theory and practice. Int. Zoo Yearb. 27: 26–41.

Foose, T. J., R. Lande, N. R. Flesness, G. Rabb and B. Read. 1986. Propagation plans. Zoo Biol. 5: 139–146.

Foose, T. J., U. S. Seal and N. R. Flesness. 1987. Captive propagation as a component of conservation strategies for endangered primates. Monographs Primatol. 9: 263–299.

Frankel, O. H. and M. E. Soulé. 1981. *Conservation and Evolution.* Cambridge University Press, Cambridge.

Gipps, J. W. H. (ed) (in press). *Beyond Captive Breeding: Reintroducing Endangered Mammals to the Wild.* Zoological Society of London, Symposium, November 1989. Oxford University Press, Oxford.

Glatston, A. R. 1986. Studbooks: the basis of breeding programmes. Int. Zoo Yearb. 24/25: 162–167.

Hedrick, P. W. and P. Miller. 1992. Conservation genetics: theory and management of captive populations. This volume.

Holt, W. V. (in press). Developments in artificial insemination and semen freezing. In: D. M. Moore, W. V. Holt, G. M. Mace and O. A. Ryder (eds) *Biotechnology and the Conservation of Genetic Diversity.* Zoological Society of London, Symposium, September 1990.

JAZGA. 1990. Report from the Japanese Association of Zoological Gardens and Aquariums. CBSG News 1(3): 23–24.

Johnston, L. E., U. S. Seal, D. L. Armstrong, R. L. Tilson, P. Wolff, K. Petrini, L. G. Simmons, T. Gross and D. E. Wildt. 1990. Successful in vitro fertilization and embryo transfer in the tiger (*Panthera tigris*). Tiger Beat, The Newsletter of the Tiger Species Survival Plan 3(2): 13–14.

Jones, S. R. 1990. Overview of goals and activities of the IUCN Captive Breeding Specialist Group and International Species Information System. Endangered Species Update 8(1): 8–9.

Kleiman, D. G., B. B. Beck, J. M. Dietz, L. A. Dietz, J. D. Ballou and A. F. Coimbra-Filho. 1985. Conservation program for the golden lion tamarin: captive research and management, ecological studies, educational strategies and reintroduction. In: K. Benirschke (ed) *Primates: the Road to Self-sustaining Populations.* Springer-Verlag, Heidelberg.

Kleiman, D. G., B. B. Beck, J. M. Dietz and L. A. Dietz (in press). Costs of reintroduction and criteria for success: accounting and accountability in the golden lion tamarin conservation program. In: J. H. W. Gipps (ed) *Beyond Captive Breeding: Reintroduction of Endangered Mammals to the Wild.* Zoological Society of London, Symposium, November 1989. Oxford University Press, Oxford.

Kydd, J., M. S. Boyle, W. R. Allen, A. Sheppard and P. M. Summers. 1985. Transfer of exotic equine embryos to domestic horse and donkey. Equine Vet. J., Suppl. 3: 80–83.

Lacy, R. C. 1989. Analysis of founder representation in pedigrees: founder equivalents and founder genome equivalents. Zoo Biol. 8: 11–123.

Lacy, R. C. and M. L. Foster. 1988. Determination of pedigrees and taxa of primates by protein electrophoresis. Int. Zoo Yearb. 27: 159–168.

Lacy, R. C., N. R. Flesness and U. S. Seal (eds) 1989. *Puerto Rican Parrot*, Amazona vittata, *Population Viability Analysis and Recommendations*. CBSG, Apple Valley, MN.

Lyles, A. M. and R. M. May. 1987. Problems in leaving the ark. Nature 326: 245–246.

MacCluer, J. W., J. L. VandeBerg, B. Read and O. A. Ryder. 1986. Pedigree analysis by computer simulation. Zoo Biol. 5: 147–160.

Nogge, G. 1988. Europäisches Erhaltungszuchtprogramm: eine neue Form der Zusammenarbeit Zoologischer Gärten zur Erhaltung bedrohter Tierarten. Zoologischer Garten N.F. 58: 40–46.

Olney, P. 1990. International studbooks and registers, pp. 15–23. In: U. S. Seal (ed) *CBSG Meeting Report, Copenhagen, August 1990*. CBSG, Apple Valley, MN.

Ounsted, M. L. (in press). Reintroducing birds: lessons to be learned for mammals. In: J. H. W. Gipps (ed) *Beyond Captive Breeding: Rreintroduction of Endangered Mammals to the Wild.* Zoological Society of London, Symposium, November 1989. Oxford University Press, Oxford.

Princée, F. P. G. 1988. Genetic variation in the zoo population of the red panda subspecies *Ailurus fulgens fulgens*. Zoo Biol. 7: 219–231.

Princée, F. P. G. 1989. *Zooresearch Studbook Management Software; User's Manual.* Stichting Nationaal Onderzoek Dierentuinen, Amsterdam.

Princée, F. P. G. (in press). Implementation of social structures in genetic management. In: J. Ballou, T. Foose and M. Gilpin (eds) *Population Management for Survival and Recovery*. University of Chicago Press, Chicago.

Puijenbroeck, B. van (ed) 1990. *Summary of the SSP/EEP Masterplan Workshop on the Okapi, November 1989*. Royal Zoological Society of Antwerp, Antwerp.

Ralls, K. and J. D. Ballou (eds) 1986. Proceedings of the Workshop on Genetic Management of Captive Populations. Zoo Biol. 5: 84–238.

Ryder, O. A. 1986. Genetic investigations: tools for supporting breeding programme goals. Int. Zoo Yearb. 24/25: 157–162.

Ryder, O. A. (in press). Some applications of PCR to studies of wildlife genetics. In: D. M. Moore, W. V. Holt, G. M. Mace and O. A. Ryder (eds) *Biotechnology and the Conservation of Genetic Diversity*. Zoological Society of London, Symposium, September 1990.

Ryder, O. A., J. H. Shaw and C. M. Wemmer. 1988. Species, subspecies and *ex situ* conservation. Int. Zoo Yearb. 27: 134–140.

Schonewald-Cox, C. M., S. M. Chambers, B. MacBryde and W. L. Thomas (eds) 1983. *Genetics and Conservation: A Reference for Managing Wild Animal and Plant Populations*. Benjamin/Cummings, Menlo Park, CA.

Scobie, P. N. and N. R. Flesness. 1987. *ARKS (Animal Records Keeping System); User Manual.* ISIS, Apple Valley, MN.

Scobie, P. N. and N. R. Flesness. 1989. *SPARKS (Single Population Analysis and Record Keeping System); User Manual.* ISIS, Apple Valley, MN.

Scott, J. M. and J. W. Carpenter. 1987. Release of captive reared or translocated endangered birds: what do we need to know? Auk 104: 544–545.

Seal, U. S. 1986. Goals of captive propagation programs for the conservation of endangered species. Int. Zoo Yearb. 24/25: 174–179.

Seal, U. S. (ed) 1990. *CBSG Meeting Report, Copenhagen, August 1990*. CBSG, Apple Valley, MN.

Seal, U. S., D. G. Makey, D. Bridgewater, L. Simmons and L. Murtfeldt. 1977. ISIS: a computerised record system for the management of wild animals in captivity. Int. Zoo Yearb. 17: 68–70.

Seal, U. S., T. Foose, R. C. Lacy, W. Zimmermann, O. A. Ryder and F. P. G. Princée (eds) 1990. *Przewalski's Horse*, Equus przewalskii, *Global Conservation Plan*. CBSG, Apple Valley, MN.

Soulé, M. E. (ed) 1986. *Conservation Biology: The Science of Scarcity and Diversity*. Sinauer, Sunderland, MA.

Soulé, M. E. (ed) 1987. *Viable Populations for Conservation*. Cambridge University Press, Cambridge.

Soulé, M. E., M. Gilpin, W. Conway and T. Foose. 1986. The millenium ark: how long a voyage, how many staterooms, how many passengers? Zoo Biol. 5: 101–103.

Stanley Price, M. R. 1989. *Animal Reintroductions: the Arabian Oryx in Oman*. Cambridge University Press, Cambridge.

Stanley Price, M. R. (in press). A review of mammal reintroductions and the role of the Reintroduction Specialist Group of IUCN/SSC. In: J. H. W. Gipps (ed) *Beyond Captive Breeding: Reintroducing Endangered Species to the Wild*. Zoological Society of London, Symposium, November 1989. Oxford University Press, Oxford.

Strover, J., J. Evans and E. P. Dolensek. 1981. Interspecies embryo transfer from the gaur to domestic Holstein. Ann. Proc. Am. Assoc. Zoo Vet. 1981: 122–124.

Stüwe, M. and K. Nievergelt. 1989. Der Alpensteinbock: Modellfall einer gelungenen Wiedereinbürgerung, pp. 237–129. In: E. Schneider, H. Oelke and H. Gross (eds) *Die Illusion der Arche Noah*. Göttingen.

Summers, P. M. 1986. Storage and use of mammalian embryos. Int. Zoo Yearb. 24/25: 131–138.

Thorne, E. T., D. R. Kwiatowski and R. Oakleaf. 1988. Black-footed ferret captive propagation: a chance for recovery, pp. 315–340. In: B. L. Dresser, R. W. Reece and E. J. Maruska (eds) *Proceedings 5th World Conference on Breeding Endangered Species in Captivity, October 1988*. Cincinnati Zoo, Cincinnati, Ohio.

Thompson, E. A. 1986. Ancestry of alleles and extinction of genes in populations with defined pedigrees. Zoo Biol. 5: 161–170.

Whittingham, D. G. (in press). Preservation of mammalian eggs and embryos. In: D. M. Moore, W. V. Holt, G. M. Mace and O. A. Ryder (eds) *Biotechnology and the Conservation of Genetic Diversity*. Zoological Society of London, Symposium, September 1990.

14 Gene banks for livestock conservation

Kalle Maijala
University of Helsinki,
SF-00780 Helsinki,
Finland

and

Nils Kolstad
Agricultural University of Norway,
N-1432 Ås-NLH,
Norway

ABSTRACT

The conservation of the genetic resources of domestic animals merits special attention. Their genetic variation has evolved over millions of years, during which different genes for many fitness traits have been tested. Under human influence in the last 10 millennia many breeds adapted to local needs. Today the potential for changing the genetic make-up of animals and concentrating on a few breeds has increased, due to new techniques. These and the similarity of breeding goals have raised concerns about losses of genetic variation. Activities for preventing these have been started in many parts of the world, to maintain the possibility of adjusting animals to future needs. Besides these economic-biological motives there are scientific and cultural-historical ones. The main methods used are live animals, frozen semen and embryos. New methods have also been suggested. Each method has its good and bad features. Hence a combination of several methods is often needed. The great variety of methods and breeds makes it important to plan the activities carefully. For the future utilization of the conserved material it is important to have reliable, many-sided information on the breeds and individuals documented and stored. Hence both global and regional data banks have been established.

GENETIC VARIATION

The genetic variation in livestock has evolved over millions of years, when the usefulness of different genes and gene combinations has been severely tested for adaptability, resistance etc. Over the last 10 mil-

Conservation of Biodiversity for Sustainable Development.
Edited by O. T. Sandlund, K. Hindar and A. H. D. Brown. 1992. Scandinavian University Press, Oslo.

lennia humans have greatly influenced this evolution, and many breeds adapted to local needs have developed. The potential for changing the genetic make-up of farm animals and of concentrating on a few breeds has increased in recent decades, due to new theories and techniques: artificial insemination, embryo transfer and data processing. These and the increased similarity of breeding goals have raised concerns about losses of breeds and of genetic variation within breeds (Maijala 1970; FAO 1981). According to a survey in 1982, 1/3 of 737 breeds of five main species were endangered in Europe (Maijala et al. 1984).

Several breeds, which had almost lost their utility, have received new importance: some draught breeds of cattle (e.g. Charolais) for beef production, sheep (e.g. Finnsheep) and pigs (e.g. Chinese pigs) for prolificacy, Nordic sheep for luxury furs, Nordic goats for cashmere wool and Cornish fighting hens for meat. On the other hand, it has been suggested that new variation useful for selection is continuously created by mutations (Hill 1982). The question is how reliably and quickly the value of this kind of variation for such complex traits as adaptability and disease resistance can be tested and evaluated in a selection programme.

MOTIVES FOR CONSERVATION (GENE BANKS)

The losses of genetic variation experienced and expected have awakened interest in conserving it, i.e. starting gene bank activities. The motives for this activity have been discussed among others by Maijala (1970), Mason (1974), Simon (1984), and Maijala et al. (1984) and are usually grouped as follows: (1) The biological-economic motives include e.g. changing production conditions and demands for animal products, need for rapid progress in present goals, need to pass selection limits and antagonisms, and possibility to apply new biotechnology in future. (2) The scientific ones include e.g. genetic variation for research, measurement of genetic change, and teaching material in biology and animal sciences. (3) The cultural-historical motives mean e.g. memorials of nature and culture, and research and education in history, ethnography etc. (4) The ethical and moral grounds concern the right of existence of each creation.

The motives of different people interested in conservation vary. Those of group 3 are most commonly understood, but the different interest groups might reach their goals faster and better, if they work together. The group 1 people with their expertise may help the group 3 in genetic problems, and the group 3 with its enthusiasm may help in financing conservation for the purposes 1 and 2. Coordination of their activities is well-founded.

CONSERVATION ACTIVITIES IN FARM ANIMALS

During the recent decades, both government and private organizations in various parts of the world have started the preservation of animal genetic resources (AGR) (Henson 1990). On both sides, the activities can be grouped into global, regional and national levels. Table 1 sketches the beginnings of these activities.

FAO has worked for more than 40 years in the field and has been an important initiator. It has worked together with UNEP. It has published studies, given recommendations to member governments, arranged training courses and organized regional cryogenic gene banks (Mason 1974; Hodges 1990). Sessions on conservation

Table 1. Summary of the beginnings of conservation activities.

Governmental organizations			Non-governmental organizations	
Global level:				
FAO[a]	publications	1953-	Rare Breeds International	1989-
	expert meetings	1966-	world conferences	
	technical consultation	1980		
	regional gene banks	1988-		
UNEP[b]	projects with FAO	1974-		
	technical consultation	1980		
Regional level:				
EAAP[c]	working party	1980-	Society for Advancement of Breeding Research in Asia and Oceania (SABRAO): conference	1979-
AGR	data bank	1987-		
ILCA[d]	breed studies	1977-		
NMC[e]	symposium	1978		
	working party	1980-		
National level:				
Poland:	national programme	1960s	France: Ethnozoo-technological Society	1971-
Hungary:	preservation on state farms	1961-	UK: Rare Breeds Survival Trust	1973-
France:	law	1966-	The Netherlands: Rare Breeds Foundation	1977-
	support to conservation programmes	1969-		
Many other countries followed			Others followed	

[a] UN Food and Agriculture Organization
[b] UN Environment Programme
[c] European Association for Animal Production
[d] International Livestock Centre of Africa
[e] Nordic Minister Council

have been included in the programmes of the four World Congresses of Genetics applied to Livestock Production since 1974.

The AGR Working Party, set up by the Genetics Commission of EAAP, has made three surveys of the AGR situation in Europe (Maijala et al. 1984; Maijala 1987b; Simon 1989) and established an AGR data bank in Hannover, Germany (Simon 1990). In 1988, this data bank became a joint EAAP/FAO Global Data Bank on AGR. Following the recommendations of the UN Environment Conference in 1972, NMC arranged a symposium and set up a working party. In Africa, ILCA has performed studies e.g. on local breeds resistant to trypanosomiasis. The Austral-Asian society (Barker and Turner 1980) arranged a symposium and planned documentation forms for local breeds.

A Polish State Programme for conservation, and a Hungarian preservation of endangered breeds on state farms were started about 30 years ago, and the state activities in France 24 years ago. The first societies for breed conservation were founded about

20 years ago. Most of the programmes have concerned conservation of endangered, indigenous breeds, but at least the EAAP and NMC programmes have concerned also that of genes in the major developed breeds. Table 2 shows activities by states and private organizations (Maijala et al. 1984).

Table 2. Examples of conservation activities (Maijala ct al. 1984).

Kind of activity	State	Private
Laws and statutes	X	
Support to conservation programmes	X	
Cryogenic semen and embryo banks	X	X
Research projects	X	
Special institutes	X	
Committees or working parties	X	
Conservation herds	X	X
Surveys or inventories	X	X
Subsidies to or establishment of breed societies	X	X
Common herdbook	X	X
Advisory service		X
Mating programmes		X

WHAT IS A GENE BANK?

In a broad sense, all conservation activities can be considered to belong to gene banks, which thus can include the following forms of conservation: (1) *In situ* conservation of live animals can take place in small holder farms or large-scale units (Bodo 1990). (2) *Ex situ* conservation can be carried out as live animals in experimental farms or zoos, or as cryogenic storage of semen, embryos or other cells, or as storage of DNA (Fitzhugh 1984; Brem et al. 1989; Pursel and Johnson 1989). (3) Documentation (data bank) concerns information of breeds, individuals and donors of cells (Maijala et al. 1984, 1990; Powell and Norman 1989; Simon 1990; Maijala 1990). In some connections, only the forms in group 2 are included in the concept of gene banks. Until now, conservation has focused mainly upon breeds, but cryoconservation of semen has been applied also to conserving genetic variation within major breeds for safeguarding the possibilities to changed breeding objectives in future. The purpose of storing DNA is to render it possible to perform genetic analyses of populations and to isolate and transfer useful genes in future.

NORDIC ANIMAL GENE BANK

The NMC working party began by defining its responsibilities in relation to those on the national level and by initiating the mapping of the national situations. The responsibilities were divided into national and Nordic levels (Maijala et al. 1990). Overall responsibility for conservation belongs to governments, because the benefits accrue mainly on the national level. The national level includes mapping out threatened breeds and conservation needs in the country, creating and maintaining gene banks, acquiring information for these, and contacting organizations involved. The national ministry of agriculture should stand for the investment and running costs, and take care of cooperation with breeding and research organizations. The materials should be stored nationally, each material at least at two places.

At the Nordic level there is an expert group, consisting of the chairmen of national groups. It initiates establishments of national gene banks, organizes and coordinates on joint level, initiates projects, and documents and maintains data banks. It mediates information of breeds and their conservation in the area on scientific and popular levels, follows up progress and applies results of research on methods, maintains research contacts (methods, animal breeding, gene technology etc.), and seeks contacts with international conservation activities, both in plants and animals.

At the beginning, the NMC working party sent a letter to the agricultural ministries of member countries, asking them to map out the national situations. Such committees were set up in each country, and the reports became ready by 1984. The Nordic data bank was placed at the Agricultural University of Norway. Since 1984 the working party has a part-time secretary. Most of the national working parties have succeeded in finding some finances for their activities, e.g. for employing a planning secretary, important for making progress.

NUMBERS OF EXISTING AND ENDANGERED BREEDS IN EUROPE

The importance of breed conservation depends greatly on the frequency of breeds endangered. The criteria for classifying a breed as endangered vary and depend, for example, on the reproduction rate of the species (Bodo 1990). In the European survey in 1982 the maximum numbers of females in breeds of cattle, sheep, goats and pigs were set at 1,000, 500, 500 and 200, respectively, or at 5,000, 1,000, 1,000 and 500, in case the numbers were declining and those of males were less than 20. With these rules and local considerations the numbers shown in Table 3 were obtained.

Table 3. Numbers of breeds and of those endangered in Europe (Maijala et al. 1984, 1990).

Species	European survey 1982			Scandinavia
	Total	Endangered		Endangered
	No.	No.	%	No.
Cattle	181	81	45	17
Horses	149	51	34	8
Pigs	66	30	45	5
Sheep	264	67	25	7
Goats	77	12	16	5
Rabbits				3
Foxes				1
Reindeer				1
Layer hens				8
Ducks				4
Geese				6
Bees				3
Minks				MT[a]
TOTAL	737	241	33	68

[a] ca. 10 mutant types

CONSERVATION METHODS AND THEIR USEFULNESS

Methods of conservation have been discussed by Fitzhugh (1984), Brem et al. (1989), Pursel and Johnson (1989) and Bodo (1990). Their use in different situations has been evaluated by Smith (1984a), Brem et al. (1984), and Maijala et al. (1984). There are several methods with varying applicability, as shown in Table 4.

Live animals (LA) is the most common and best known method, but is being more and more supplemented with frozen semen (FS) and frozen embryos (FE). Live animals and frozen embryos have high initial costs, but they conserve full functional genotypes, rendering rapid utilization possible at any time. Both initial and maintenance

Table 4. Available methods of conservation and their usefulness.

Method	Abbreviation	Availability
Present methods:		
Live animals	LA	In all species
* pure breeds	LAP	
– random mating	LAR	
– divergent selection	LAS	
* gene pools	LAG	
Frozen semen	FS	In most species [a]
Frozen embryos	FE	In many species [b]
New methods:		
Frozen oocytes	FO	In cattle, but
Embryo vitrification	EV	possible in
Stored cells	SC	most
Stored chromosomes	CH	species [c]
DNA or gene libraries	GL	In all species [c]

[a] nearly all domestic mammals and birds (Pursel and Johnson 1989)
[b] in most domestic ruminants (Pursel and Johnson 1989)
[c] technically possible in most species (Brem et al. 1989)

costs for frozen semen are low, but regenerating a breed takes a long time. Disease and accident risks are highest for live animals. There is no risk for genetic drift, inbreeding and genetic contamination from other populations in frozen semen and frozen embryos, but these methods do not make it possible to develop the adaptability or other valuable traits during conservation. Only live animals satisfy also the cultural-historical and emotional interests and render it possible to investigate traits which become topical.

Since no method fulfills all requirements, combined use of live animals, frozen semen and frozen embryos are recommended in the species in which they are possible. In species like pigs (frozen embryos not possible), live animals and frozen semen deserve to be combined. The use of embryo vitrification, frozen oocytes, stored cells, stored chromosomes and gene libraries deserves consideration, since techniques for utilizing genetic material stored in these ways are developing. They have almost the same advantages and disadvantages as frozen embryos. The gene library method is important for estimating genetic distances between breeds, and for finding and isolating important genes in the future.

PROBLEMS IN MAINTAINING SMALL POPULATIONS

The importance of alternative or supplementary methods of conservation becomes understood, when the following problems involved in maintaining small populations are considered: (1) Genetic problems are caused by the risks from inbreeding and random drift, and by limited possibilities to artificial selection, while undesired responses to natural selection occur (Bodo 1990; Rochambeau and Chevalet 1990). (2) Hygienic problems cause a need to protect gene bank animals from disastrous diseases and ensure that they do not cause a health risk to other animals at the time of utilization (Wierzbowski 1984). (3) Organizational problems concern information, mating planning, exchange of animals etc. (Maijala 1986; Bodo 1990; Henson 1990). (4) The risk of accidents is reason for maintaining the material at least at two locations (Maijala et al. 1984). (5) Increased costs are very likely, since the investments have to be made well before returns, which in turn accrue to other people, and hence governmental participation is needed (Smith 1984b).

The problems mean that many kinds of expertise are needed for rational and effective conservation. A central problem is the avoidance of genetic drift and inbreeding. Sensible utilization of different methods can help to a certain extent: Frozen storage methods help in minimizing genetic drift and inbreeding, and in conserving genes. Choice between frozen semen and frozen embryos depends on the costs of collection and on the number of progeny required. With frozen embryos the original stock can be regenerated at any time, and frozen semen needs a small permanent female breeding group or a store of frozen embryos.

An essential concept in estimating the numbers of animals needed for conservation is the effective population size, N_e (Wright 1931), which determines the random drift of gene frequencies (Figure 1). It is important to remember that N_e depends on the number of breeding males, not only on females (Figure 2). An idea of the numbers of animals needed in live animal conservation without frozen semen and frozen embryos is given in Table 5.

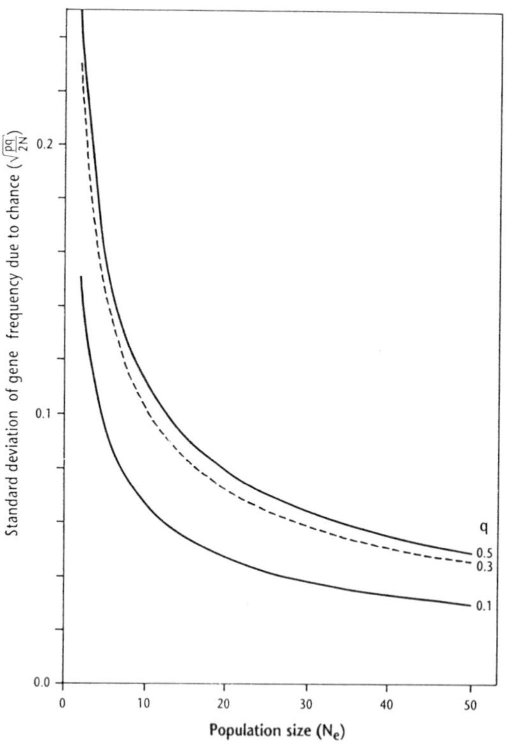

Figure 1. Random drift of gene frequency at different population sizes (N_e) and gene frequencies (q) (according to Wright 1931).

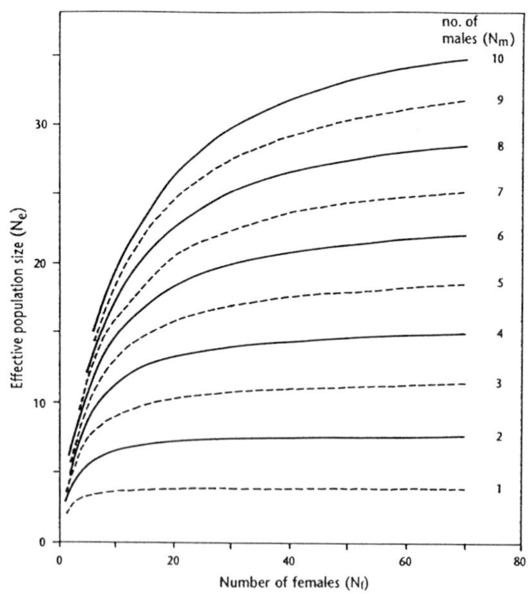

Figure 2. Effective population size $N_e = 4N_m \cdot N_f/(N_m + N_f)$ of varying numbers of males (N_m) and females (N_f).

When lower inbreeding rates are required, much larger numbers of animals are needed. Hence supplementation with frozen semen and frozen embryos is important. In the case of cattle, the number of embryo donors needed for getting a satisfactory number of progeny in the future depends on several factors (Brem et al. 1989): (1) Biologically determinable figures include per cent cows suitable to be donors, donors able to superovulate, lost during pregnancy, peri- and postnatal mortality, and sex ratio. (2) Biotechnically determinable figures are per cent embryos suitable for freezing, lost during freezing and storage, and pregnant after freezing and thawing. (3) Other components such as selection intensity. For example, 616 donors are needed for producing 25 unrelated females, if the post-thawing embryo survival rate is 50 % and the pregnancy rate 20 %, while only 88 are needed, if the corresponding values are 100 % and 70 %.

Table 5. Required minimum numbers in a conservation herd, when inbreeding rate is kept under 2 % per year (after Smith 1984b).

Species	Total no. of animals		No. annual replacements	
	Males	Females	Males	Females
Cattle	10	26	10	5
Sheep	22	60	22	12
Pig	44	44	44	18
Poultry	72	72	72	72

ECONOMIC ASPECTS OF CONSERVATION

Smith (1984b, 1985) has tried to estimate both the costs and possible benefits of conservation and arrived at the following conclusions: (1) Even small improvements of production efficiency pay on the national level. (2) Even low proportions of genes used may be profitable. (3) The maintenance of strains with low probabilities of utilization pays. (4) The costs are low as compared to national benefits. (5) The required probabilities of utilization are lowest for frozen semen. (6) The benefits accrue primarily at the national level. (7) There is space for and many advantages of creating and manipulating genetic diversity in maximizing future economic efficiency of farm animals.

Because of the high costs of live animal conservation, ways of decreasing them are of interest. Such ways are listed in Table 6.

ANIMAL GENETIC RESOURCES DATA BANKS

EAAP/FAO Global Data Bank

In 1988, EAAP and FAO agreed that the EAAP Data Bank established in Hannover Veterinary University, Germany, for breeds of five species (cattle, horses, pigs, sheep, and goats) be developed to a joint EAAP/FAO Global Data Bank on animal genetic resources and to include buffaloes. To collect information for it, the following questionnaires have been prepared (Simon 1990):

1. General information (country, species, breed, location, responsible organizations and informers etc.).
2. Origin and development of breed (origin, immigration, numbers of animals and their trends, herd size, average age of breeding animals etc.).
3. Breed description (colours, appearance, genetic peculiarities etc.).
4. Qualification of breed (various uses, special qualities).
5. Management conditions.
6. Performance records (relative to standard breeds).

Table 6. Possible alternative uses of breeds of some farm animal species for economic production or for pleasure and recreation (based mainly on Maijala 1987a).

Way of using	Possible in				
	Cattle	Goats	Horses	Pigs	Sheep
Economic production:					
Tractive power in difficult conditions	x		x		
Production of "biological food"	x	x			x
Production in school or prison farms	x	x	x		x
Pasture or lawn management	x		x	x	
Forest management, underbrush clearing					x
Production of sera for research and health	x		x	x	
Production of unallergic milk		x			
Production of other medicines	x	x	x	x	x
Dam line in crossbreeding for meat	x			x	x
Utilization in harsh environments	x	x		x	x
Utilization in marginal areas	x	x		x	x
Experimental animals in research	x	x	x	x	x
Production of luxury furs/wool for handicraft					x
Animals in part-time farming		x	x	x	
Pleasure and recreation:					
Animals in national parks	x				x
Farm animal parks and museums	x	x	x	x	x
Trotting matches, riding for hobby and racing			x		
Agricultural and native place museums	x	x	x	x	x
Social company of humans, pet-keeping		x	x	x	x
Aid in bringing up children, 4 H-farms	x	x	x	x	
Maintenance of local culture and tradition	x	x	x	x	x
Exhibition in zoos/tourist attraction	x	x	x	x	x
Folk art, ceremonial purposes	x		x		x

7. Additional information (genetic distances, DNA storage, live animal conservation programmes, possible information sources).

The information is planned to be up-dated every third year and to be utilized, for example, for establishing a World Watch List for endangered breeds (Maijala 1990).

Nordic Data Bank (NDB)

The NCM working party started to plan the Nordic AGR Data Bank in 1985, but postponed the realization, in order to coordinate the forms with the EAAP-forms as far as possible. The database is to be located at the Agricultural University of Norway, Ås. It can exchange data on floppy disks and is independent of language, so that outputs can be sent in six languages. It has seven different kinds of forms: Forms 1–2 are like in the EAAP/FAO system. Form 3 is for breed description (as in EAAP/FAO, but with special details for poultry, rabbits, and fur animals). Form 4 is for qualification of breed (as in EAAP/FAO, but includes some additional traits, and special details for poultry, rabbits and fur animals). Forms 5–6 are as in the EAAP/FAO system. Form

7 contains individual qualities of live animals, and of semen and embryo donors, and is of two kinds: one for ruminants, horses, pigs, sheep, goats and reindeers, and another for poultry, rabbits and fur animals. Each piece of information is at a given place, and numerical or coded information is used, if possible. Both the forms and the software are coordinated with the EAAP/FAO database, but there are forms for breeds of 18 species instead of six as in the latter.

SPREADING OF INFORMATION (NORDIC COUNTRIES)

The NMC-working party has realized that effective spreading of information on breeds and their conservation is important for success. Hence, the Nordic and the national working parties have organized seminars, participated in agricultural shows, prepared films and TV-programmes, published articles in popular journals and given talks in various meetings. In Finland, an enthusiastic artist arranged a "cattle wandering" of ca. 500 km with her East-Finnish cows, stopping in four cities in 1988.

CHOICE OF BREEDS FOR CONSERVATION

Because of the large number of indigenous breeds endangered and of the limited resources available it is not possible to conserve all breeds. Thus, it is necessary to choose them for conservation on a basis of sensible criteria (Smith 1984a; Maijala 1986, 1990; Bodo 1990). These are linked with both objectives and methods of conservation. It has sometimes been suggested that only those breeds, which have been properly evaluated, deserve to be conserved. However, evaluation is often a time-consuming and difficult task, and it is not possible for traits that may become recognized and topical only in the future. Conservation is often needed before evaluation. Many kinds of compromises are obviously needed, as for example between ideals and possibilities, between motives and methods, and between different breeding objectives (Maijala 1986). In considering the various aspects, both practical experience and theoretical knowledge are needed.

Table 7 shows several aspects which should be considered in choosing breeds for conservation.

UTILIZATION OF CONSERVED BREEDS

Whether and when certain conserved genetic resources will be utilized cannot be predicted. However, it is highly probable that some of them will be used, as change is a typical feature of our needs, techniques and living conditions. It is already possible to know, how the material can best be regenerated provided it has been stored with proper documentation. The utilization may require regeneration of the breed for use as a pure breed or as a component in a cross. In both cases rapid regeneration is required. This is possible only, if material stored as live animal (pure breed) or frozen embryos is available. The embryos can be transferred to females of the same species, and thus purebred progeny of both sexes can be produced. These can then be inseminated with frozen semen-stores of the breed.

Frozen semen and live animal (gene pool) can be used as sources of genes for special traits. Marker genes linked closely to the desired gene may in future assist the introgression of genes into a population. In the

Table 7. Aspects for choosing breeds for conservation (Maijala 1987a).

Viewpoint	Examples of aspects included
Value as biological material	Performance, Adaptation, Disease resistance, Special traits, Complementarity and heterosis in crosses
Genetic status and distinctiveness	History, Age as separate breed, Breed purity, Breed structure, Genetic distance to other breeds, Population size and its trends
Ecological importance	Landscape management
Cultural-historical importance	National traditions
Social-political importance	Employment, Leisure time, Hobbies
Possibilities for evaluation and maintenance	Personal and financial resources
Availability of information	Production records, Parentages
Conservation with or without simultaneous selection	Live vs. cryogenic conservation
Possibility for international collaboration	Related populations in other countries
Possibilities for other uses	See Table 6

more distant future, it may be possible to isolate a gene determining an important trait and transfer it to a population. Examples of traits that are determined by major genes, are resistance to some diseases (e.g. trypanosomiasis), adaptability, milk protein types, fecundity (e.g. fertility gene in Booroola-Merino), egg size/body size-ratio and shell strength in Jær-hens.

Additional possibilities for utilization will follow from developments in methods and techniques of conservation and gene manipulation. Hence, there is intensive research going on concerning various biotechnologies as e.g. semen-, embryo- and DNA-techniques (Brem et al. 1989). In the DNA-technology, studies on gene mapping and linkage analyses, estimation of genetic distances with the aid of DNA-markers and gene transfer will serve to utilize gene bank materials. On the other hand, development of statistical techniques to evaluate breeds and individuals more accurately and to find associations between markers and economic traits is also important and is being continued in many countries.

CONCLUSIONS

1. Many economic-biological, scientific and cultural-historical motives speak for conservation of animal breeds and genetic variation.
2. Even small improvements in efficiency and small proportions of genes from conserved stocks bring profit to the nation.
3. Each conservation method has its advantages and disadvantages. A combination of several methods is often well-founded.
4. Conservation of frozen semen and embryos besides live animals makes conservation less expensive, and better than live animals alone.
5. The high costs for live animal conservation can be decreased by using gene bank animals for alternative production forms, human pleasure and recreation.
6. National, regional and global gene bank activities for livestock have increased during the last two decades.
7. Gene banks in the narrow sense consist of *ex situ* conservation, in the wide sense

also of *in situ* conservation, data banks etc.
8. Adequate documentation is necessary for rational conservation and utilization.
9. Coordination of activities among people, organizations and countries is important.
10. International gene and data banks have been established.
11. For cost reasons it is important to choose material for gene banks, using several criteria.
12. Information of and for gene and data banks is important.

LITERATURE CITED

Barker, J. S. F. and H. N. Turner (eds) 1980. *Proc. SABRAO Workshop on Genetic Resources in Asia and Oceania.* Tropical Agricultural Research Centre, Tsukuba, Publ. No. 47.

Bodo, I. 1990. Methods and experiences with in situ preservation of farm animals, pp. 85–102. In: *Animal Genetic Resources. A Global Programme for Sustainable Development*. FAO Anim. Prod. Health Paper 80.

Brem, G., B. Brenig, M. Müller and K. Springmann. 1989. Ex situ cryoconservation of genomes and genes of endangered cattle breeds by means of modern biotechnological methods. FAO Anim. Prod. Health Paper 76: 1 – 123.

Brem, G., F. Graf and H. Kräusslich. 1984. Genetic and economic differences among methods of gene conservation in farm animals. Livestock Prod. Sci. 11: 65–68.

Food and Agriculture Organization. 1981. Animal Genetic Resources Conservation and Management. FAO Anim. Prod. Health Paper 24: 1–388.

Fitzhugh, H. A. 1984. Genetic aspects of germplasm storage and genetic engineering, pp. 21–42. In: *Animal Genetic Resources: Cryogenic Storage of Germplasm and Molecular Engineering*. FAO Anim. Prod. Health Paper 44/2.

Hill, W. G. 1982. Predictions of response to artificial selection from new mutations. Genet. Rcs. 40: 255–278.

Henson, E.L. 1990. The organization of live animal preservation programmes, pp. 103–117. In: *Animal Genetic Resources. A Global Programme for Sustainable Development*. FAO Anim. Prod. Health Paper 80.

Hodges, J. 1990. Review of regional animal gene banks and recommendations from Hannover Workshop on associated topics raised by the Tenth Committee on Agriculture, pp. 51–57. In: *Animal Genetic Resources. A Global Programme for Sustainable Development*. FAO Anim. Prod. Health Paper 80.

Maijala, K. 1970. Need and methods of gene conservation in animal breeding. Ann. Génét. Sél. Anim. 2: 403–415.

Maijala, K. 1986. Motives, possibilities and methods of maintaining numerically small cattle breeds. World Rev. Anim. Prod. 22 (1): 43–50.

Maijala, K. 1987a. Possible role of animal gene resource in production, natural environment conservation, human pleasure and recreation, pp. 205–215. In: *Animal Genetic Resources. Strategies for Improved Use and Conservation*. FAO Anim. Prod. Health Paper 66.

Maijala, K. (in collaboration with D. L. Simon and D. Steane) 1987b. Surveying animal breed resources in Europe, pp. 208–218. In: *Research in Cattle Production.*

Danish Status and Perspectives. Contribution in Honour of A. Neimann-Sörensen. Det Kgl. Danske Landhusholdningsselskab.

Maijala, K. 1990. Establishment of a World Watch List for endangered livestock breeds, pp. 167–184. In: *Animal Genetic Resources. A Global Programme for Sustainable Development.* FAO Anim. Prod. Health Paper 80.

Maijala, K., A. V. Cherekaev, J. M. Devillard, Z. Reklewski, G. Rognoni, D. L. Simon and D. E. Steane. 1984. Conservation of animal genetic resources in Europe. Final report of an E. A. A. P. working party. Livestock Prod. Sci. 11: 3–22.

Maijala, K., A. Neimann-Sörensen, S. Adalsteinsson, N. Kolstad, B. Danell and B. Gjelstad. 1990. Conservation of animal gene resources in the Nordic countries. Proc. 4th World Congr. Genet. Appl. Livestock Prod. 14: 459–464.

Mason, I. L. 1974. Moderator's introduction for Round Table A – The conservation of animal genetic resources. Proc. 1st World Congr. Genet. Appl. Livestock Prod. 2: 13–21.

Powell, R. L. and H. D. Norman. 1989. Animal germplasm information systems, pp. 427–443. In: L. Knutson and A. K. Stoner (eds) *Biotic Diversity and Germplasm Preservation, Global Imperatives.* Kluwer, Dordrecht.

Pursel, V. G. and L. A. Johnson. 1989. Cryopreservation of animal germplasm resources, pp. 337–353. In: L. Knutson and A. K. Stoner (eds) *Biotic Diversity and Germplasm Preservation, Global Imperatives.* Kluwer, Dordrecht.

Rochambeau, H. de and C. Chevalet. 1990. Genetic principles of conservation. Proc. 4th World Congr. Genet. Appl. Livestock Prod. 14: 434–442.

Simon, D. L. 1984. Conservation of animal genetic resources – a review. Livestock Prod. Sci. 11: 23–36.

Simon, D. L. 1989. Interim report of the Working Party on Animal Genetic Resources on the survey 1988 in EAAP-member countries. 40th Ann. Meet. EAAP, Dublin. August 1989.

Simon, D. L. 1990. The global animal genetic data bank, pp. 153–166. In: *Animal Genetic Resources. A Global Programme for Sustainable Development.* FAO Anim. Prod. Health Paper 80.

Smith, C. 1984a. Genetic aspects of conservation in farm livestock. Livestock Prod. Sci. 11: 37–48.

Smith, C. 1984b. Estimated costs of genetic conservation in farm animals. FAO Anim. Prod. Health Paper 44/1: 21–30.

Smith, C. 1985. Scope for selecting many breeding stocks of possible economic value in the future. Anim. Prod. 41: 403–412.

Wierzbowski, S. 1984. Cryo-preservation of genetic material: disease risk and control, pp. 49–65. In: *Animal Genetic Resources: Cryogenic Storage of Germplasm and Molecular Engineering.* FAO Anim. Prod. Health Paper 44/2.

Wright, S. 1931. Evolution in Mendelian populations. Genetics 16: 97–159.

SECTION 5

Biotechnological challenges

Recent advances in biotechnology, and in particular gene technology, are being used worldwide to address problems of basic biology, human health, developed-world agriculture and developing-world hunger. The practice of gene technology, however, raises a number of concerns regarding the environmental, ethical, proprietary, and socio-economic consequences of these activities.

Gene technology provides conservation biology with a powerful tool for analyzing genetic variation at the ultimate level, the DNA molecule. Levin (Chapter 15) argues that this novel genetic information will generate a better understanding of the importance of genetic variation, improve the delineation of endangered populations and species, and aid in the design and implementation of captive breeding programs for those populations and species.

Gene technology challenges conservation biology with the release of genetically engineered organisms into the environment. Whereas the effects of these releases on biodiversity may be no different from those of releasing exotic species or domesticated organisms, it must be kept in mind that introduced species are one of the major causes for loss of biodiversity. Levin notes the shortcomings of genetic and ecological theory to make precise predictions of the environmental effects of releases. He recommends that any large-scale release of novel organisms – whether transformed by gene technological or traditional procedures – be preceded by microcosm and small-scale field experiments and "worst-case" risk assessments. Moreover, the utility to society of any release must be identified and weighed against potential environmental hazard.

Is the manipulation of genetic material ethically acceptable? Brøgger (Chapter 16) answers this question in the affirmative because it is ethically acceptable for humans to interact with other life structures in nature, for example in medicine and plant breeding. Limitations to the practice of gene technology must be set by making concessions to the intrinsic value of living organisms, and to nature's utility value for humans. Brøgger concludes by asking for a declaration of ethics in biological research along the lines of the Helsinki declaration for biomedical research.

The advances in gene technology are based on the diversity of genes and organisms, human innovation and ample economic funding. Patent systems have been developed in the industrialized countries to secure that any discovery of significant economic value may be treated as a property for the benefit of the scientist or company making the discovery. New agricultural crops produced by gene technology are now patentable matter in some countries. But how should Third World countries, where most agricultural crops originated, be compensated for years of past innovation by local farmers in shaping the plant varieties that form the basis for present-day agriculture? Fowler (Chapter 17) argues that concrete recognition of "farmers' rights" through a mandatory gene fund is imperative not only to reward innovation, but also to reward and finance conservation. The establishment of such a fund will soften the negative impacts of patents on the international apportionment of power and profit, and address in a farsighted way the link between biodiversity conservation, economic development and justice.

Shiva (Chapter 17) places the socioeconomic consequences of biotechnology in a broader context of different human needs in rich and poor countries. Western science and technology may be out of place in non-western contexts. Shiva uses the symbol of Mahatma Gandhi's spinning wheel – which sustained more jobs than did mechanized textile industry in India – to argue that diverse, low-technology uses of biological resources are better for the livelihoods of most Third World communities than western, high-technology monocultures. The merits of these different production technologies, she maintains, must also be evaluated with respect to their effects on biodiversity.

15 DNA technology and the release of genetically engineered organisms: some implications for the conservation of genetic resources

Bruce R. Levin
Department of Zoology,
University of Massachusetts,
Amherst, Massachusetts 01003,
USA

ABSTRACT

Genomic data obtained through DNA restriction fragment and sequence analysis is having a major impact on the study of the genetic diversity, structure and evolution of populations. The development of "genetically engineered" organisms by *in vitro* DNA procedures is playing an increasingly important role in agriculture, medicine and a number of industries. In this essay, I consider the implications and effects of these applications of "DNA technology" to the conservation of genetic resources. I review those elements of population and evolutionary genetics that are relevant to the conservation genetics, how DNA technology is now being employed by population geneticists and consider the ways this technology may be useful to conservation genetics. I describe some of the applications for which the genetically engineered organisms are being developed for release into the environment, the possible effects of these novel organisms on our genetic resources, the procedures used to assess the environmental hazards of releasing genetically engineered organisms and the limitations of this "risk assessment" enterprise. I conclude with a few critical observation and recommendations about the utility of DNA technology for conservation genetics and the consequences of the release of genetically engineered organisms on the genetic diversity of natural populations.

INTRODUCTION

The development of procedures to extract, cut, splice, and sequence DNA, to introduce DNA manipulated *in vitro* into organisms and clone those genetic molecules have been a boon to basic and applied biology. In addition to the utility of this "DNA technology" for research in molecular, cellular and developmental biology, its application is becoming increasingly important to agriculture, medicine, epidemiology, waste treatment and the chemical and drug industry. These tools are also proving valuable for studies of the genetics and evolution of natural populations. On the other side, there have been considerable anxieties about the health and environmental consequences of the inadvertent or intentional

Conservation of Biodiversity for Sustainable Development.
Edited by O. T. Sandlund, K. Hindar and A. H. D. Brown. 1992. Scandinavian University Press, Oslo.

release of the organisms whose genomes have been modified *in vitro*, "genetically engineered organisms", GEOs.

This essay is divided into two primary sections. The first is an excursion into those elements of population genetics that are relevant to the conservation of genetic resources; (a) the study of genetic variability and ideas about the evolutionary role of standing genetic diversity, (b) how DNA technology is being used for studies of the genetics of natural populations and, (c) the implications of this technological "revolution" for applied conservation genetics. The second part considers the release of genetically engineered organisms; (a) some of the applications proposed for releasing GEOs, (b) potential deleterious consequences of these releases on our genetic resources, and (c) the assessment of these risks. I conclude this, admittedly personal, treatment of these topics with some critical observations and recommendations.

POPULATION GENETICS, DNA TECHNOLOGY, AND THE CONSERVATION OF GENETIC RESOURCES

Traditionally, the academic discipline of population genetics has been called upon to provide answers to three questions that are fundamental to the design of programs for the conservation of genetic resources: (1) How much evolutionarily relevant genetic variability is maintained in natural populations? (By "evolutionarily relevant" genetic variability, I am considering that needed for populations to adapt to their current environment and to the continuous and inevitable changes that occur in that environment.) (2) How is this genetic variability maintained in natural populations? (3) What are ˙ ways to maintain evolutionarily netic variability in threatened and managed populations? In its recent incarnation as "molecular population genetics" (Nei 1987), this academic discipline is also being called upon to help define the taxonomic bounds of endangered species.

Measuring genetic variability and ideas about the role and maintenance of standing genetic variation

Ancient history

In the old days, roughly 1908 through 1966, population genetics was rich in theory and poor in the data needed to evaluate the validity of that theory. It was clear that there was standing genetic variation in natural populations. By artificial selection one could readily obtain all kinds of changes in quantitative characters in organisms isolated from natural sources (Robertson and Reeve 1952; Thoday and Boam 1961; Dobzhansky and Spassky 1967). However, there was no way to estimate the absolute magnitude of the genetic variation responsible for these selected changes in phenotypes. Studies of the genetics of natural populations relied on either variation in chromosomes (ploidy, inversions, transpositions) or in individual genes recognized by their phenotypic effects (colors, spots, stripes, antigens, viability, etc.). With these sorts of data, it was not possible to directly estimate the proportion of genes represented by more than one allele in measurable frequency, **polymorphic**, or the probability that a pair of randomly chosen genes would be different alleles, **genic diversity**, (for a clever indirect approach to this estimation, see Lewontin 1967).

In the absence of these critical data, there were "schools of thought". One school, the "classical", championed the view that natural selection favored unique alleles at individual loci. Mutations to new allelic states were almost invariably deleterious and purged by purifying (stabilizing) selection (Müller 1950). Implicit in this classical interpretation is the proposition that natural populations maintain little or no evolutionarily relevant genetic variability and evolution relies primarily on variability generated by mutation. In the view of the second or "balanced" school, an unspecified, but substantial, fraction of genes were polymorphic and these polymorphisms were maintained by some form of balancing natural selection, such as heterosis (overdominance), stabilizing frequency-dependent selection, or selection favoring different alleles in different geographic regions or at different times. In this interpretation, natural populations maintain a fair amount of evolutionarily relevant genetic diversity, including much of that needed to adapt to environmental changes.

The classical and balanced schools had their advocates and articulate proponents. Although there were enterprising methods to distinguish between these two hypotheses (e.g. see Schull and Neel 1965), these procedures were not terribly sensitive and the data being generated were insufficient to distinguish between them (Lewontin 1974). Thus the discussion and argument went on.

The era of protein electrophoresis

This scholastic period of population genetics ended abruptly in the mid 1960s, with a technical and a conceptual advance. Gel electrophoresis, a workhorse of biochemistry, was applied to studies of the genetics of natural populations. There was evidence for variation in the electrophoretic mobility of enzymes in a variety of species (Shaw 1965), which suggested that protein electrophoresis could be employed to identify variation in structural genes in natural populations. And, in 1966 two groups independently published the results of studies employing gel electrophoresis to estimate the quantity and distribution of structural gene variation in natural populations of humans (Harris 1966), and *Drosophila pseudoobscura* (Lewontin and Hubby 1966). The idea was straight forward, by selecting proteins (mostly enzymes) because they could be identified by staining gels, rather than by their biological function, one would have a sample of structural genes that were chosen without regard to how natural selection may act on them. Estimates of the number and relative frequencies of different electrophoretic mobility classes of this "random" sample of structural gene products would provide an estimate of variation in the structural genes of the species under study.

The results of the initial electrophoretic surveys, and the thousands of kilowatts of protein variation data obtained since, indicated that most natural populations of eukaryotes abound with allelic variation at structural gene loci (for a review see Nevo et al. 1984). On average, the genic diversity (heterozygosity) of natural populations of eukaryotic species was on the order of 10 %, and for bacterial "species" like *E. coli* and *Haemophilus influenzae* on the order of 50 % (Selander and Levin 1980; Porras et al. 1986).

The neutrality – selection controversy

The application of protein electrophoresis to investigate genetic variability within and between populations and the parallel use of amino acid sequence data for studies of the phylogenetic relationship in the 1960s ushered in a new romantic period in population genetics. At one level, the protein polymorphism data led to the demise of the classical school, or at least the traditional curriculum of that school. At the same time, these data indicated that there was too much variability to be maintained by traditional balancing selection schemes, such as heterosis (overdominance) operating independently at many loci (Lewontin and Hubby 1966). New models were constructed, old ones resurrected, and procedures were developed to test hypotheses generated by this theory.

Within short order, the answers were in. But alas, these answers were conflicting. In theory, all of the variability observed could be maintained by some modified form of balancing selection in which loci were not acting independently (Sved et al. 1967; King 1967; Milkman 1967). Also, in theory, all of the variability observed at the protein loci could be selectively neutral and maintained by a balance between recurrent mutation and random genetic drift (Kimura and Ohta 1971).

Thus, population genetics entered the 1970s with two new schools, selectionist and neutralist and what became known as the "neutral gene" or "neutrality – selection" controversy.

Molecular population genetics

With the advent of DNA manipulation technology, it became possible to use genomic data for direct inferences about genetic variability, genetic relationships between populations of the same and different species, and the genetic structure of populations. These techniques also made it practical to study genetic variation in DNA-bearing organelles, mitochondria and chloroplasts, as well as the chromosomal regions flanking genes, spacers, intervening sequences, pseudogenes, regulatory regions, and the vast amounts of repeated DNA sequences that abound in the genomes of eukaryotes.

For retrospective survey studies of genetic variation and population genetic structure, the most important of these DNA procedures are: (1) The enzymatic cutting of DNA in positions of known sequences (of from 4 to 8 base pairs) with restriction endonucleases, "restriction sites". (2) Gel electrophoresis to determine the size (molecular weight) of DNA. (3) *In vitro* hybridization of DNA on membranes with radioactively labeled DNA probes, "Southern" and "dot" blots. (4) Cloning to increase the concentrations of specific sequences of DNA, in living hosts like bacteria, or *in vitro*, using a polymerase chain reaction, PCR. (5) DNA sequencing, to determine the primary, nucleotide, structure.

The protocols for these procedure are well described in a number of excellent manuals (e.g. Sambrook et al. 1989; Innis et al. 1990) and the equipment, supplies, and ingredients needed are readily available commercially.

Along with these biochemical and molecular procedures, a variety of statistical and other tools were developed to analyze the genomic variation data. With these methods it is possible to: (1) Use restriction fragment analysis to estimate the magnitude of nucleotide sequence variation in populations (Schaeffer et al. 1988). (2) Align DNA sequences and determine the nature as well as quantity of nucleotide sequence diversity (Nei 1987; Hudson et al. 1987; Li and Graur 1991). (3) Use DNA data to estimate genetic distances within and between populations and construct trees to specify their phylogenetic relationships and population genetic structure (Felsenstein 1988; Nei 1987; Slatkin 1989; Slatkin and Maddison 1989).

With the use of organelle and chromosomal DNA restriction fragment and sequence data it has been possible to generate far more accurate pictures of the genetic structure of natural populations than those obtained by the use of morphological characters or protein electrophoresis. Questions about; (1) the origin and geographic bounds of populations; (2) breeding structure, extent of inbreeding, migration and introgression between populations; (3) the degree to which populations have passed through variability-reducing bottlenecks, can now be answered with fair precision, albeit with some effort (e.g. Smith et al. 1989; Avise et al. 1988; Meylan et al. 1990; Reeve et al. 1990; Flint et al. 1989).

The application of DNA manipulation techniques has also made it possible to obtain far more detailed and accurate descriptions about the magnitude and nature of genetic variation. Through the use of these molecular tools, we have estimates of the amount of genetic variability at the ultimate, level, DNA (Kreitman 1983; Milkman and Crawford 1983; Dykhuizen and Green 1986; Milkman and Stolzfus 1988; Aquadro et al. 1988; DuBose et al. 1988).

Finally, DNA data have proven extremely valuable for resolving taxonomic and phylogenetic problems; identifying species and determining the relationships between species (see Li and Graur 1991 for a recent review).

Current ideas about the maintenance and role of standing genetic variation

For those interested in a general discourse on the neutrality – selection controversy, I suggest one, or preferably more than one, of the following: Lewontin (1974), Clarke (1979), Kimura (1983), Crow (1985), Hartl and Dykhuizen (1985), Nei (1987), Berry (1989), Li and Graur (1991), and, for sure, the chapter by Berry in this volume. In my own opinion, the neutrality – selection controversy mellowed in the 1980s, but was never resolved, save possibly in the minds of some of the combatants. At this time it is not at all clear what fraction of the genetic variability observed at the morphological level, as amino acid sequence variation in proteins, or nucleotide sequence variation in DNA, is maintained by natural selection and critical to the present and future adaptation of populations, and how much of this genetic variation is selectively neutral, noise or parasitic genetic molecules.

That is not to say there is no evidence that some standing genetic variation is maintained by natural selection. For specific genes, like that coding for the oft cited sickle cell haemoglobin polymorphism in humans, there is overwhelming evidence that allelic variation is maintained by selection favoring heterozygotes. There are also a number of other compelling examples of morphological or protein variation at specific loci being maintained by some form of balancing selection, stabilizing frequency-dependent selection, or selection due to environmental heterogeneity (Endler 1986; Berry, this volume, and the references therein). DNA data have also provided compelling evidence that variation in specific genes, like the ADH locus of *Drosophila pseudoobscura* (Kreitman 1983, 1991), and the major histocompatibility, MHC, loci of humans (Nei and Hughes 1991) is maintained by some form of selection.

To reiterate, the questions that remain unanswered are the old ones. How much of the standing genetic variation in populations is maintained by natural selection? And, how much of the variation maintained in populations is essential for their survival and adaptation to environmental change?

Implications for conservation genetics

It is clear that DNA manipulation technology will continue to be of considerable value to academic population genetics and evolutionary biology. However I believe that utility of these procedures to applied conservation genetics is relatively modest.

One application of DNA technology for conservation genetics is for the design and implementation of breeding programs to maintain endangered species. There is little question about the negative consequences of inbreeding organisms from normally outbred populations due to the segregation of recessive deleterious alleles. DNA restriction fragment data ("finger printing") is an ideal tool for determining or checking the genetic relationships of mating pairs and unequivocally identify individuals (as long as there is DNA sequence variability and the technique is performed correctly). It is also possible to use DNA restriction and blotting techniques to identify carriers of previously identified deleterious alleles. Finally, in theory at least, it should be possible to use DNA restriction and blotting procedures to determine the extent of outbreeding and whether outbreeding depression is anticipated (Templeton 1986).

Another useful application of DNA technology, as well as classical protein electrophoresis, for applied conservation biology is as tools for the unambiguous identification of species (Avise 1989). This is particularly critical, since the taxonomic designation of endangered species is often fundamental to the laws governing their protection and programs managing their populations. The New Zealand tuatara, *Sphenodon*, the only surviving genus of an ancient order of reptiles, provides an excellent illustration of this. The taxonomy of tuatara had been in question for some time and only a single species of this genus, *S. punctatus*, was recognized by many biologist using purely morphological criteria, as well as by the New Zealand legislation protecting this reptile. However a combination of morphological and protein electrophoresis data clearly indicate that there are at least two extant species of this genus, *S. punctatus*, and a species without official protection, *S. guntheri* (Daugherty et al. 1990; May 1990).

With a sufficient amount of DNA sequence data, we may be able to actually determine the proportion of genetic variation that is evolutionarily relevant. (Personally, I don't believe that by themselves purely retrospective studies of DNA sequence variation will be sufficient for this. We will have to learn a great deal more about the mechanisms of gene action and regulation and the constraints natural selection imposes on genes and their products, i.e. find out just how myopic natural selection is.) It is clear that the acquisition of these data and this information is now a major focus of population, developmental and molecular genetics. However, I expect it will be some time before we know enough to actually estimate the proportion of standing genetic variation that is evolutionarily relevant and before we really understand the relationship between the magnitude of standing genetic variation and the ability of a species to adapt to environmental changes. Furthermore, even when those data and that information do become available, they will be obtained primarily from model species, like *E. coli, Drosophila*, and humans. As a result, this information may be of little relevance for endangered species. Thus, for now and I expect the more immediate future, we will have to operate without even rough estimates of the proportion of sustained genetic variation in populations that is "evolutionarily relevant" and has to be maintained for the survival and future adaptation of species.

RELEASE OF GENETICALLY ENGINEERED ORGANISMS

The development and application of *in vitro* procedures to produce "hybrid" (recombinant) organisms have generated more concern about potentially deleterious health and environmental problems than any single contribution of the biological sciences. During its early days, the mid 1970s, this, "recombinant DNA", technology was limited almost exclusively to bacterial hosts and vectors (plasmids and bacteriophages). And, during this period, concerns about potential hazards of this technology were dealt with by restricting its application to physically and biologically contained situations; low volume laboratory and industrial vessels and hosts and vectors that were "disarmed" (of low fitness), to prevent their establishment in natural communities (Krimsky 1984).

By the early 1980s the range of potential hosts for genetic engineering was broadened considerably. By using vectors such as the Ti plasmid of *Agrobacterium tumerfaciens* for plants or transposons like the P-elements of *Drosophila melanogaster* it became possible to introduce DNA manipulated *in vitro* into higher plants and animals. Also by the early 1980s, there was pressure to expand the commercial applications of recombinant DNA technology. Although the containment requirement had little inhibitory effect on the application of these methods for basic research or production of drugs, hormones and other compounds (Davies 1988), this constraint precluded applications which required the release of engineered organisms into the environment.

Potential applications for released genetically engineered organisms

At this time, there are a number of enterprises for which the release of genetically engineered organisms into the environment has been proposed and, in some cases, actually implemented: agriculture, medicine (epidemiology) and waste treatment.

For agriculture, recombinant DNA procedures are being used for crop and domestic animal improvement, pest and pathogen control, and extending growing seasons. By using *in vitro* recombination with DNA from different plants, bacteria and even animals it is (or soon will be) possible to produce recombinant crop plants with; (1) altered nutritional values, e.g. improved amino acid compositions; (2) resistance to viral, fungal and bacterial diseases, herbivorous animals, and chemical herbicides; and (3) modified requirements for added nutrients (e.g. nitrogen). In theory, and surely soon in practice, it is also possible to use *in vitro* procedures to genetically modify domestic animals to alter their rates of development, disease resistance, efficiency, diet, and nutritional qualities. By *in vitro* genetic modification, the plant host-range of existing nitrogen-fixing bacteria could be expanded and other root-inhabiting bacteria can be made into nitrogen-fixers. In the same way the host-range or effectiveness of insect-pathogenic or plant-pathogenic bacteria, viruses and fungi can be modified to improve their effectiveness as biological control agents (Bishop 1988). Expanding the season length of crop plants was the motivation of the first release of a GEO, *Pseudomonas syringae* with altered ice nucleation properties (Lindow 1985). For a more extensive discourse on these and other agricultural applications of GEOs see Gaertner and Kim (1988) and Gould (1988).

Currently, vaccine production is the most promising general application of released GEOs for epidemiology and medicine. The basic idea is to introduce genes responsible for the synthesis of antigens (protein subunits) from pathogenic microorganisms (bacteria, viruses, and protozoa) into a non-pathogenic organism, like the Vaccinia (cow pox) virus (Perkus et al. 1985; Hilleman 1985). Vaccination with these genetically altered organisms could confer protection against infections by different pathogens without the danger of the vaccinating microorganism becoming virulent, as would be the case for attenuated bacteria and viruses. These engineered viruses would also have the advantage of being live vaccines and this provide longer-term protection than nonproliferating vaccines, like free antigen or killed organisms.

Decomposing wastes and detoxifying pollutants are an environmentally attractive and promising application for released genetically engineered bacteria and fungi. Using conventional (*in vivo*) genetic procedures hydrocarbon decomposing strains of *Pseudomonas* have been developed to clean-up oil spills (Chakrabarty 1979, 1986). There have also been suggestions to use *in vitro* gene manipulation procedures to manufacture microorganisms to detoxify pesticides and other toxic compounds (Malik 1989).

Implications for the conservation of genetic resources

With recombinant DNA technology one can produce organisms with truly novel properties. However, from the perspective of the conservation of genetic resources, novelty itself is not a problem. I believe that from this conservationist perspective, the problems of releasing genetically engineered organisms are essentially the same (at least in kind) as those of releasing exotic species or indigenous or domestic species that are genetically modified by conventional *in vivo* procedures (mutation, hybridization and selection). Through competition, predation (parasitism), habitat modification or encroachment (as in agriculture), these released organisms could reduce the size of established natural populations, even to the point of extinction. They could also change the genetic composition of natural populations, by introgression (mating or horizontal transfer of genes or accessory genetic elements) or by acting as selective agents.

Assessing the biological risks of releasing genetically engineered organisms

Ideally the assessment of the potential biological consequences of releasing GEOs would be quantitative and provide answers to a series of questions: (1) What is the likelihood (probability) that a released population of GEOs will become established and be maintained in a natural habitat? (2) To what extent will established populations of GEOs spread beyond their intended target area or host (prey) species? (3) What effect will released GEOs have on the densities and range of existing natural populations? (4) To what extent will GEOs exchange genes with members of natural populations and what are the consequences of those gene exchanges?

While this list may appear straightforward, a quantitative assessment of the potential biological hazards of released GEOs is a formidable endeavor. Each of these questions has many elements. Answering them requires the efforts of scientists in a number of different areas, ecology, molecular biology, genetics and evolutionary biology and a substantial amount of resources and personnel (Halversen et al. 1985; Tiedje et al. 1989).

One approach to answering these questions is to identify the primary parameters affecting the ecology and genetics of the released GEOs and indigenous populations and separately estimate their values. Mathematical models and computer simulations could then be used to study and provide answers to the above questions for the system at large, a protocol analogous to that used to assess the likelihood of breakdown of machinery or meltdown of nuclear power plants. Unfortunately even if this enterprise was restricted to the dominant parameters, the number of such parameters would be large and the possible interactions between them astronomical. For the released GEOs alone, it would be necessary to have at least information about; (1) rates of population growth and mortality; (2) physical conditions for growth and survival (temperatures, humidity ...); (3) densities of the indigenous populations (hosts, prey, parasites, pathogens, predators and competitors) with which it interacts ecologically and genetically; (4) and the nature and values of the parameters governing the population dynamics of these interactions. Even if all the presumed major parameters were identified and their values estimated, our current knowledge of the ecology and genetics of natural communities is not sufficient to produce accurate models and computer simulations to study the systems at large.

One interpretation of this is that in practice it is not possible to produce comprehensive, first-principle assessments of the environmental risks of releasing GEOs. There are alternatives. One is microcosm experiments; releasing GEOs into contained synthetic environments that mimic the natural habitat and community for which these organisms are intended. Another is small-scale field experiments; monitoring the consequences of releases in controlled and contained natural habitats. Finally, there is "worst-case assessment"; evaluating only the likelihood and reality of the worst conceived of consequences of particular releases.

These more qualitative approaches have problems and limitations of their own. The accuracy with which microcosms can be made to mimic complex natural ecosystems is, at best, modest. Even if it were possible to do truly contained, small-scale release experiments, the risks for repeated large-scale releases may be very much greater. While imaginative investigators with knowledge and experience with the ecology and genetics of particular groups of organisms and their habitats could generate realistic worst-case scenarios for particular releases, there is no way to know to what extent these scenarios are exhaustive (Simonsen and Levin 1988). Furthermore, even worst-case risk assessment is a formidable enterprise and, unless the criteria (experimental outcomes) for terminating a release program are established in advance, it is an enterprise of little utility. For an illustration of these problems with worst-case risk assessment, I suggest a perusal of the history of the polyoma – cancer risk experiments of the late 1970s (Krimsky 1984; or, more briefly in Simonsen and Levin 1988).

Thus, it seems that no matter how we approach it, our ability to quantitatively assess the potential consequences of releasing GEOs is necessarily limited. I don't see this as an argument against assessing the biological risks of releasing GEOs. Quite the contrary, positive results in these studies, evidence for hazards, could be obtained in advance of an actual, large-scale release. While negative results in a single study are of little value, as this "no-problem" evidence accumulates for a particular application, so does our confidence in its safety (Simonsen and Levin 1988). These risk assessment studies will also add to our general understanding of the genetics and ecology of natural populations and communities, worthy information, independently of the motivation for gathering it.

The main implication, as I see it, is that we have to appreciate the limitations of *a priori* assessments of the biological risks of releasing GEOs and deal with them accordingly. Most importantly, we have to accept the possibility that in some cases these releases will engender environmental problems that cannot be readily rectified, including irreversible reductions in the genetic diversity of natural populations and possibly the extinction of some of these population. I don't see this potential argument against the release of GEOs, but rather as argument for us to question whether the utility of a specific release of a GEO is worth the risk (Simonsen and Levin 1988).

Other risks of releasing genetically engineered organisms

We would be seriously remiss if we maintain the fiction that the potential biological hazards of this or other biotechnology are the only hazards to consider, or even the most important. There are ethical considerations that have to be taken into account. There are also a plethora of social, political and economic problems associated with this technology, some of which may seriously impinge on the future of our genetic resources (see the chapters in this volume by Brøgger, Fowler, and Shiva).

I believe the most significant of these other potential hazards involves the genetic "improvement" of crop plants and domestic animals. By expanding the geographic range in which crops and animals are grown or maintained, there is the very real possibility of additional encroachment of agriculture into natural habitats. This expansion of the growing range of crops may also induce farmers to reduce the diversity of species under cultivation, especially in the Third World where multi-species cultivation is common and perhaps essential for a sustained agriculture (McNeely 1989). It would seem that this replacement would be especially likely, if the "improved" crops are of high cash value. Finally, there is a good possibility that genetically "improved" and, in the short-term more productive, crops of a particular species, will replace the more genetically diverse plants of that species in current use.

To be sure, none of these problems are unique to genetically engineered organisms. Most of them were manifest in the first "Green Revolution", long before the "invention" of *in vitro* genetic engineering. On the other hand, the techniques of contemporary biotechnology can certainly exacerbate the situation.

CONCLUSIONS AND RECOMMENDATIONS

The application of DNA technology for conservation genetics

Eventually genomic data obtained through the application of DNA technology will add to our understanding of the magnitude and nature of maintained genetic variability needed for the adaptation and survival of species. However, for now and the near future, I believe we will have to rely on existing, phenotypic, evidence as support for the fundamental premise of conservation genetics, to wit; "genetic variability is a good thing, the more the better". Existing evidence certainly supports the proposition that a substantial, if unknown, fraction of the genetic variability maintained in eukaryotic populations is "evolutionarily relevant" (Allendorf and Leary 1986; Ledig 1986; Berry, this volume). Furthermore, save for cases of outbreeding depression, there is no evidence against this interpretation. Finally, there is no choice. Once evolutionarily relevant genetic variability is lost, it will be virtually impossible to regenerate.

At this time, I believe there are two major applications for which DNA technology is useful for the conservation of genetic resources. First, for the design and implementation of breeding programs; to determine and/or confirm genetic relationships among breeding pairs and to identify deleterious recessive genes. Second, for the unambiguous identification of endangered species and populations. The information and materials needed for the use of DNA procedures for these purposes are available and there is little justification for not employing them when their application can avoid inbreeding or outbreeding depression or the extinction of species due to bad taxonomy.

The release of genetically engineered organisms

I believe that the potential deleterious effects of the release of genetically engineered organisms on the genetic diversity of natural communities are no different in kind than those of the release of exotic species or organisms whose genomes have been modified by conventional, *in vivo*, procedures. However, for reasons that are historical, political and sociological rather than scientific, public concern about the health and environmental consequences of releasing genetically engineered organisms has provided us with an unprecedented opportunity to evaluate a new technology before its widespread implementation. I believe that we have both the need and obligation to exploit this opportunity.

It seems clear that we will not be able to provide precise estimates of the likelihood of deleterious ecological and genetic effects of released genetically engineered organisms or the nature of those effects. Nevertheless, I believe that it is critical to establish and enforce formal protocols to assess the genetic and ecological risks of all intentional releases of novel organisms into the environment. Furthermore, because the potential problems are of the same in kind, there is no logical justification to not include all intentional releases of novel (exotic) organisms, independently of the process by which their genome was modified. Finally, these protocols should be developed and administered at a global, rather than local (national) level and should consider the utility, and potential social and economic hazards of this technology as well as the biological.

ACKNOWLEDGMENTS

I wish to thank, Anton Brøgger, Ralphael Falk, Lone Simonsen and an anonymous reviewer for comments on this manuscript and their various, but only occasionally successful, attempts to "keep me honest". I also wish to thank the sponsors, organizers and participants of the International Conference on the "Conservation of Genetic Resources for Sustainable Development" for stimulating discussions and ideas, many of which are incorporated in this report.

LITERATURE CITED

Allendorf, F. W. and R. F. Leary. 1986. Heterozygosity and fitness in natural populations of animals, pp. 57–76. In: M. E. Soulé (ed) *Conservation Biology: The Science of Scarcity and Diversity*. Sinauer, Sunderland, MA.

Aquadro, C. F., K. M. Lado and W. A. Noon. 1988. The rosy region of *Drosophila melanogaster* and *Drosophila simulans* I. Contrasting levels of naturally occurring DNA restriction map variation and divergence. Genetics 119: 875–888.

Avise, J. C. 1989. A role for molecular genetics in the recognition and conservation of endangered species. Trends Ecol. Evol. 4: 279–281.

Avise, J. C., R. M. Ball and J. Arnold. 1988. Current versus historical population sizes in vertebrate species with high gene flow: a comparison based on mitochondrial lineages and inbreeding theory for neutral genes. Mol. Biol. Evol. 5: 331–344.

Berry, R. J. 1989. Electrophoretic studies: perspectives for population biology, pp. 7–16. In: H. D. Loxdale and J. Den Hollander (eds) *Electrophoretic Studies on Agricultural Pests*. Clarendon Press, Oxford.

Berry, R. J. 1992. The role of ecological genetics in biological conservation. This volume.

Bishop, D. H. L. 1988. The release into the environment of genetically engineered viruses, vaccines and viral pesticides. Trends Ecol. Evol 3: 12–14.

Brøgger, A. 1992. Ethical concerns regarding gene technology. This volume.

Chakrabarty, A. M. 1979. Microbial genetic engineering by natural plasmid transfer – Some representative benefits and biohazards, pp. 185–193. In: A. M. Chakrabarty (ed) *Genetic Engineering*. CRC Press, Boca Raton, FLA.

Chakrabarty, A. M. 1986. Genetic engineering and problems of environmental pollution, pp. 515–530 In: W. Schonborn (ed) *Biotechnology*, Vol. 8. *Microbial Degradations*. Weinheim, FRG.

Clarke, B. C. 1979. The evolution of genetic diversity. Proc. R. Soc. Lond. B 205: 453–474.

Crow, J. F. 1985. The neutrality – selection controversy in the history of evolution and population genetics, pp. 1–18. In: T. Ohta and K. Aoki (eds) *Population Genetics and Molecular Evolution*. Japan Sci. Soc. Press, Tokyo.

Daugherty, C. H., A. Cree, J. M. Hay and M. B. Thompson. 1990. Neglected taxonomy and continuing extinctions of tuatara (*Sphenodon*). Nature 347: 177–179.

Davies, J. 1988. Genetic engineering: process and product. Trends Ecol. Evol. 3: 7–11.

Dobzhansky, T. and B. Spassky. 1967. Geotactic and phototactic behavior in *Drosophila* I. Proc. R. Soc. Lond. B 168: 27–47.

DuBose, R. F., D. E. Dykhuizen, and D. L. Hartl. 1988. Genetic exchange among isolates of bacteria: Recombination within the *phoA* gene of *Escherichia coli*. Proc. Natl. Acad. Sci. USA 85: 7036–7040.

Dykhuizen, D. E. and L. Green. 1986. DNA sequence variation, DNA phylogeny, and recombination in *E. coli*. Genetics 113: 71.

Endler, J. A. 1986. *Natural Selection in the Wild*. Princeton University Press, Princeton, NJ.

Felsenstein, J. 1988. Phylogenies from molecular sequences. Inferences and reliability. Ann. Rev. Genet. 22: 521–565.

Flint, J., A. J. Boyce, J. J. Martison and J. B. Clegg. 1989. Population bottle necks in Polynesia revealed by minisatellites. Hum. Genet. 83: 257–263.

Fowler, C. 1992. Biotechnology, patents and the Third World. This volume.

Gaertner, F. and L. Kim. 1988. Current applied recombinant DNA projects. Trends Ecol. Evol. 3: 4–7.

Gould, F. 1988. Genetic engineering, integrated pest management and the evolution of pests. Trends Ecol. Evol. 3: 15–18.

Halversen, H. O., D. Pramer and M. Rogul (eds) 1986. *Engineered Organisms in the Environment: Scientific Issues*. Amer. Soc. Microb., Washington, DC.

Harris, H. 1966. Enzyme polymorphism in man. Proc. R. Soc. Lond. B 164: 298–310.

Hartl, D. L. and D. E. Dykhuizen. 1985. The neutral theory and the molecular basis of preadapation, pp. 107–124 In: T. Ohta and K. Aoki (eds) *Population Genetics and Molecular Evolution*. Japan Sci. Soc. Press, Tokyo.

Hilleman, M. R. 1985. Newer directions in vaccine development and utilization. J. Infect. Dis. 151: 407–419.

Hudson, R. R., M. Kreitman and M. Aquade. 1987. A test of neutral molecular evolution based on nucleotide data. Genetics 116: 153–159.

Innis, M. A., D. H. Gelfand, J. J. Snomsky and T. J. White. 1990. *PCR Protocols: A Guide to Methods and Applications*. Academic Press, San Diego.

Kimura, M. 1983. *The Neutral Theory of Molecular Evolution.* Cambridge University Press, Cambridge.

Kimura, M. and T. Ohta. 1971. Protein polymorphism as a phase in molecular evolution. Nature 229: 467–469.

King, J. L. 1967. Continuously distributed factors affecting fitness. Genetics 55: 483–492.

Kreitman, M. 1983. Nucleotide polymorphism at the alcohol dehydrogenase locus of *Drosophila melanogaster.* Nature 304: 412–417.

Kreitman, M. 1991. Detecting selection at the level of DNA, pp. 204–221. In: R. K. Selander, A. G. Clark, and T. S. Whittam (eds) *Evolution at the Molecular Level.* Sinauer, Sunderland, MA.

Krimsky, S. 1984. *Genetic Alchemy.* MIT Press, Cambridge, MA.

Ledig, T. F. 1986. Heterozygosity, heterosis and fitness in outbreeding plants, pp. 77–104. In: M.E. Soulé (ed) *Conservation Biology: The Science of Scarcity and Diversity.* Sinauer, Sunderland, MA.

Lewontin, R. C. 1967. An estimate of average heterozygosity in man. Amer. J. Hum. Genet. 19: 681–685.

Lewontin, R. C. 1974. *The Genetic Basis of Evolutionary Change.* Columbia University Press, New York.

Lewontin, R. C. and J. L. Hubby. 1966. A molecular approach to the study of genic heterozygosity in natural populations. II. Amount of variation and degree of heterozygosity in natural populations of *Drosophila pseudoobscura.* Genetics 54: 595–609.

Li, W.-H., and D. Graur. 1991. *Fundamentals of Molecular Evolution.* Sinauer, Sunderland, MA.

Lindow, S. E. 1986. Construction of isogenic ice strains of *Pseudomonas syringae* for evaluation of specificity of competition on leaf surfaces, pp. 509–515. In: F. Megusar and M. Ganter (eds) *Perspectives in Microbial Ecology.* Slov. Soc. Microbiol., Ljubljana, Yugoslavia.

Malik, V. S. 1989. Biotechnology – the golden age, pp. 263–306. In: S. L. Neidleman (ed) *Advances in Applied Microbiology,* Vol. 34.

May, R. M. 1990. Taxonomy as destiny. Nature 347: 129–130.

McNeely, J. 1989. Conserving genetic resources at the farm level. Ileia News Letter 5: 3–6.

Meylan, A. B., B. W. Bowen and J. C. Avise. 1990. A genetic test of the natal homing versus social facilitation models for green turtle migration. Science 248: 724–727.

Milkman, R. 1967. Heterosis as a major cause of heterozygosity in nature. Genetics 55: 493–496.

Milkman, R. and I. Crawford. 1983. Clustered third-base substitutions among wild strains of *Escherichia coli.* Science 221: 378–380.

Milkman, R. and A. Stolzfus. 1988. Molecular evolution of the *Escherichia coli* chromosome. II. Clonal segments. Genetics 120: 359–366.

Müller, H. J. 1950. Our load of mutations. Amer. J. Hum. Genet. 2: 111–176.

Nei, M. 1987. *Molecular Evolutionary Genetics*. Columbia University Press, New York.

Nei, M. and A. L. Hughes. 1991. Polymorphism and evolution of the major histocompatibility complex loci in mammals, pp. 222–247. In: R. K. Selander, A. G. Clark, and T. S. Whittam (eds) *Evolution at the Molecular Level*. Sinauer, Sunderland, MA.

Nevo, E., A. Beiles and R. Ben-Shlomo. 1984. The evolutionary significance of genetic diversity: ecological, demographic and life history correlates. Lect. Notes Biomath. 53: 13–213.

Perkus, M. E., A. Piccini, B. R. Lipinskas and E. Paoletti. 1985. Recombinant vaccinia virus: immunization against multiple pathogens. Science 229: 981–984.

Porras, O., D. A. Caugant, B. Gray, T. Lagerard, B. R. Levin and C. Svanborg Edén. 1986. Difference in structure between type b and nontypable *Haemophilus influenzae* populations. Infect. Immun. 53: 79–89.

Reeve, H. K., D. F. Westneat, W. A. Noon, P. W. Sherman and C. F. Aquadro. 1990. DNA fingerprinting reveals high levels of inbreeding in colonies of the eusocial naked mole-rat. Proc. Natl. Acad. Sci. USA 87: 2496–2500.

Robertson, F. W. and E. C. R. Reeve. 1952. Studies in quantitative inheritance. I. The effects of selection on wing and thorax length in *Drosophila melanogaster*. J. Genet. 50: 414–448.

Sambrook, J., E. F. Fritsch and T. Maniatis. 1989. *Molecular Cloning: A Laboratory Manual*. 2nd Edition. Cold Spring Harbor Press, Cold Spring Harbor.

Schaeffer, S. W., C. F. Aquadro and C. H. Langley. 1988. Restriction-map variation in the notch region of *Drosophila melanogaster*. Mol. Biol. Evol. 5: 30–40.

Schull, W. J. and J. V. Neel. 1965. *The Effect of Inbreeding on Japanese Children*. Harper & Row, New York.

Selander, R. K., and B. R. Levin. 1980. Genetic diversity and structure in populations of *Escherichia coli*. Science 210: 545–547.

Shaw, C. R. 1965. Electrophoretic variation in enzymes. Science 149: 936–943.

Shiva, V. 1992. The seed and the spinning wheel: technology development and biodiversity conservation. This volume.

Simonsen, L. and B. R. Levin. 1988. Assessing the risks of releasing genetically engineered organisms. Trends Ecol. Evol. 3: 27–30.

Slatkin, M. 1989. Detecting small amounts of gene flow from phylogenies of alleles. Genetics 121: 609–612.

Slatkin, M. and W. P. Maddison. 1989. A cladistic measure of gene flow inferred from the phylogenies of alleles. Genetics 123: 603–613.

Smith, D. R., O. R. Taylor, and W. M. Brown. 1989. Neotropical africanized honey bees have African mitochondrial DNA. Nature 339: 213–215.

Sved, J. A., T. E. Reed and W. F. Bodmer. 1967. The number of balanced polymorphisms that can be maintained in a natural population. Genetics 55: 469–481.

Templeton, A. R. 1986. Coadaptation and outbreeding depression, pp. 105–116 In: M. E. Soulé (ed) *Conservation Biology: The*

Science of Scarcity and Diversity. Sinauer. Sunderland, MA.

Thoday, J. M. and T. B. Boam. 1961. Regular responses to selection. I. Description of responses. Genet. Res. 2: 161–176.

Tiedje, J. M., R. K. Colwell, Y. L. Grossman, R. E. Hudson, R. E. Lenski, R. Mack and P. Regal. 1989. The planned introduction of genetically engineered organisms: ecological considerations and recommendations. Ecology 70: 298–315.

16 Ethical concerns regarding gene technology

Anton Brøgger
Department of Genetics,
Institute for Cancer Research,
The Norwegian Radium Hospital,
N-0310 Oslo 3,
Norway

ABSTRACT

Ethical theory distinguishes between two basic arguments. According to the deontological argument an action is right or wrong in itself. With the consequentialist argument it is the ends, the result of the action which carries an intrinsic value, and not the action in itself. Gene technology implies the isolation and analysis of genetic material, the combination of genetic material from different organisms and the eventual insertion of the recombination into an organism leading to genetic change. This presentation focuses on the ethical aspects of (1) isolation and manipulation of genetic material in general; (2) transfer of genetic material from one species to another; (3) transfer of human genetic material to any other species; (4) the release of genetically engineered organisms into the environment; and (5) the conservation of genetic resources. The view is maintained that genetic material is not in principle different from other life structures in nature. Since it is ethically acceptable for man to interact with nature in general, genetic manipulation must also be acceptable. The limitations must be governed from two sets of values: (1) living nature has a value in itself; (2) living nature has a utility value for man. A declaration on ethics in biological research comparable to the Helsinki declaration for biomedical research could form a basis for the regulation of human conduct in relation to nature in general.

REMARKS ON ETHICS IN GENERAL

During my school days nobody talked about **ethics**, but we were much concerned with matters of **morals**. Morals may be defined as our opinions of the right and wrong, the good and bad; the base upon which one wants to govern one's own behaviour and judge the behaviour of others. Ethics is the more or less systematic reflection concerning our own and others' morals. Ethics has been called the theory of morals, or moral philosophy. At present this distinction between morals and ethics seems to have been lost. In the ongoing

Conservation of Biodiversity for Sustainable Development.
Edited by O. T. Sandlund, K. Hindar and A. H. D. Brown. 1992. Scandinavian University Press, Oslo.

debate on scientific research, nobody speaks about morals, but we are full of ethical concerns. According to current practice I shall also use the word ethics both as regards theory and in relation to the practical application of rules of behaviour. Ethics is not only dealing with observation and description of human conduct (descriptive ethics), but is concerned with **values**, with estimates. It is a normative science. Ethics is not primarily occupied with the actual character of human behaviour, but with its ideal; not so much with how we behave, but with how we ought to behave.

Ethical theory distinguishes between two basic approaches. According to a **deontological** argument a certain action is right or wrong in itself – independent of the eventual good or bad consequences. A right action has a positive value in itself, and therefore it is a **duty** to perform the act. A bad action has a negative value in itself, and it is a duty to refrain from it. According to a **consequentialist** argument an action is right or wrong depending on what it is leading up to. It is the ends which carry the intrinsic value – good or bad – and not the action in itself. The action will have a positive or negative instrumental value in so far it has good or bad consequences. Within this type of ethical reasoning the **utilitarianism** is the best known principle: the highest good is the greatest amount of happiness for the largest number of people. Within ethics theory these two arguments may be clearly distinguished and analysed separately. In practice both arguments may serve at the same time governing our reasoning about good and bad, although it seems that in the Western culture today the utilitarian argument is dominating.

Generally, the ethical concerns have been focused on the negative aspects of the use of gene technology. In the ongoing debate much less attention has been paid to the positive aspects. Ethics does not only tell us what we should not do. Ethics certainly carries imperatives about what we should do, about our obligations to do the right and the good. Through gene technology the science of genetics gives us an increasingly better understanding of the development and functioning of living organisms including ourselves. We are becoming much better armed to fight genetic disease as well as other diseases. In agriculture and fish farming important gains will be made in favour of a qualitative as well as quantitative production of food for an increasing world population.

ETHICAL ASPECTS OF GENETIC KNOWLEDGE IN ITSELF

As regards knowledge, we may ask: is there any aspect of scientific knowledge which is unethical, which is bad in itself? I believe that our traditional view is that knowledge in itself is in a way ethically neutral; it is neither good nor bad. Ethical concern is connected either to how the knowledge is acquired or to how the knowledge is used.

One ethical argument against the acquirement of knowledge is that the knowledge may be used in such an extremely bad way that it would be better to refrain from having the knowledge in itself, even if it is ethically neutral. I shall not claim that this is never the case. However, in most cases scientific knowledge may be used both for the good and for the bad. Then, to stop scientific research is unethical because this will lead to the lack of a certain knowledge, which may be used for a good purpose. I believe most people agree that this reasoning certainly is valid as regards the science of genetics. I have problems in imagining an area of genetic knowledge, which is unethical to possess in itself.

The question then comes as to how the knowledge is acquired. Genetic knowledge is obtained in several ways: (1) through the biochemical and cell biological analysis of the genetic material itself, how it is organised into chromosomes, and how it functions in the organism; (2) through the study of the genotypes and phenotypes of organisms and their offspring, either in nature or after experimental crossing. Gene technology has become a very potent instrument for this type of research. Genetic research implies the interference with the integrity of animals and plants. This seems to be ethically acceptable provided that the rules for the treatment of animals in research are strictly followed. In fact, other scientific disciplines than genetics represent ethical problems as regards the handling of experimental animals. In comparison breeding experiments are fairly innocent.

GENE TECHNOLOGY, THE CHALLENGE

Gene technology consists of a battery of different methods. First comes the isolation of the genetic material – the DNA molecules – from the organism in question. Needed is a sample of cells containing nuclei with the chromosomes, which contain the DNA. Such a sample of cells is easy to obtain from any organism including man. The procedure is now purely biochemical: lysing of the cells and extraction of DNA by using specific chemicals. DNA molecules are extremely long and difficult to analyse biochemically unless they are cut into shorter pieces. This is done with specific so-called restriction enzymes, which cut the DNA at specific points defined by a short sequence of base pairs. The shorter DNA pieces may now be separated by electrophoresis, because they move differently according to their size in an electric field. The pieces may be further identified by means of hybridisation with radioactive probes and autoradiography. Or they may be subject to sequence analysis, revealing the sequence of base pairs in the molecule.

One very important aspect is that DNA fragments from different sources cut with the same restriction enzyme may be fused together into new combinations. In this way fragments of DNA from one organism may be inserted into a bacterial plasmid. Plasmids are ring-formed DNA-molecules able to enter bacteria and replicate there. The inserted fragments will also be replicated together with the plasmid DNA in the bacteria. This is one way of producing more DNA of a specific type. DNA may be inserted into cells – either by micromanipulation, by direct uptake or by means of plasmids or viruses which enter the cells. In principle DNA from any organism may be transferred by gene technological methods to any other organism. The organism receiving the foreign DNA is called transgenic. There are now methods available for the production of transgenic animals as well as transgenic plants (Watson et al. 1983; Emery 1985).

ISOLATION AND MANIPULATION OF GENETIC MATERIAL IN ITSELF

It would put an end to all gene technology if we were not allowed to isolate and work with the genetic material in general. It would end the extention of genetic knowledge and deprive us of all the benefits from the application of such knowledge, although it would also deprive us of any misuse of this knowledge. The basic argument against gene technology is that the genetic material is sacred in some way and different from everything else in nature and therefore should not be subject to human intervention. To me, this reasoning is not acceptable. I strongly maintain the view that the genetic material is not in principle different from other life structures in nature. We generally accept that man may interfere, interact and manipulate with living nature. I can see no specific reason why it should be ethically unacceptable in itself for man to examine and manipulate any genetic material. In this case my reasoning is based on a deontological, ethical principle. It is the process of genetic manipulation in itself I am considering and not any of the consequences such manipulation may have in specific cases. That aspect will be considered separately later.

TRANSFER OF GENETIC MATERIAL BETWEEN SPECIES

The transfer of genetic material from one species to another will add a new character to the transgenic organism. The other way around, deletion of genetic material may take away a certain character. It may be said that the integrity of a species is dependent on its total genetic material, its genome. Genetic manipulation will interfere with integrity and change the identity of the species. The species on Earth are not given once and for all. Evolutionary processes lead to a constant change. I cannot see why it should be ethically unacceptable for man to contribute to this process. In fact man has been changing organisms by means of breeding before the days of genetics. Actually, we can distinguish three stages in our breeding of plants and animals: (1) the pre-genetics stage which lasted from the beginning of agriculture up to this century; (2) the conventional genetics stage in which genetic knowledge was used to create highly efficient breeding programs; (3) the gene technological stage which allows genetic manipulation of organisms in a much more precise and efficient way than conventional breeding.

I am unable to see any ethical difference in principle between these three ways of breeding plants and animals. The two first mentioned have never been disputed and raised any ethical concern. In itself it is ethically acceptable to change the genetic constitution of cultivated plants and animals. That is not to say that any such genetical change is ethically acceptable. Each case must be considered with regard to the characters in question and in relation to the wellbeing of the animals and to any consequences for human welfare. My point is that the ethical aspects of agriculture have been with us for centuries, and that gene technology does not represent any entirely new situation in this respect.

A recent example is the transgenic sheep, which is now producing coagulation factor IX in its milk. The protein is extracted from the milk and used in the treatment of hemophilia. The animal is not suffering in any way from this new character. But it interferes with the integrity of sheep as a species. As already stated above I find such microevolutionary changes ethically acceptable. In addition the result is a better treatment of a serious human disease, since treatment with blood transfusions or factor IX produced from human blood carries the risk of HIV and other infections.

TRANSFER OF HUMAN GENETIC MATERIAL

Each generation of man may be said to contain a certain amount of information, partly the genetic information inherited through the DNA of the germ cells, partly the cultural information transmitted by various signal systems of which language is the most important. Accordingly, there are two lines of information between generations, one molecular and one cultural. In what ways these two types of information are interrelated is not known. But somehow the interaction between the molecular and the cultural information results in a human being. At the moment it seems as if the ethical discussion puts most emphasis on the molecules. It is claimed that it is fundamentally wrong to change the genetic constitution of the next generation. Any interference with our molecular constitution is considered to be an indecent tampering with life. One may ask that, if it is not only the DNA molecules which make us human, but also the cultural heritage, why is it more serious to change the genetic than the cultural constitution of the next generation? Cultural evolution is also an irreversible process.

The transfer of human genetic material to other species has raised the question whether nucleotide sequences of human origin represent a certain human value, which is in jeopardy when transferred to another organism. I find it hard to accept that a human value is connected to a molecule. The human value is connected to the human subject. The development of a human subject is dependent on a certain total combination of genetic material, not on any humanity resting in each single gene. The amino acid sequence in the beta chain of hemoglobin is identical in man and gorilla with the exception of number 104, which is arginine in man and lysine in gorilla (Zuckerkandl 1965). But it is not arginine which makes us human, and lysine is not responsible for the development of a gorilla. There is no specific humanity connected to the arginine, nor is there any "gorillity" dependent on the lysine.

Therefore, I find it ethically acceptable to transfer any human gene to any other organism. On the other hand, there is a difference between single genes coding for certain proteins and the whole combination of genes leading to development of a human subject. The use of such a combination of genes in making a hybrid between man and for instance a gorilla would certainly be ethically unacceptable. I maintain that there is a difference between single genes for specific proteins and a total genome.

PATENTABILITY OF LIVING MATTER

Traditionally, a patent is issued to protect the economic interests of the person who has made an invention. If an invention is making money, it seems reasonable that some of the profit should go to the person who made the invention. Basically it is ethically acceptable to earn money, but clearly within reason. We distinguish between a reasonable and an unreasonable profit. The question is then how would the patentability of living matter lead to an unreasonable and therefore an ethically unacceptable profit? One other aspect of profit must also be taken into account. The demand for a quick and high profit may lead to pressure upon the evaluation process regarding the safety of releasing genetically altered organisms into the environment.

Patents are issued for inventions and not for discoveries. However, we have seen a proposal for the patentability of living matter isolated from nature, analysed and described (CEEC 1988a,b). This means that plasmids, DNA sequences, proteins, peptides and enzymes may be patented. In other words, scientific discoveries would be patentable. This is against the principle of free communication of knowledge as a common property of mankind. Patentability of scientific discoveries is clearly ethically unacceptable.

There are two areas of patentability as regards inventions, namely the process and the product. It may be argued that the true invention as regards new organisms produced by gene technology is the process – the way in which the organism has been genetically altered. It seems reasonable, then, that processes in gene technology are patentable. On the other hand, the patentability of the product – the genetically altered organism in itself – or the use of the product is highly questionable. A biological product is self-replicating. It seems therefore to lead to a quite unacceptable profit if the patent should cover the product and its offspring for generations to come.

GENE TECHNOLOGY AND THE PRESERVATION OF NATURE

On Earth today some 430,000 plant species and more than one million animal species have been described. On the one hand, each of these species has a value in itself. On the other, each of these species represents a potential value for man as a resource of genetic information. Thus, there are two ethical bases for the conservation of nature: the value of nature itself and the possible utility value for man.

Each single organism playing its role in the whole ecosystem has its value regardless of whether it is used in any way by man. It is a difficult philosophical question whether nature has a value in itself regardless of the presence of man. Values depend upon a subject to appreciate them. On the other hand, it gives one a certain uneasiness to say that nature – and the whole universe for that matter – was worthless until man arrived on the scene to appreciate its existence (cf. Næss, this volume).

In order to protect the value of nature, man has an obligation to live and interfere with nature in such a way that this value is not compromised. It follows that we must take responsibility for the preservation of nature as it is today. Of course, it may be argued that nature is changing anyway. The evolution of life on Earth is an irreversible process. New species will develop, and some older species will become extinct. By preservation of nature we do not mean that man should counteract natural evolution. But our responsibility is to minimize the man-made changes in this process. In the efforts to preserve nature in its present state there is clearly no place for gene technology. In this narrow context the use of gene technology for changing existing species would be ethically unacceptable. But perhaps not in all respects. The introduction of genes providing resistance against disease and pests in wild animals and plants would certainly be a matter for discussion. Furthermore, it could be argued that the man-made development of new species by means of gene technology would enrich the living world and add to its value rather than the opposite.

We are depending upon nature for our existence. Therefore, a utilitarian ethical argument is also a foundation for how we are allowed to interfere and interact with nature. On this background it is not only ethically acceptable to cultivate, breed and improve plants and animals. It is an ethical imperative. In order to feed an increasing human population any means for sustainably increasing qualitatively or quantitatively the production of food must be most welcome also from an ethical point of view and not only an economical. We expect that the potential in gene technology is great in many respects – increasing production, improve quality, and introduce resistance against pests and diseases. I can see only one serious restriction for the use of gene technology in this area, and that is the possible harm of introducing new microorganisms, plants and animals into the environment. Any risk for unwanted effects must be carefully assessed in each single case.

GENE TECHNOLOGY AND FARM ANIMALS

Clearly there are certain limits on the extent to which we are allowed to interfere with the integrity and dignity of the animals we keep for farming purposes. Any change in characters which may lead to suffering and pain for the animal is certainly ethically unacceptable. The farm animals have a right to live a decent life. Much of the modern industrialised agriculture has already interfered with this. The way hens are kept in cages and cattle kept indoors is already ethically questionable. Any use of gene technology which continues this development is not acceptable. Farm animals should be allowed to feed and move outdoors when the climate allows this. We will only accept minor changes in anatomical and physiological characteristics. The fusion of embryos to produce chimaeras (although not gene technology) such as the sheep-goat is a major insult to the integrity of each species. On the other hand, a minor transgenic change allowing the animal to produce a foreign but useful protein in the milk seems to me to be acceptable. Furthermore, genetic manipulation to increase resistance against diseases and to decrease the occurrence of genetic disease is not only acceptable but desirable.

THE RELEASE OF GENETICALLY ENGINEERED ORGANISMS INTO THE ENVIRONMENT

During the discussion of the new methods of gene technology at The Gordon Conference on Nucleic Acids in New Haven in June 1973 the concern was raised that this research represented a risk for serious unknown dangers for people. One possibility would be the production of genetically changed bacteria leading to unknown diseases. The following year The Committe on Recombinant DNA Molecules chaired by Paul Berg recommended that scientists should refrain from further recombinant DNA research until the risks were properly considered and the necessary safeguards could be taken. This is the first time in history that scientists restricted their work in such a way. The expert conference in February 1975 in Asilomar, California, laid the foundations for further work in the evaluation of the hazards of gene technology (Berg et al. 1975). National committees in many countries have considered the problems and issued reports and guidelines. Some of the dangers first envisioned have proved not to be serious. Today, a major concern is the production of genetically modified organisms and their use in agriculture or release into the environment in general (Levin 1990).

There are risks associated with the introduction of any new organism – a microorganism, plant or animal – into an environment regardless whether the new organism has been produced by means of gene technology or in a conventional way (Crawley 1990). The new organism may change the ecological balance between the already existing organisms in a harmful way. It may occupy the same ecological niche as some old species and lead to an eventual extinction of the old species. The new organism may find an ecological niche as a new weed, which eventually carries resistance against pesticides used against old weeds. We must consider the possibility that genes are transferred from the new organism to old organisms – either by a viral vector or by hybridisation – leading to an unwanted change of the recipients.

It is quite clear that the ethical concern regarding the use of gene technology has resulted in important measures taken by both national and international bodies. Directories have been issued both concerning the use of geneticially modified microorganisms and the release of genetically modified organisms into the environment (CEEC 1988c, COECD 1986). With reference to agricultural and environmental applications the member countries are recommended to (1) use the existing considerable data on the environmental and human health effects of living organisms to guide risk assessments; (2) ensure that recombinant DNA organisms are evaluated for potential risk, prior to applications in agriculture and the environment by means of an independent review of potential risks on a case-by-case basis; (3) conduct the development of recombinant DNA organisms for agricultural or environmental applications in a stepwise fashion, moving from the laboratory to the growth chamber and greenhouse, to limited field testing, and finally to large-scale field testing; (4) encourage further research to improve the prediction, evaluation, and monitoring of the outcome of applications of recombinant DNA organisms.

HANDLING THE MATTER – AN EXAMPLE FROM DENMARK

Denmark was the first country to issue a law on gene technology and the environment (Danish Law, 4 June 1986, No. 288). The law is generally stating that organisms or cells produced by means of gene technology must not be produced, sold, imported or used, except under certain conditions, which in each separate case requires a specific permission from the Minister of the Environment. The Danish law has missed an important point, namely that there is biologically no difference between an organism produced by conventional breeding and an organism produced by gene technology. It would have been logical to consider the use and particularly the release into the environment of any organism modified by human intervention in the same way, regardless of the techniques applied in the production of the organism.

On 27 December 1988 De Danske Sukkerfabrikker A/S filed an application with the Ministry of the Environment for the release of two types of genetically modified sugar beets. One type was resistant against the pesticide glyphosate. Since this was the first case involving the release of a genetically modified organism in Denmark, the Minister of the Environment had promised to present the case to the Danish Parliament before a final decision was taken about the permission.

The release of the plants was planned in a stepwise manner. In the first place the plants should not be allowed to carry flowers. Any flowers would be picked off to inhibit the spreading of pollen and seed. Later, flowering plants would be covered by a tent in order to inhibit the spreading of pollen and seed. Furthermore, the plants should first be tested in a physical containment in a greenhouse, then in the open in a very restricted area and later in somewhat larger areas in different localities before the eventual final release for general agricultural use. Each step would imply the careful collection of observations as a basis for the decision whether to go to the next step.

Separate governmental authorities considered the ecological and toxicological aspects. The conventional sugar beet does not spread in nature in Denmark. There is no reason to assume that the genetically modified beet would behave differently in this respect. The sugar beet may hybridise with the wild beet (the Danish strandbete, *Beta vulgaris maritima*), but such hybrids are very rare. The genetically new factors are not considered to give the wild beet any advantages under natural conditions. The resistance against glyphosate depends on an enzyme which is slightly different from a naturally occurring protein. From a toxicological point of view there is no reason to assume that this protein will harm the final product, the sugar, and represent a hazard to the consumer. Glyphosate is already permitted as a pesticide in Denmark. The total use of pesticides in Denmark will not be considerably changed if glyphosate is used in sugar beet cultures, neither as regards quantity or frequency of treatment. The possibility of a selection in favour of weeds resistant against glyphosate has been considered to be minimal. The Minister of the Environment presented a report to the Danish Parliament with the recommendation that the project should be approved. After a heated debate the Parliament also recommended the project.

CONCLUDING REMARKS

My general view is that genetic material is not in principle different from other structures in nature. Since it is ethically acceptable for man to interact with nature in general, genetic manipulation must also be acceptable. The limitations must be governed from two sets of values, which form the bases for the ethical concerns: (1) the living nature has a value in itself; (2) living nature has a utility value for man. The most serious question as regards the use of gene technology concerns the introduction of new microorganisms, plants or animals into the environment. Any risk for unwanted effects must be carefully assessed in each single case.

Farm animals have a right to live a decent life. Any change in characters which may lead to suffering and pain for the animal is certainly ethically unacceptable. On the other hand, genetic manipulation to increase their resistance against diseases and to decrease the occurrence of genetic disease is not only acceptable but desirable.

A declaration on ethics in biological research comparable to the Helsinki declaration for biomedical research (WMA 1964) could form a basis for the regulation of human conduct in relation to nature in general.

ACKNOWLEDGMENTS

I am grateful to Kai Dramer, Center for Medical Ethics, Oslo for valuable discussions during the preparation of the manuscript.

LITERATURE CITED

Berg, P., D. Baltimore, S. Brenner, R. O. Roblin and M. F. Singer. 1975. Asilomar conference on recombinant DNA molecules. Science 188: 991–994.

Council of the European Economic Community. 1988a. Proposal for a council directive on the legal protection of biotechnological inventions. CEEC, Brussels, COM (88) 496 final – syn 159.

Council of the European Economic Community. 1988b. Proposal for a council directive on the deliberate release to the environment of genetically modified organisms. CEEC, Brussels, COM (88) 160 final.

Council of the European Economic Community. 1988c. Proposal for a council directive on the contained use of genetically modified organisms. CEEC, Brussels, COM (88) 160 final.

Council of the Organisation for Economic Co-operation and Development. 1986. Recommendations concerning safety considerations for applications of recombinant DNA-organisms in industry, agriculture and the environment. COECD, Paris, 16 July 1986.

Crawley, M. J. 1990. The ecology of genetically engineered organisms: assessing the environmental risks, pp. 133–150. In: H. A. Mooney and G. Bernardi (eds) *Introduction of Genetically Modified Organisms into the Environment*. Wiley, Chichester, UK.

Emery, A. E. H. 1985. *An Introduction to Recombinant DNA*. Wiley, Chichester, UK.

Levin, S. A. 1990. Ecological issues related to the release of genetically modified organisms into the environment, pp. 151–159. In: H. A. Mooney and G. Bernardi (eds) *Introduction of Genetically Modified Organisms into the Environment*. Wiley, Chichester, UK.

Næss, A. 1992. Sustainability! The integral approach. This volume.

Watson, J. D., J. Tooze and D. T. Kurtz. 1983. *Recombinant DNA. A Short Course*. Scientific American Books. Freeman, New York.

World Medical Association. 1964. Declaration of Helsinki. Recommendations guiding physicians in biomedical research involving human subjects. Cited in J. Norw. Med. Assoc. 99: 1377–1380, 1979.

Zuckerkandl, E. 1965. The evolution of hemoglobin. Scient. Amer. 212: 110–118.

17 Biotechnology, patents and the Third World

Cary Fowler
Norwegian Centre for International Agricultural Development (NORAGRIC),
Agricultural University of Norway,
N-1432 Ås-NLH,
Norway

ABSTRACT

Who owns, controls and benefits from biological diversity has been a driving force behind global geopolitics for hundreds of years. These questions are intimately connected with the success and failure of conservation programs. Much of the economically important diversity for agricultural crops originated in the Third World. Patent systems under pressure from new biotechnologies to expand their scope may provide the mechanism for ownership and benefits of biological materials to flow into private hands in industrialized countries. Third World farming communities, however, also engage in innovation and should receive some form of recognition and compensation for their creativity and work. Recognition of these "farmers' rights" and more effective conservation efforts could be combined in a mandatory fund to support conservation and utilization of genetic resources in the Third World.

INTRODUCTION

This paper is intended to provide a very basic introduction to some of the controversies currently surrounding biological patents and an explanation as to why these debates are important to industry, to Third World development aspirations and ultimately to genetic conservation efforts.

In modern industrialized societies, the purposes and effects of patents are taken for granted. So uncritical has our thinking about them become, that indeed they are seen as a civil right, something above questioning and examination. In this atmosphere it is difficult to see patent systems as they really are – as human-created institutions designed for human purposes, as the offspring not of religion or philosophy, but of politics and economics.

Early precursors to modern patents system date back to the seventh century B. C. when the Greeks began granting short-term monopolies to cooks for new recipes. The city-state of Venice instituted a patent law in 1474. England adopted its first patent

Conservation of Biodiversity for Sustainable Development.
Edited by O. T. Sandlund, K. Hindar and A. H. D. Brown. 1992. Scandinavian University Press, Oslo.

law in 1623, the US in 1790, France in 1791 and Sweden in 1819. Amid great controversy, patent laws were passed, defeated or repealed by many countries during the 1800s. No clear concensus existed for establishing government-granted rights to inventors to exclude others from imitating, manufacturing, using or selling a product (without permission) or utilizing a patented matter or process, as patents are usually defined. So defined, patents may be seen by some as a civil right, but it would be more appropriate to view them as a legal mechanism of control in the marketplace.

Until recently, patents were not needed for biological materials. There were more efficient mechanisms of control. A brief look at some of these methods will be useful.

The voyage of Columbus to the New World marked the beginning of what is called the "Columbian Exchange." Columbus returned from his first successful trip to the New World with maize, a crop of Central and South American origins. Future trips by others would bring back potatoes, squash, cassava, peanuts and common beans. But the "exchange" was really more acquisition than exchange. Crops moved in both directions. New crops were introduced to the New World. The crops, however, moved under colonial control. This was particularly true for the industrial/plantation crops which were the cash crops of the day, those traded commercially.

Rubber was smuggled out of Brazil and shipped to Kew Gardens. From Kew it made its way to the botanical gardens in Singapore where it was distributed to initiate the rubber industry of Southeast Asia. The Southeast Asia rubber boom caused a rubber bust in Brazil. The economy of Northeast Brazil collapsed and hundreds of thousands of people perished as a result. In "exchange", sugarcane travelled from Southeast Asia to the New World where systems of human slavery were instituted to tend the cane on lands clear-cut of trees just for that purpose. The boom-bust-famine cycles of sugar production are well known. Coffee came from Africa and Arabia. And bananas, which were to lend their name to a particularly unbalanced form of development in the so-called "Banana Republics," were brought from Southeast Asia.

Many native crops of the Americas were obliterated. Amaranth, an important grain crop of South America, was outlawed by the Spanish due to its association with "pagan" religious ceremonies. In North America, many domesticated food crops disappeared when the cultures which valued them and the people who raised them were wiped out. In some cases we no longer know the original names of these plants (today found only in a wild state), for some 200 languages were lost when the Indians were hunted down in North America.

Colonial powers attempted to control biological materials through control of production. The Dutch limited production of cloves and nutmeg to three islands in the Moluccas. The French brandished the guillotine to anyone who dared take indigo off of Antigua (Fowler and Mooney 1990).

Explorers were not simply looking for gold and silver or shorter trading routes. They had an eye for plants. In the 16th century fewer than 100 new plants were introduced into England. Some 1,000 were introduced in the 17th century and 9,000 were brought in during the 18th century (Lemmon 1968). These were the plants which were to form the basis of European dye, chemical and pharmaceutical industries. These were the plants which made possible the plantation economies in the Third World that brought such wealth to Europe.

Europe established its first botanical gardens in the 1600s. By the 1800s, most colonial powers had such gardens which effectively helped to channel biodiversity to these countries for their use and even control (Brockway 1979; Smith 1985). For the most part the search for plants was a search for new and valuable species of plants. Intra-species diversity was not terribly important, even though the Dutch were looking for clove and nutmeg diversity in the late 1700s to combat problems they were having in their plantations.

The economic importance of the botanical garden as a source of material for introduction into agricultural systems began to diminish with the rise of plant breeding. The United States built a system of "plant introduction stations" and finally a gene bank, the modern-day equivalent of the botanical garden, for the collection and storage of diversity within species of economically important crops. The Columbian exchange continued to work. Biodiversity was still flowing from South to North under the economic and political influence of commercial interests.

FROM PHYSICAL TO LEGAL MEANS OF CONTROL

With the rise of political independence in the Third World, the decline of plantation economies and the growing importance of plant breeding, new methods of control became necessary. Old methods of coercion and of physical control of production sites were no longer viable. Slowly, the idea of plant patents was broached after the turn of the century. Patents were still quite controversial. Most of the Third World countries that had been brought along by the major powers to sign the Paris Convention had long since repealed their patent statutes. Chemical, pharmaceutical and food patents were routinely prohibited. (Chemicals were not patentable in the Nordic countries until 1968 and in Switzerland until 1978. Many countries still reject pharmaceutical patents and Spain has no provision for chemical patents even today.) The selling of patents on "life" was not easy. But two powerful forces were combining to make it easier.

The first was the consolidation and internationalization of the seed industry. Approximately 1,000 mergers and acquisitions have taken place in the seed industry in the past 20 years as once family-owned, locally-oriented companies have been integrated into the multinational petrochemical and drug industry. As seed companies have expanded, they have become national and international in the scope of their business operations, their needs and their vision.

Old systems of control were either unavailable or increasingly inappropriate. This was particularly true for the seed industry where new seed varieties could easily be taken and reproduced by a competitor. With much value being added to seed through plant breeding, a healthy premium could be won by anyone who could gain exclusive rights to market a new seed variety. The introduction of hybrids solved that problem for some crops, especially maize. Farmers wanting to grow maize hybrids were forced to return to the seed company yearly. Biology itself offered the companies a method of control. But for grain crops, soybeans, cotton and many vegetables, no such control was possible.

In 1961 the Union for the Protection of New Varieties of Plants (UPOV) was organized to coordinate and promote "plant breeders rights" (PBR), a nobel term for the granting of patent-like certificates for plants. The Union was based in Europe, though by 1970 the US had passed its own "Plant Variety Protection Act", which provided similar protection. Typically, these laws permitted the protection of discoveries. And they provided for "research" and "farmer" exemptions, meaning that protected varieties could be used in research and in the breeding of new varieties and that farmers were allowed to save the seed of protected varieties and replant that seed the following year. With these concessions, PBR laws were adopted in a number of industrialized countries. (Notably, Canada approved its law only in 1990 and Norway has yet to adopt such a system.)

PBR was a significant departure from traditional patenting systems where inventions have to be new, useful, inventive and described sufficiently to enable someone skilled in the art to reproduce the invention. In the 1960s and 1970s, it was not thought that plants could meet such strict criteria. Thus the trade which society makes with the inventor, namely the grant of monopoly rights in exchange for the contribution of the knowledge (inventive step) was compromised. To meet PBR requirements, a variety need only be distinct, uniform and stable.

The second factor encouraging the spread of life-patenting laws was the development of biotechnology. The needs of biotechnology were met neither by traditional patent systems nor by plant breeders rights.

Are the "inventions" of biotechnology new and inventive in the legal sense? Can they be described adequately? If not, traditional patent systems offer no protection. In the US a series of court and Patent Office decisions have ameliorated some of these problems. In contrast, in Europe, the European Patent Convention expressly prohibits the patenting of plant varieties and animal breeds – a formidable obstacle.

Plant breeders rights offered no hope of protection for genes, characteristics or processes. And in any case, given the ease with which organisms multiply, PBR with its research and farmer exemptions provided too little control for the new industry.

Constructing new patent laws to accommodate biotechnology might be the most logical thing to do, but political opposition to this would be far too great in most countries. Instead, there have been and continue to be a multitude of efforts to modify existing patent laws through regulatory, administrative and judicial decisions. Space does not allow us to do more than generalize. In various countries, these accomodations are taking a number of forms: New microorganisms are being redefined so that they are no longer considered as products of nature (and thus excludable from many patent laws), but as products of human inventiveness. Thus, invention has been interpreted as including discovery and attempts are being made to give such an interpretation the force of law. The scope of coverage is broadening to include genes and characteristics (and plants and animals). Descriptive requirements are being loosened. And, it seems that the burden of proof is in the process of being reversed. The European Commission, for example, is proposing that defendants prove their innocence by proving that they have not used a plaintiff's patented microorganism (Newman 1989). The result of these developments is to give more power and more control to the patent holder. The effect is to maintain the centralization of control over biodiversity which has existed since colonial times. The mode of control has simply shifted from physical to legal.

In summary, with biotechnology we have a new, lucrative technology, easily copied, coming at a time when old methods of control and even old patent systems have become inadequate. The response has been one of accommodation.

The importance of this accommodation cannot easily be overestimated. The US patent office currently has a backlog of 8,000 "biotechnology" patents pending, though there are only a dozen products actually on the market now. Genentech, the premier American biotech company, has four times as many lawsuits to protect its patents as it has products. And in the US, patent appeal cases have been consolidated into one court system with the result that the rate of patents being upheld has jumped from 30 % to 80 % almost overnight (Dwyer 1989). Increasingly, research is being guided by the potential for patent acquisition and enforcement rather than by the results of experiments and tests (Dwyer 1989). Indeed, at least one company was created in the US whose "main business" according to the Wall Street Journal "is buying up broad patents and then suing other companies for alleged infringements..." (Lambert and Hayes 1990).

MOVING TO THE INTERNATIONAL PATENT ARENA

Patents appear to be the only viable form of assuring control and ownership of biotechnological innovations (though trade secrets are also of importance). Yet, the new techniques can be applied internationally. In order to be effective, patents must be internationally recognized. At present they are not. Patents granted in one country are not necessarily recognized by any other country. Some countries, such as India, maintain a distinctly hostile attitude towards patents.

The Third World critique of the industrial patent system is an old one. It found expression in and gained strength from work done at the UN Conference on Trade and Development in the 1960s. According to this analysis, most patents granted in the Third World simply function as a reverse system of preferences, giving legal preference, protection and monopoly rights to foreign imports. Fewer than 5 % of the patents granted in the Third World are used there in the production process. Fewer than 1 % of the patents issued in the Third World go to Third World nationals (Patel 1989). Third World governments then see no evidence that patents encourage research, a benefit often cited in industrialized countries. Instead they wonder how imported patent systems could serve their own development needs. Indeed, "development" seems to be the question. Whereas industrialized countries are more comfortable talking about patents as "rights," Third World governments seem to see patents in terms of development and transfer of technology policies. If a patent is to be granted, the typical Third World country will want that patent to be "worked" inside the country. It will want the technology imported and used inside the country. If only the product is imported, its contribution to development will be slight.

The conflict of interests over patents in recent years has made international patent reform difficult to accomplish in the traditional arenas like the World Intellectual Property Organization. Given this, American industry has pressed for a change of arenas. This may explain the US initiative in the General Agreement on Tariffs and Trade (GATT) in Geneva where the US can use its trading power to bargain for intellectual property rights.

At the beginning of the current Uruguay Round of negotiations agreement had been reached on the scope of negotiations. In the "General Principles Governing Negotiations" (paragraph v of Section B) developed countries formally agreed that they did not expect developing countries "to make contributions which are inconsistent with their individual development, financial and trade needs. Developed contracting parties shall therefore not seek, neither shall less-developed contracting parties be required to make, concessions that are inconsistent with the latter's development, financial and trade needs."

Nevertheless, the challenge was made. The US put forth a novel position: the absence of intellectual property rights protection (at a mimimum level similar to that offered by the US) constitutes an unfair trade barrier and should be subject to trade sanctions under GATT. Strategically, the threat of trade restrictions is being offered to persuade countries into adopting foreign levels of patent protection through the only forum in which such a tactic might work across the board. As outlined by Gadbaw and Richards (1988), the strategy was fourfold:

1. attempt to persuade Third World countries of the value of intellectual property rights;
2. threaten loss of access to US markets;
3. offer increased or decreased research and development funding through the private sector, and;
4. pursue agreements on minimum patent standards through GATT.

"The primary thrust of the Uruguay Round efforts," according to Gadbaw and Gwynn (1988), "is to draw the developing countries into a new set of rights and obligations".

FARMERS AS INNOVATORS?

Thus far we have examined the accommodation of northern, industrialized country patent systems to changes in technology and advances in the biological sciences. And we have briefly looked at the efforts of those systems to "clone" themselves in the Third World. What about the interests of the Third World amidst these changes? Does the fact that only 1 % of the patents go to Third World nationals mean that only 1 % of the innovation on this planet is done in Asia, Africa and Latin America? Or, might it reflect a bias in the system, an indication of who has written patent conventions and for whom?

While it may seem impolite to say this, it can hardly be questioned that industrialized countries need protection for their inventions and desire free access to the biological resources of the South. Nowhere is this more evident than with agricultural crops. Modern plant breeding and biotechnology have made plant genetic resources extremely valuable. These are the resources which make evolution in our crops possible. Yet, note the curious way in which we speak of this genetic diversity, these raw materials which are the basis of our agricultural system. We hear these resources described as "primitive varieties," "Stone Age varieties," as heirloom or traditional varieties or as landraces. The first two terms are blatantly unscientific, inaccurate and demeaning. The other terms imply that these resources are something from the past, something which is not implicitly the product of human creativity. It is probably true that in using the term "landrace", scientists mean no harm or disrespect. But language often has a way of revealing unconscious feelings and hidden social relationships. Are these genetic resources in the form of "varieties" being grown by Third World peoples really just raw materials or are they improved materials? Should these priceless materials remain priceless?

Such questions provided the backdrop for a remarkable series of debates in the United Nations Food and Agriculture Organization (FAO) in Rome in the 1980s. A number of Third World nations (particularly Latin American) had begun to sense inequities in patterns of seed collection and storage. Patterns of germplasm flows and usage under modern seed bank systems were similar to those under the botanical garden system of colonial days. A modern-day Columbian exchange was taking place, once again one that was less exchange than acquisition from their vantage point.

Between 1974 and 1984, over 125,000 duplicate samples were drawn from collecting expeditions under the International Board for Plant Genetic Resources (IBPGR). According to their data, over 90 % of these came from the Third World. Only 15 % of the duplicates of this material were stored in the Third World under the control of their governments. And only

25 % of the regional and 16 % of the global repositories (so designated by IBPGR) for individual crops were in the Third World (Fowler and Mooney 1990). Clearly the North had acted on former US Secretary of Agriculture Richard Lyng's observation that the US (and other industrialized countries) are "totally at the mercy" of the Third World for crop germplasm.

After a number of meetings and many hours of often acrimonious debate, the FAO established:

1. A Commission on Plant Genetic Resources where governments, as governments could meet to discuss questions about the conservation and use of these resources.
2. An International Network of Gene Banks, under the auspices of FAO.
3. An "International Undertaking," a low-level, voluntary agreement which called for the free exchange of all classifications of germplasm, from wild varieties, mutants and the "folk varieties" of the Third World farmer to the breeding lines and commercial varieties of the seed company.
4. The concept of Farmers' Rights as a means of claiming that plant genetic resources are not just a raw material divorced from human creativity, and as a way of pressing forward the demand that if plant breeders rights are to be observed and rewarded, then farmers rights must be treated similarly.
5. An International Gene Fund to finance genetic conservation and utilization programs and to serve possibly as an instrument for the concrete manifestation and recognition of the concept of farmers rights.

Third World delegates to FAO claim that their part of the globe has a more informal, less visible, less recognized and totally unrewarded system of innovation. If it could be shown that Third World farmers are plant breeders, the argument for a parallel system of compensation to PBR would be strengthened politically. Let us examine, briefly, what is known about the steps of the "plant breeding" process at the farm level in the Third World.

In the Andes, Dr. Steve Brush (1986) has described folk taxonomy systems for potatoes among that region's farmers. Similar studies have shown up to five levels of classification for rice among farmers in Asia (H. Conklin, cited in Brush 1986). Farmers may be able to describe and classify their plants, but do they know about heredity? The Pima and Tepehuan people in Mexico have ways of describing introgression of wild and weedy genes into cultivated crops and "a sophisticated sense of the relationship between domesticated and spontaneous species," according to Dr. Gary Nabhan, an ethnobotanist (Nabhan 1989).

Can they use such knowledge of heredity? In the 1960s Dr. Garrison Wilkes observed harvests of hybrids of Indian maize with teosinte, a wild relative of maize. In the Nabogame Valley in Mexico, he found farmers allowing and even encouraging the introgression of maize and teosinte. According to them the result was "stronger" and flintier kernels (Wilkes 1977).

In the most far-reaching study of its kind to date, Dr. Paul Richards (Richards 1986) found that farmers in Sierra Leone can distinguish between 70 different rices. These are differentiated according to length to maturity, ease of husking, proportion of husk size to grain size and weight, susceptibility to bird and insect attack, appropriateness in different soils and with different levels of moisture, color of plant parts, cooking time and qualities, taste, etc.

When harvesting, these farmers stay clear of the boundaries between different "varieties." They harvest panicle by panicle saving the interesting material for future experiments. They also rogue off-types when they desire to keep a variety "pure." They have trial plots. They record data. They test germination rates. They try to match their rice to the local ecological niches. (Do the traits uncovered in this process constitute discovery or invention in the legal sense? Does it matter?) They multiply promising new varieties and they conduct field tests.

In summary, Third World farmers have been found to employ taxonomic systems, encourage introgression, use selection, make efforts to see that varieties are adapted, multiply seeds, field test, record data and even name their varieties. In short, they do what many northern plant breeders do, except that they do not apply for legal protection – the one step in the process which requires skills (and money) which they do not have. Do these actions deserve any less recognition or reward than those of the 13 year-old boy who discovered the world's first thornless rose and the company which eventually got the patent for this "innovation" (Kneen 1948)?

The actions of Third World farmers may meet the requirements of patent or plant breeders rights laws, but the plants themselves do not. These plants may be new, distinct and useful. They are rarely uniform. The obvious question is, why should they be uniform? Uniformity is a quality sometimes needed by industrialized agricultural systems and always needed by patent laws, but its agronomic appropriateness in the Third World is highly suspect. In any case, the Third World farmer's field is at once a seed bank, a test plot and a production site. How could it be uniform? How can we ask the Third World farmer to be innovative and to produce uniformity simultaneously? We might better ask what it is about our politics, economic needs or the nature of our own scientific establishment which makes it so difficult to acknowledge the skills, creativity and contributions of others – especially in the Third World. Could it be that our lack of understanding blinds us to the reality of a different style of innovation? Dr. Melaku Worede, director of Ethiopia's renowned gene bank, tells the story of an American scientist who collected a high-lysine sorghum there some years ago. After testing in the laboratory this scientist announced his discovery – a high-lysine sorghum. The locals had a name for this variety, which the scientist had not bothered to learn. It was called "Why Bother With Wheat?" They already knew the sorghum's qualities. Who made the discovery?

It is readily acknowledged that collected seeds in gene banks are poorly documented. Often there is little "passport" information. There is almost no information on local uses, names, etc. In such a situation, attributes bred by folk science can become anonymous attributes, a gift from nature, disconnected from the labor and creativity of real people – the common heritage of mankind. In the extreme, this material can even become patentable. In this regard one conclusion of an exclusive meeting co-sponsored by the American Society of Agronomy, the Crop Science Society of America, the American Agricultural Economics Association and the American Society of Horticultural Science is particularly interesting: "It is important to support better characterization of material in gene banks so that characterization is known and in the public domain, available, and therefore not subject to intellectual property rights" (Anonymous 1989). This statement serves to highlight the function patents may come to play in relation to Third World genetic resources. Are these organizations not suggesting that material (potentially the "Why Bother With Wheat?" sorghum) might be patented if efforts are not soon made to catalog what is contained in gene banks?

CONCLUSION

We are now entering a period of intensified conflict over questions of ownership and control of biological materials. Patents and patent-like protection will be at the center of controversies, which will spill over into questions of access to and use of biological diversity. Industrialized countries can no longer expect the Third World to donate graciously genetic material for the improvement of northern crops and the enrichment of foreign corporations only to be told that this material is the "common heritage of mankind," while corporate material is private property.

One cannot expect Madagascar, for example, to donate vanilla germplasm only to watch biotech companies begin commercial production – without vanilla plants, without Madagascar's 75,000 vanilla producers and without compensation either directly or indirectly to the farmers or the country.

Patents are a means of allocating ownership, assigning control, regulating access and apportioning benefits. In the field of biology they are the latest in a series of mechanisms which date back hundreds of years which includes colonialism, physical control of production and markets, hybrids, trade secrets, etc. Today, the Third World is beginning to develop and employ its own mechanisms of control strikingly similiar to ones that have been employed against them – physical control, in this case over the genes themselves.

In this context, the debate at GATT is part of the same debate going on at FAO (over farmers' rights and the International Fund), at UNEP (over biodiversity treaties and access to genetic diversity), at UNESCO (over cultural heritage and protection of folklore) and over the agenda and outcome of the 1992 Conference on Environment and Development scheduled for Rio de Janeiro, Brazil.

One cannot pretend that the profound changes underway in patent law exist outside a larger milieu. Patents reflect power relationships and as such, they change. Patent systems which recognize the contributions of some but not of others are based on injustice and will be unstable, if not unsustainable. This, perhaps, is the lesson of the patent controversies of the last decade. Means should therefore be employed to soften the negative impacts of patents and provide systems of protection and compensation for forms of innovation not now covered by traditional patent systems. Concrete recognition of "farmers' rights" through a mandatory gene fund would be one such method. Such a fund would not only reward innovation, but reward and finance conservation as well. And as the first Keystone Dialogue report notes: "The value of a resource to a country depends on the country's present and future ability to utilize them. It follows that support for genetic resources conservation should be supplemented by support for plant breeding activities" (Keystone 1988). Short of totally reconceptualizing the patent system – an interesting but unrealistic goal – a fund might serve to recognize and redress some of inequalities of biological patent systems.

Decisions made regarding patents, ownership and control will inevitably affect our ability to conserve genetic resources and have access to them. The link between environmental conservation, economic development and justice is a link which cannot and should not be severed.

Finally, the decisions which are made concerning the ownership and control of the planet's biological diversity will also say much about the kind of people we want to become and the kind of relationship with the world and its people we want to bequeath to future generations. As the 500th anniversary of the voyage of Columbus is observed, many will watch to see if anything has really changed in the relationship between the centers of world economic power and the periphery.

LITERATURE CITED

Anonymous. 1989. Quandry over plant patenting brings diverse group of experts together. Diversity 5: 36.

Brockway, L. H. 1979. *Science and Colonial Expansion: The Role of the British Royal Botanical Gardens.* Academic Press, New York.

Brush, S. B. 1986. Genetic diversity and conservation in traditional farming systems. J. Ethnobiol. 6: 151-167.

Dwyer, P. 1989. The battle raging over intellectual property. Business Week; 22 May 1989.

Fowler, C. and P. Mooney. 1990. *Shattering: Food, Politics and the Loss of Genetic Diversity.* University of Arizona Press, Tucson.

Gadbaw, R. M. and R. E. Gwynn. 1988. Intellectual property rights in the new GATT round, pp. 38–88. In: R. M. Gadbaw and T. J. Richards (eds) *Intellectual Property Rights: Global Concensus, Global Conflict?* Westview Press, Boulder, CO.

Gadbaw, R. M. and T. J. Richards. 1988. Introduction, pp. 1–37. In: R. M. Gadbaw and T. J. Richards (eds) *Intellectual Property Rights: Global Concensus, Global Conflict?* Westview Press, Boulder, CO.

Keystone Center. 1988. *Final Report of the Keystone International Dialogue on Plant Genetic Resources*, Session I: Ex situ *Conservation of Plant Genetic Resources.* Keystone Center, Keystone, CO.

Kneen, O. H. 1948. Plant patents enrich our lives. National Geographic, March, 1948.

Lambert, W. and A. S. Hayes. 1990. Investing in patents to file suits is curbed. Wall Street Journal. 30 May 1990.

Lemmon, K. 1968. *Golden Age of Plant Hunters.* Barnes, Cranbury, UK.

Nabhan, G. P. 1989. *Enduring Seeds: Native American Agriculture and Wild Plant Conservation.* North Point Press, San Francisco, CA.

Newman, P. 1989. A modest proposal for European patents. Bio/Technol. 7: 26.

Patel, S. J. 1989. Intellectual Property Rights in the Uruguay Round: A disaster for the South? Economic and Political Weekly. 6 May 1989.

Richards, P. 1986. *Coping with Hunger: Hazard and Experiment in an African Rice-farming System.* Allen & Unwin, London.

Smith, N. J. H. 1985. *Botanic Gardens and Germplasm Conservation.* Harold L. Lyon Arboretum Lecture Number Fourteen. University of Hawaii Press, Honolulu.

Wilkes, H. G. 1977. Hybridization of maize and teosinte, in Mexico and Guatemala and the improvement of maize. Econ. Bot. 31: 254-293.

18 The seed and the spinning wheel: technology development and biodiversity conservation

Vandana Shiva
Research Foundation for Science and Ecology,
105, Rajpur Road,
Dehra Dun 248001,
India

ABSTRACT

Biodiversity conservation is generally conceived of independent of the technologies of production which use and transform biological resources. This paper demonstrates the interdependence of these two factors. In Third World countries where most of the world's biodiversity is concentrated, many communities of tribals and peasants derive their livelihoods and satisfy multiple needs directly from the rich diversity of biological resources. Production technologies based on uniform monocultures of trees, crops or livestock threaten livelihoods while they displace biodiversity. The common misconception that diversity is linked to low productivity, and that uniformity is essential for high productivity is addressed. It is shown that when multiple yields, values and outputs of biological systems are taken fully into account, diversity does not preclude high productivity. Using the symbol of Gandhi's spinning wheel, the paper urges for a deeper consideration of the social and ecological context in which technology development takes place.

INTRODUCTION

Biological richness is not uniformly distributed across the world. It is concentrated in the tropical countries of the Third World. Most current conservation plans for biodiversity, however, originate in the North, and they carry within them the social categories of development and planning that are characteristic of industrialized and affluent countries. For example in *Conserving the World's Biological Diversity*, McNeely et al. (1990) have divided the economic values of biological resources into three categories:

1. "Consumptive use value" – value of products consumed directly without passing through a market such as firewood, fodder and game meat.
2. "Productive use value" – value of products commercially exploited.
3. "Non-consumptive use value" – indirect values of ecosystem functions, such as watershed protection, photo-synthesis, regulation of climate and production of soil.

Conservation of Biodiversity for Sustainable Development.
Edited by O. T. Sandlund, K. Hindar and A. H. D. Brown. 1992. Scandinavian University Press, Oslo.

An interesting value framevork has thus been constructed which predetermines analysis and options. If the Third World poor, who derive their livelihoods directly from nature only "consume", and the trading and commercial interests are the only "producers", it follows quite naturally, that the Third World is responsible for the destruction of its biological wealth, and the North alone has the capacity to conserve it.

This ideologically constructed divide between consumption, production and conservation hides the political economy of the processes which underlie the destruction of biological diversity.

According to the dominant paradigm of production, diversity goes against productivity, which creates an imperative for uniformity and monocultures. This has generated the paradoxical situation in which modern plant and animal improvement has been based on the destruction of the biodiversity which it uses as raw material. The irony of plant and animal breeding is that it destroys the very building blocks on which the technology depends. Forestry development schemes introduce monocultures of industrial species like eucalyptus and push into extinction the diversity of local species which fulfill local needs. Agricultural modernization schemes introduce new and uniform crops into farmers' fields and destroy the diversity of local varieties. In the words of Professor Garrison Wilkes of the University of Massachusetts, this is analogous to taking stones from a building's foundations to repair the roof (Myers 1985).

In agriculture and forestry, in fisheries and animal husbandry, production is being incessantly pushed in the direction of destruction of diversity. Production based on uniformity thus becomes the primary threat to biodiversity conservation and to sustainability.

However, this threat to biodiversity from technology development has been little understood and analysed. This paper is an attempt to fill this lacunae; and to enrich the understanding of the relation between technology, natural resources and human needs. In particular it focuses on the social and ecological impact of technology change in areas related to biological resources.

TECHNOLOGY, DEVELOPMENT AND SUSTAINABILITY

Science and technology are conventionally accepted as what scientists and technologists produce, and development is accepted as what science and technology produce. Scientists and technologists are in turn taken to be that sociological category formally trained in western science and technology, either in institutions or organisations in the West, or in Third World institutions mimicking the paradigms of the West. These tautological definitions are unproblematic if one leaves out people, especially poor people; if one ignores ecological and cultural diversity and distinct civilisational and natural histories of our planet which have created diverse and distinctive cultures and ecosystems. Development in this view is taken as synonymous with the introduction of western science and technology in non-western contexts. The magical identity is development = modernisation = westernisation.

In a wider context, where science is viewed as "ways of knowing" and technology as "ways of doing", all societies, in all their diversity, have had science and technology systems on which their distinct and diverse development has been based. Technologies or systems of technologies bridge the gap between nature's resources and human needs. Systems of knowledge and culture provide the framework for the perception and utilisation of natural resources. Two changes occur in this shift of definition of science and technology. Science and technology are no longer viewed as uniquely western but as a plurality associated with all cultures and civilisations. And a particular science and technology do not automatically translate into development everywhere. Ecologically and economically inappropriate science and technology can become causes of underdevelopment and poverty, not solutions to underdevelopment and impoverishment. Ecological inappropriateness is a mismatch between the ecological processes of nature which renew life support systems and the resource demands and impacts of technological processes. Technological processes can lead to higher withdrawals and consumption of natural resources or higher additions of pollutants than ecological limits allow. In such cases they contribute to underdevelopment through destruction of ecosystems.

Economic inappropriateness is the mismatch between the needs of society and the requirements of a technological system. Technological processes create demands for raw materials and markets, and control over both raw materials and markets becomes an essential part of the politics of technological change.

The lack of the theoretical cognition of the two ends of technological processes, their beginning in natural resources and their end in basic human needs, has created the current paradigm for economic and technological development which demands increasing withdrawals of natural resources and generates increasing addition of pollutants while marginalizing and dispossessing an increasing number of people from the productive process. These characteristics of contemporary scientific industrial development are the primary causes for the ecological, political and economic crisis. The combination of ecologically disruptive scientific and technological modes and the absence of the criteria for evaluating scientific and technological systems, in terms of efficient resource use and capability of satisfying basic needs, has created conditions where society is increasingly propelled towards ecological and economic instability and has no rational and organised response to arrest and curtail these destructive tendencies (Shiva et al. 1991).

In Third World economies, many communities depend on biological resources for their sustenance and well being. In society, biodiversity is simultaneously a means of production, and an object of consumption. It is the survival base that has to be conserved. Sustainability of livelihoods is ultimately connected to the conservation and sustainable use of biological resources in all their diversity.

However, biodiversity based technologies of tribal and peasant societies have been viewed as backward and primitive and have been displaced by technologies which use biological resources in such a way that they destroy diversity and people's livelihoods.

DIVERSITY AND PRODUCTIVITY

There is a general misconception that diversity based production systems are low productivity systems. However, the high productivity of uniform and homogeneous systems is a contextual and theoretically constructed category; based on taking only one-dimensional yield and outputs into account. On this same construct, the multidimensional yields of diversity based systems of production are excluded and ignored. Were the "miracle" seeds of the Green Revolution inherently superior and "advanced" in comparison to the diversity of indigenous crops and varieties that they displaced? The miracle of the new seeds has most often been communicated through the term "high yielding varieties" (HYV). The HYV category is a central category of the Green Revolution paradigm. Unlike what the term suggests, there is no neutral or objective measure of "yield" on the basis of which the cropping systems based on miracle seeds can be established to be higher yielding than the cropping systems they replace. It is now commonly accepted that even in the most rigorous of scientific disciplines such as physics, there are no neutral observational terms. All terms are theory laden (Kuhn 1972).

The HYV category is similarly not a neutral observational concept. Its meaning and measure is determined by the theory and paradigm of the Green Revolution. And this meaning is not easily and directly translatable for comparison with the agricultural concepts of indigenous farming systems for a number of reasons. The Green Revolution category of the HYV is essentially a reductionist category which decontextualizes properties of both the native and the new varieties. Through the process of decontextualization, costs and impacts are externalized and systematic comparison with alternatives is precluded.

Cropping systems, in general, involve an interaction between soil, water and plant genetic resources. In indigenous agriculture, for example, cropping systems include a symbiotic relationship between soil, water, farm animals and plants. Green Revolution agriculture replaces this integration at the level of the farm with the integration of inputs such as seeds and chemicals. The seed/chemical package sets up its own interactions with soils and water systems, which are, however, not taken into account on the assessment of yields.

Modern plant breeding concepts like HYV reduce farming systems to individual crops and parts of crops. Crop components of one system are then measured with crop components of another. Since the Green Revolution strategy is aimed at increasing the output of a single component from a farm, at the cost of decreasing other components and increasing external inputs, such a partial comparison is by definition, biased to make the new varieties "high yielding" even when at the systems level, they may not be. Traditional farming systems are based on mixed and rotational cropping systems of cereals, pulses, oilseeds with diverse varieties of each crop, while the Green Revolution package is based on genetically uniform monocultures. No realistic assessment has ever been made of the yield of the diverse crop outputs in the mixed and rotational systems. Usually the yield of a single crop like wheat or maize is singled out and compared to yields of new varieties.

As Dr. Palmer concluded in the United Nations Research Institute for Social Development's 15 nation study of the impact of the seeds, the term "high yielding varieties" is a misnomer because it implies that the new seeds are high-yielding in and of themselves (Pearse 1980). The distinguishing feature of the seeds, however, is that they are highly responsive to certain key inputs such as fertilizers and irrigation. Palmer therefore suggested the term "high responsive varieties" (HRV) in place of the "high yielding varieties" (HYV). In the absence of additional inputs of fertilizers and irrigation, the new seeds perform worse than indigenous varieties. With the additional inputs, the gain in output is insignificant compared to the increase in inputs. The measurement of output is also biased by restricting it to the marketable part of crops. However, in a country like India, crops have traditionally been bred and cultivated to produce not just food for man but fodder for animals, and organic fertilizer for soils. According to A. K. Yegna Iyengar, a leading authority on agriculture, "as an important fodder for cattle and in fact as the sole fodder in many tracts, the quantity of straw obtainable per acre is important in this country. Some varieties which are good yielders of grains suffer from the drawback of being low in respect to straw" (Yegna Iyengar 1944). He illustrated the variation in the grain-straw ratio with yields from the Hebbal farm (Table 1).

Table 1. Grain and straw production of various rice varieties.

Name of Variety	Grain (in lb. per acre)	Straw (in lb. per acre)
Chintamani sanna	1,663	3,333
Budume	1,820	2,430
Halubbalu	1,700	2,740
Gidda Byra	1,595	2,850
Chandragutti	2,424	3,580
Putta Bhatta	1,695	3,120
Kavada Bhatta	2,150	2,940
Garike Sanna	2,065	2,300
Alur Sanna	1,220	3,580
Bangarkaddi	1,420	1,760
Banku (rainy season 1925–26)	1,540	1,700
G.E.B. (rainy season 1925–26)	1,900	1,540

In the breeding strategy for the Green Revolution, multiple uses of plant biomass seem to have been consciously sacrificed for a single use, with non-sustainable consumption of fertilizer and water. The increase in marketable output of grain has been achieved at the cost of decrease of biomass for animals and soils and the decrease of ecosystem productivity due to the over-use of resources. The increase in production of grain for marketing was achieved in the Green Revolution strategy by reducing the biomass for internal use on the farm. Similarly indigenous breeds of Indian cattle have been described as "unproductive", and all animals breeding programmes are aimed at replacing them with exotic strains of the Jersey, Holstein, Fiesian, Red Dane and Brown Swiss. However, Indian cattle do not only yield milk. They provide energy and fertilizer, which are crucial to farming systems.

It should be noted that two-thirds and more of the power requirements of Indian villages are met by some 80 million work animals of which 70 million are the male progeny of what the western perspective sees as "useless" low milk-yielding cows. It has been calculated that to replace animal power in agriculture, India would have to spend about a thousand million US dollars annually on petrol. Indian cattle excrete 700 million tonnes a year of recoverable manure: half of this is used as fuel, liberating the thermal equivalent of 27 million tonnes of kerosens, 35 million tonnes of coal or 68 million tonnes of wood, all of which are scarce resources in India; the remaining half is used as fertilizer. As for other livestock produce, it may be sufficient to mention that the export of hides, skins, etc. brings in 150 million dollars annually into the national coffers. With limited resources, indigenous cattle produce a multiplicity of uses (George 1985).

Indian cattle provide food in excess of the edible food consumed, in contrast to the US where six times as much edible food is fed to cattle as is obtained from them (Leon 1975) (Table 2). Yet this highly efficient food system, based on the multiple uses of cattle, has been dismantled in the name of efficiency and "development" by the reductionist strategies of the green and white revolutions splitting and dichotomising an integrated system of crop production and animal husbandry, necessary for maintaining each other sustainably (Shiva 1988, 1991).

Table 2. Inputs and useful outputs from US cattle and Indian cattle and buffalo (1972). Reprinted from Leon (1975).

Inputs and outputs	Matter (10^{10} kg)		Energy (10^{12} calories)		Protein (10^9 kg)	
	USA	India	USA	India	USA	India
Inputs						
Edible by man	11.9	0.68	38.6	1.7	16.0	2.1
Inedible by man	22.2	40.00	88.0	120.5	25.1	33.3
Total	34.1	40.68	126.6	122.2	41.1	35.4
Outputs						
Work	–	–	–	6.50	–	–
Milk	1.12	0.51	5.04	2.09	2.06	0.88
Meat	0.90	0.50	4.40	2.23	0.17	0.11
Hides	0.11	0.07	–	–	–	–
Manure	0.87	10.81	–	16.16	–	–
Total	3.00	11.89	9.44	26.98	2.23	0.99
Efficiency (%)	9	29	7	22	5	3

The low productivity of diverse, multiple dimensional systems and the high productivity of uniform, one dimensional systems of agriculture, forestry and livestock is therefore not a neutral, scientific measure but is biased towards the commercial interests for whom maximising of one dimensional output is an economic imperative.

This push towards uniformity, however, undermines the diversity of biological systems which form the production system. It also undermines the livelihoods of the people whose work is associated with diverse and multiple use systems of forestry, agriculture and animal husbandry. As an example, in the State of Kerala in India, which derives its name from the coconut palm, coconut is cultivated in a multistoreyed, high intensity cropping system along with betel and pepper vines, bananas, tapioca, drumstick, papaya, jackfruit, mango and vegetables. Compared to annual labour requirement of 157 man-days ha^{-1}, in a monoculture of coconut palm, the mixed cropping system increases employment to 960 man-days ha^{-1} (Government of Kerala 1964).

In the dry land farming systems of the Deccan, the shift from mixed-cropping of millets with pulses and oilseeds to eucalyptus monocultures has led to a loss of employment of 250 man-days ha^{-1} yr^{-1} (Shiva and Bandyopadhyay 1987).

When labour is scarce and costly, labour displacing technologies are productive and efficient. When labour is abundant, labour displacement is unproductive because it leads to poverty, dispossession and destruction of livelihoods.

In Third World situations, sustainability has therefore to be achieved at two levels simultaneously – the sustainability of natural resources and the sustainability of livelihoods. Biodiversity conservation has therefore to be linked to conservation of livelihoods derived from biodiversity.

CONSERVATION OF THE SEED AND THE SPINNING WHEEL

The conservation of livelihoods along with the conservation of resources has been a special concern for us in India. It was the basis of our freedom movement and the struggle against colonialism. Mahatma Gandhi had recognised that poverty and underdevelopment in India was rooted in the destruction of jobs linked to our rich textile industry. The regeneration of livelihoods was central to the process of regaining independence. Gandhi categorically stated that what is good for one nation situated in one condition is not necessarily good for another differently situated. One man's food is often another man's poison. Mechanization is good when hands are too few for the work intended to be accomplished. But according to Gandhi, it is an evil where there are more hands than required for the work as is the case in India (Pyarelal 1959).

The spinning wheel became for Gandhi and India a symbol of a technology that conserves resources, people's livelihoods and people's control over their livelihoods. In contrast to the imperialism of the British textile industry, which had destroyed India's industrial base, the "charkha" was decentred and labour generating, not labour displacing. It needed people's hands and minds, instead of treating them as surplus, or as mere inputs into an industrial process.

This critical mixture of decentralization, livelihood generation, resource conservation and strengthening of self-reliance were essential to undo the waste of centralization, livelihood destruction, resource depletion and creation of economic and political dependence that had been engendered by the industrialization associated with colonialism.

Gandhi's spinning wheel is a challenge to notions of progress and obsolescence that arise from absolutism and false universalism in concepts of science and technology development. Obsolescence and waste are social constructs that have both a political and ecological component. Politically, the notion of obsolescence gets rid of people's control over their lives and livelihoods by defining productive work as unproductive and removing people's control over production in the name of progress. It would rather waste hands than waste time. Obsolescence also destroys the regenerative capacity of nature by substituting manufactured uniformity in place of nature's diversity. Technological obsolescence translates into obsolescence of biodiversity. This induced dispensability of poorer people on the one hand and diversity on the other constitutes the political ecology of technological development guided by narrow and reductionist notions of productivity. Parochial notions of productivity, perceived as universal, rob people of control over their means of reproducing life and rob nature of her capacity to regenerate diversity.

Ecological erosion and destruction of livelihoods are linked to one another. Displacement of diversity and displacement of people's sources of sustenance both arise from a view of development and growth based on uniformity created through centralized control. In this process of control, reductionist science and technology act as handmaidens for economically powerful interests. The struggle between the factory and the spinning wheel continues as new technologies emerge for the manipulation of biological resources.

Like the spinning wheel was rendered backward and obsolete by the industrialization of the manufacture of textiles, farmers' seeds are being rendered obsolete and valueless by technological change associated with the industrialization of seed production.

The indigenous varieties or landraces in agriculture have evolved through millenia of natural and human selection, these varieties produced and used by Third World farmers worldwide are called "primitive cultivars". Those varieties created by modern plant breeders in international research centres or by transnational seed corporations are called "advanced" or "elite". The hierarchy in words like "primitive" and "elite" has strong cultural roots, even when these words are used in scientific domains. Underlying these categorisations is an inherent bias which assumes that technologies that emerge in the industrialized North are superior in an absolute sense. However, the experience of the Green Revolution informs us that in the domain of biodiversity, technology development can lead to progress for one interest group while creating underdevelopment for others.

The change in the nature of seed is justified by creating a framework that treats self-regenerative seed as "primitive" and as "raw" germplasm and the seed that is inert without inputs and non-reproducible as a finished product. The whole is rendered partial, the partial is rendered whole. The commoditised seed is ecologically incomplete and ruptured at two levels:

1. It does not **reproduce** itself, while by definition, seed is a regenerative resource. Genetic resources are thus, through technology, transformed from a renewable into a non-renewable resource.
2. It does not **produce** by itself. It needs the help of inputs to produce. As the seed and chemical companies merge, the dependence on inputs will increase, not decrease. And ecologically, whether a chemical is added externally or internally, it remains an external input in the ecological cycle of the reproduction of seed.

It is the shift from the ecological processes of reproduction to the technological processes of production that underlies both the problem of dispossession of farmers and of genetic erosion.

The new plant biotechnologies will follow the path of the earlier HYVs of the Green Revolution in pushing farmers onto a technological treadmill. Biotechnology can be expected to increase the reliance of farmers on purchased inputs even as it accelerates the process of polarization. It will even increase the use of chemicals instead of decreasing it. The dominant focus of research in genetic engineering is not on fertilizer-free and pest-free crops, but pesticide and herbicide-resistant varieties. For the seed-chemical multinational companies, this might make commercial sense, since it is cheaper to adopt the plant to the chemical than to adopt the chemical to the plant. The cost of developing a new crop variety rarely reaches USD 2 million, whereas the cost of a new herbicide exceeds USD 40 million (Fowler et al. 1988).

Like Green Revolution technologies, biotechnology in agriculture can become an instrument for dispossessing the farmer of seed as a means of production. The relocation of seed production from the farm to the corporate laboratory relocates power and value between the North and South; and between corporations and farmers. It is estimated that the elimination of home grown seed would dramatically increase the farmers' dependence on biotech industries by about USD 6,000 million annually (Kloppenburg 1988).

It can also become an instrument of dispossession by selectively removing those plants or parts of plants that do not serve the commercial interest but are essential for survival of nature and people. "Improvement" of a selected characteristic in a plant, is also a selection **against** other characteristics which are useful to nature, or for local consumption. Improvement is not a class or gender neutral concept. Improvement of partitioning efficiency is based on the enhancement of the yield of the desired product at the expense of unwanted plant parts. The desired product is, however, not the same for rich people and poor people, or rich countries and poor countries, nor is efficiency. On the input side, richer people and richer countries are short of labour and poorer people and poorer countries are short of capital and land. Most agricultural development, however, increase capital input while displacing labour, thus destroying livelihoods. On the output side, which parts of a farming system or a plant will be treated as "unwanted" depend on what class and gender one is. What is unwanted for the better off may be the wanted part for the poor. The plants or "plant parts" which serve the poor are the ones whose supply is squeezed by the normal priorities of improvement in response to commercial forces.

CONCLUSION

The destruction of people's livelihood and sustenance goes hand in hand with the erosion of biological resources and their capacity to fulfill diverse human needs while regenerating and renewing themselves. Attempts to increase commodity flows in one direction generate multiple levels of scarcities in related outputs. Increase of grain leads to decrease of fodder and fertilizer. Increase of cereals leads to decrease of pulses and oilseeds. The increase is measured. The decrease goes unnoticed, except by those who are deprived through the creation of new scarcity. Both people and nature are impoverished; their needs are no longer met by the one dimensional production systems which replace biologically rich and diverse ecosystems, and put added burdens on remaining pockets of biodiversity to satisfy their needs.

The extinction of people's livelihoods and sustenance is closely connected with the erosion of biodiversity. Protection of biodiversity can only be ensured by regenerating diversity as a basis of production in agriculture, forestry and animal husbandry. The practice of diversity is the key to its conservation.

Biodiversity cannot be conserved until diversity is made the logic of production. If production continues to be based on the logic of uniformity and homogenisation, uniformity will continue to displace diversity. "Improvement" from the corporate viewpoint, or from the viewpoint of western agricultural research is often a loss for the Third World, especially the poor in the Third World. There is therefore no inevitability that production acts against diversity. Uniformity as a pattern of production becomes inevitable only in a context of control and profitability.

All systems of sustainable agriculture, whether of the past or the future, work on the basis of the perennial principles of diversity and reciprocity. The two principles are not independent but interrelated. Diversity gives rise to the ecological space for give and take, for mutuality and reciprocity. Destruction of diversity is linked to the creation of monocultures, and with the creation of monocultures, the self-regulation and decentred organisation of diverse systems gives way to external inputs and external and centralized control. Sustainability and diversity are ecologically linked because diversity offers the multiplicity of interactions which can heal ecological disturbance to any part of the system. Nonsustainability and uniformity means that disturbance to one part is translated into a disturbance to all other parts. Instead of being contained, ecological destabilisation tends to be amplified. Closely linked to the issue of diversity and uniformity is the issue of productivity. Higher yields and higher production have been the main push for the introduction of uniformity and the logic of the assembly line. The imperative of growth generates the imperative for monocultures. Yet this growth is, in large measure, a socially constructed, valueladen category. It exists as a "fact" by excluding and erasing the facts of diversity and production through diversity. Sustainability, diversity and decentred self-organisation are therefore linked, as are unsustainability, uniformity and centralization.

Diversity as a pattern of production, not merely of conservation, ensures pluralism and decentralization. It prevents the dichotomising of biological systems into "primitive" and "advanced", or of knowledge systems into "primitive" and "advanced". Like Gandhi challenged the false concepts of obsolescence and productivity in the production of textiles by his search for the spinning wheel, groups across the Third World are challenging the false concepts of obsolescence in agricultural production which necessarily generate non sustainability. They are searching for the diversity of seeds used by farmers over centuries and making them the basis of a futuristic, self-reliant, resilient and sustainable agriculture (Altieri 1991; Shiva 1991).

LITERATURE CITED

Altieri, M. 1991. Traditional farming in Latin America. Ecologist 21: 93–96.

Fowler, C., E. Lachkovics, P. Mooney and H. Shand. 1988. The laws of life: Another development and the New Biotechnologies. Development Dialogue 1988 (1–2): 1–350.

George, S. 1985. *Operation Flood.* Oxford University Press, New Delhi.

Government of Kerala. 1984. Report of High Level Committee on Land and Water Resources. Trivandrum, India.

Kloppenburg, J. 1988. *First the Seed.* Cambridge University Press, Cambridge.

Kuhn, T. 1972. *The Structure of Scientific Revolution.* University of Chicago Press, Chicago.

Leon, B. 1975. Agriculture: A sacred cow. Environment 17: 38.

McNeely, J. A., K. R. Miller, W. V. Reid, R. A. Mittermeier and T. B. Werner. 1990. *Conserving the World's Biological Diversity.* IUCN, Gland, Switzerland.

Myers, N. (ed) 1985. *The Gaia Atlas of Planet Management.* PAN Books, London.

Pearse, A. 1980. *Seeds of Plenty, Seeds of Want.* UNRISD, Geneva.

Pyarelal. 1959. *Towards New Horizons.* Navjivan Press, Ahmedabad.

Shiva, V. 1988. *Staying Alive: Women, Ecology and Development.* Kali for Women, New Delhi and Zed Books, London.

Shiva, V. 1991. *Violence of the Green Revolution.* Third World Network, Penang and Zed Books, London.

Shiva, V. and J. Bandyopadhyay. 1987. *Ecological Audit of Eucalyptus Cultivation.* Research Foundation for Science, Technology and Natural Resource Policy, Dehra Dun.

Shiva, V., J. Bandyopadhyay, P. Hegde, B. V. Krishnamurthy, J. Kurien, G. Narendranath, V. Ramprasad and S. T. S. Reddy. 1991. *Ecology and the Politics of Survival: Conflicts Over Natural Resources in India.* UNU, Tokyo and SAGE, New Delhi/Newbury Park, London.

Yegna Iyengar, A. K. 1944. *Field Crops of India.* BAPPCO, Bangalore (reprinted 1980).

SECTION 6

Biodiversity and sustainable development

This section discusses the constraints upon long-term conservation of biodiversity by political, social and ideological factors.

Agarwal (Chapter 19) reviews the unique biodiversity of the Indian subcontinent and outlines India's work of biodiversity conservation. He demonstrates how the central political authorities, partly through political structures and thinking inherited from the British, often act to protect biodiversity in ways that exclude the local communities. He argues that one precondition for any long-term conservation is that the local communities assume the management responsibilities for the resources in their natural environment. Similar conclusions are drawn by other authors in this volume, e.g. Janzen (Chapter 2) and Shiva (Chapter 18). Agarwal also indicates how large scale economic activities, e.g. mining, hydropower developments, and forestry, destroy biodiversity and the livelihood of local people. Agarwal brings two main messages. First, the local communities must be given control over their lands, to reestablish traditional management structures. Second, a global policy of fair distribution of the world's economical opportunities must be developed to enable the developing countries to run a policy for long-term conservation of biodiversity. As long as the developing countries are kept at an economic level where they are forced to choose the short term solutions just to avoid disaster, there is no hope for a long-term conservation of living natural resources.

Næss (Chapter 20) discusses the conflict between the consumption of resources caused by the material standard of living in the rich countries and the conservation of global natural resources. Assuming that we all, as humans, have a special ethical obligation towards other human beings, the fulfilment of the vital needs of humanity requires some priority. Reducing the gap in the material standard of living between the rich and the poor countries will most probably necessitate a reduced consumption in the rich countries. Næss argues that no countries, and least of all the industrialized countries, are developed, in the sense that they are at a globally sustainable consumption level.

Næss also discusses the intrinsic value

of non-human life as a political premise for protection, referring to the report of the World Commission on Environment and Development. He argues that the intrinsic value of life forms should be accepted as a basic truth, just as we accept, without opposition, that two plus two is four.

19 Sociological and political constraints to biodiversity conservation: a case study from India

Anil Agarwal
Centre for Science and Environment,
807, Vishal Bhawan, 95 Nehru Place,
F-6, Kailash Colony,
New Delhi 110048,
India

ABSTRACT

Biodiversity in the Indian subcontinent is threatened by modern economic programmes. Numerous projects and schemes have been launched by the Indian government to protect the country's biodiversity. These programmes are being opposed by both the country's poor and promoters of economic schemes. Failure to take into account the needs of the poor is leading to tensions around protected areas. Programmes to protect biodiversity will succeed only if programmes can be launched to help the poor to regenerate degraded lands to meet their subsistence needs. International programmes to protect biodiversity which do not simultaneously address the problem of poverty will be viewed with suspicion in the Third World.

INTRODUCTION

The Indian subcontinent is one of the most fascinating ecological and geographic regions in the world. Within the confines of this one country exists the cold arid desert of Ladakh, the hot arid desert of Rajasthan, the world's rainiest areas of Meghalaya and Kerala, the high mountain region of the Himalaya, and the deep, alluvial flood plains of the Indus and Ganga valleys. The ecological diversity sustains a tremendous biological diversity. India lies at the confluences of the Afrotropical, Palaeoarctic and Indomalayan biogeographical regions. It possesses interesting components from each of these realms as well as numerous endemic species of flora and fauna. With just a fortieth of the world's land area, India possesses around a twentieth of the known living organisms (Gadgil 1982).

The total Indian flora consists of about 45,000 species. The known flowering trees and plants of India make up about 15,000 species, but several areas of the country remain unexplored or underexplored. About 7,000 endemic plant species have been found. The Himalaya and the Khasi Hills in Meghalaya account for about 3,000 endemic species and the Deccan peninsula

Conservation of Biodiversity for Sustainable Development.
Edited by O. T. Sandlund, K. Hindar and A. H. D. Brown. 1992. Scandinavian University Press, Oslo.

for about 2,000 species (Jain and Sastry 1980). The area of maximum diversity in India is the northeastern region, followed by the lower region of the Western Ghats around the Nilgiri Hills. India has a rich diversity of animal life, including that of domesticated animals like cattle, buffaloes, goats, pigs, sheep, donkeys and poultry (Bhat et al. 1981).

India is known especially for its rich genetic wealth of bamboo, fruit trees (mango, banana and citrus), vegetables like eggplant, okra and cucurbits (melons and gourds), grain legumes (mung beans, black gram and cowpeas), rice, pepper, cardamom, ginger, turmeric, sugarcane and jute. There is also a rich diversity in wheat, maize, millets, oilseeds and various spices, fodder and plantation crops. At least 2,500 plants are used for medical purposes. About 2,000 species of forage legumes and grasses occur, of which two-thirds are grasses. It is estimated that about 250 Indian species of wild plants have agricultural and horticultural importance and of these about 60 are rare or endemic (Mehra and Arora 1982). Rice has attracted the maximum interest amongst Indian scientists working with plant genetic resources. The areas with maximum diversity of rice are the northeastern tracts and the hilly regions of Orissa, Bihar, Madhya Pradesh and Andhra Pradesh, which are mostly inhabited by tribal populations. Numerous local rice varieties have also been found in the Malabar region and in the cold Himalayan tracts (Mehra and Arora 1982; IRRI/IBPGR 1983).

THREATS TO BIODIVERSITY

Human beings have coexisted with plants and animals on the Indian subcontinent for at least 50,000 years. For as much as 90 % of this time they were hunters and gatherers, influencing life around them only mildly. The impact of human activity increased dramatically when people began to cultivate plants and maintain herds of domestic animals, beginning with the river valleys of northwestern India some 5,000 to 8,000 years ago. Agricultural civilization spread gradually over much of India and, in the process, forests disappeared over the vast Indo-Gangetic plains and other parts of the country. The hilly and dry tracts, however, retained a good forest cover with tribal populations living in them depending on a hunting-gathering way of life, which they continue to do until today. With such natural habitats scattered widely over the land, a fairly representative sample of India's plant and animal life remained intact until the British times. The consolidation of the British Raj however, began a process of decimation of these habitats.

The hunting activities of the British wiped out large populations of animals. The British also took over, without compensation, vast tracts of communally managed forests throughout the country – an act of robbery surpassed only by the usurping of entire continents like North America and Australia by white settlers (Gadgil 1982).

The post-independence period since 1947 has seen major technological interventions to speed up the process of economic development which have further diminished the natural wealth of India. With the growing need for communications, roads were built into the interior areas of India, including those hilly and dry tracts where the natural areas still remained. With the opening up of these areas, the forests were ruthlessly overexploited. Like the British, post-independence governments in India have continued to claim ownership of forests and exploit them heavily for industrial and urban growth. Wood-based industries like rayon, paper and plywood mills have been a major cause of deforestation in India under the auspices of government-sponsored logging programmes.

India is a federal country with numerous state governments. To encourage industrialization in their own states, these governments sought to attract industries and, in the case of **wood-based industries**, encouraged the state forest departments to enter into contract with these industries to ensure timber and pulpwood supplies at very low rates. The low prices left no incentive to harvest the resource in a sustainable manner even though that was always the stated policy of forest management. Developments in commercial forestry have not only wiped out large areas but also put immense pressure on specific trees which have acquired an economic value.

River dams have emerged as a major threat to the rich forest areas in the hilly tracts. Numerous dams have been proposed in the Himalaya, Western Ghats and the Central Highlands. These are the very regions where most of the forested areas of the country now exist. Some of the remaining forests in the deep and remote valleys are also genetically very rich. The Silent Valley forest reserve in Kerala which was going to be inundated by a hydroelectric project, has been an important source of genes with pest resistant qualities when crossbred with domestic plants. The campaign to prevent the Silent Valley from being inundated was one of the earliest and most successful environmental campaigns in India (IRRI/IBPGR 1983). The ongoing dam building activities pose a serious threat to the country's remaining biodiversity (Agarwal and Narain 1990).

Mining activities pose yet another serious threat. Since independence, India has witnessed a phenomenal growth in mineral production. The geographical distribution of fossil fuels and metallic mineral reserves is highly uneven. About 40 contiguous districts in central and eastern India account for more than half of all the national mining output. Unfortunately, these very districts are extremely rich in forests and support large tribal populations (Ghotge and Chopra 1985).

Crop and livestock development programmes have posed a serious threat to the diversity in domesticated plants and animals. About 50,000 local rice varieties are estimated to be still in cultivation but they are being rapidly replaced by new high yielding varieties, particularly in the irrigated areas (IRRI/IBPGR 1983). Growing deforestation also poses a serious threat to many wild varieties of rice.

Animal crossbreeding programmes are posing a serious threat to India's natural animal wealth. Some of the finest milk breeds of cattle in India are facing constant genetic erosion because of indiscriminate crossbreeding and neglect. Mass artificial insemination and import of exotic stock for cross breeding have now made it impossible even to locate genetically superior animals of Sahiwal, Red Sindhi and Tharparker breeds in India. Urban centres attract cows and buffaloes with the highest milk yield from the rural areas. But from there they go to slaughterhouses after completion of lactation. Thus, high producing animals (genotypes) are being regularly removed from breeding stocks, leaving inferior genotypes for breeding. The genetic quality of the Murrah breed of buffaloes has deteriorated (Bhat et al. 1981; Bhat 1981).

India has a rich variety of fish fauna though only a few species have ever been cultivated. The growing pollution of India's rivers, streams and lakes poses a serious threat to the country's fish populations. The dams and canals that are being built to harness the country's water resources, apart from preventing migration of fish, have serious implictions for fish genetic resources. They permit intermixing and breeding of different strains and species which hitherto enjoyed isolation and preservation due to geographical and ecological boundaries.

Together with all the above pressures exerted by modern economic development programmes, there are additional pressures being exerted by human populations through subsistence activities, the most common of which are **grazing**, **firewood collection** and **shifting agriculture**. Grazing is common in the forests of dry and sub-humid regions and plays a major role in suppressing the regeneration of ground vegetation. Firewood collection is not a major cause of deforestation in sub-humid and humid areas unless it is undertaken through organized logging operations to meet urban needs (Agarwal and Bhatt 1985). In arid and semi-arid areas, however, the remaining tree populations are under intense pressure to meet firewood and fodder needs (Palmberg 1981; Armitage et al. 1980).

Shifting agriculture is not a common practice in India and is restricted mainly to the northeastern region and certain districts of Orissa, Madhya Pradesh and Andhra Pradesh. Unfortunately, these areas also contain much of the country's remaining forest wealth. Shifting agriculture, therefore, puts a certain pressure on the country's biodiversity in specific areas.

NATIONAL BIODIVERSITY RELATED PROGRAMMES

Indian efforts to protect biodiversity from further erosion can be classified into three categories:

1. collection and preservation of plant and animal germplasm through *in situ* and *ex situ* programmes;
2. evaluation and control of modern economic programmes like road and dam building to prevent further deforestation and erosion of genetic resources; and,
3. optimal utilization of the country's land and water resources and regeneration of degraded areas to meet the basic biomass needs of the people for food, fuel, fodder and timber so that the pressure on remaining areas of rich natural wealth can be reduced.

Gene banks and protected areas

To collect and store germplasm, the government has set up separate national Bureaus of Plant, Animal and Fish Genetic Resources under the aegis of the Indian Council of Agricultural Research. These institutions are recent and still have to take on full-fledged operations. However, there are already several gene banks in the country specializing in crop germplasm conservation.

An extensive system of protected areas has also been established. From a modest number of 65 national parks and sanctuaries in 1960, they grew to 65 national parks and 380 wildlife sanctuaries in early 1989 (Kothari et al. 1989). An expert committee set up by the Indian Board of Wildlife has recommended that at least four per cent of the total land area of the country should be under nature reserves. At the moment, about three per cent of the country's land is protected.

The establishment of these protected areas has been at an enormous human and social cost. Numerous villages have been displaced and many more stand to be displaced as more wildlife sanctuaries get converted into national parks which have stricter regulations. No official figures have been published on the number of people displaced by protected areas and subsequently resettled elsewhere. The living conditions of people displaced by protected areas are abysmal. They live in constant fear of animals and remain deprived of basic facilities like communications, access to medical care and education. Conservationists often lament the fate of those displaced by dams, but totally neglect the fate of those displaced by proteced areas (Fernandes et al. 1989).

Since there is an acute biomass scarcity in areas around wildlife sancturies and national parks, local people look upon them as biomass reserves which can meet their basic needs, especially during distress periods like drought years. Since the park authorities want to keep the people out, this leads to tensions, and, in a few cases, police have had to fire upon protesting crowds. The park authorities do not have a clear policy for culling the growing populations of animals inside the protected areas. These animals often raid the surrounding fields of the farmers. Human killings by tigers around tiger reserves have been another major problem. The local people see these protected areas as a clear symbol of an anti-people government which wants to throw the poor out and open up nature's bounties to middle and upper class tourists. As things stand, Indian conservationists have not developed joint management systems for protected areas in which local populations living inside or on the fringe of the protected areas can participate.

Environmental impact assessment (EIA) and forest conservation

The Ministry of Environment and Forests has developed an elaborate structure for evaluating the environmental impact of development projects. Though there is no law on environmental impact assessment, public sector authorities have to submit all major projects for assessment by the Ministry, under an executive order of the government. Since the government is itself the major development agency in the country, the Ministry of Environment's EIA system is able to cover a major sector of the Indian economy. Major private sector projects need a clearance from the pollution control board of their respective state before they can get an industrial licence. The existing procedures are, however, weak with respect to the private sector and the small-scale industrial sector.

No projects are permitted inside sanctuaries and national parks. Moreover, the Indian Board of Wildlife has set out rules under which no major industrial project can be set up within 20 km of thc perimeter of a national park or wildlife sanctuary, so that population pressures do not build up around them. In several states which have a large number of protected areas, these rules preclude a large area from industrial development. Numerous projects under construction have had to be abandoned.

In 1980, the Central Government passed the Forest (Conservation) Act, which stipulated that no forest land or any portion of it, may be used for non-forest purposes without the permission of the Central government. With developmental activities growing, there is growing pressure on forest authorities to divert forest land for construction and siting of dams, roads and industrial projects. The act applies even to those lands which are designated as forest land but may be without any tree cover at the moment. The Forest (Conservation) Act tries to make room for the expected increase in forest cover as India begins to invest in afforestation.

Both the EIA procedures and the Forest (Conservation) Act have been strongly opposed by development lobbies and their implementation has demanded considerable political will. As pointed out earlier, a large part of the country's mining output today comes from those areas which have a good forest cover. As a result, numerous mining projects have been refused or delayed. Dam builders also have to wait for a long time to get clearances. This has caused inter-ministerial tensions within the central government and a worsening of Centre-State relations. Since forests are owned by the state governments, they see tl Forest (Conservation) Act as an intrusion into their Constitutional rights.

It is obvious that in such a situation, non-governmental organizations play an important role as watchdog bodies to ensure that internal pressures within the government do not permit environmental harmful projects to be cleared. It is also vital that conservationists do not overstate conservation needs which would lead to unnecessary tensions with development agencies on one hand and local populations living around protected areas on the other.

In order to conserve the country's forest wealth, import of pulp and timber for paper and plywood factories has been liberalized by reducing import duties. Unfortunately, the import prices of these commodities are now so low that wood prices have gone down considerably and farm forestry, which had picked up in the early 1980s, has been on a decline.

The central government has been urging all State Governments to ban logging completely. But several states are reluctant to do so because they earn substantial revenue from timber sales. There are proposals to compensate various forest-rich state governments for the loss of revenue they would incur through a ban on logging but no definite action has yet been taken.

Ecological restoration of degraded lands

Between one-third to one-half of India's land is estimated to be **wastelands** which are producing far below their optimum biological potential. The restoration of these lands can yield large quantities of fuelwood and fodder for rural needs. In 1985, the government of India set up a National Wastelands Development Board to undertake a massive afforestation programme. Between 1980 and 1987, the government claims to have afforested 7.9 million hectares of land.

Government afforestation programmes have been criticized for exaggerated claims and promotion of exotic species on private farmlands. These are meant more for urban and commercial timber needs than local needs of fuelwood and fodder. Moreover, trees have mainly been planted on private farmlands in the fertile plains, whereas afforestation is more urgently needed on community and government lands in the hill and dry areas.

During the 1980s numerous small experiments were carried out both by government agencies and non-governmental organizations for environmental regeneration. Several of these projects show that environmental regeneration is possible and it can greatly improve and stabilize the subsistence economy of the Indian villages. Cost-benefit ratios on investments are not particularly high. These projects also show that environmental regeneration is possible only if there is community control over the local environment and village democracy is strengthened.

India is a country with a high population density and a high dependence on biomass sources, including non-private biomass sources obtained on common lands for meeting basic needs of the people, especially the poor. Indian farmers practice an integrated system of farming and animal husbandry to diversify their occupational base and reduce the economic impact of rainfall variations. As a result, they have transformed their local environment into a complex agrosylvopastoral ecosystem with interacting components of croplands, grazing lands and forest and treelands. Traditionally, croplands have been private property while grazing lands and treelands have been common property. When the British colonialists nationalized the grazing and forest lands, they converted these **managed common property resources** into **free access common property resources**, which steadily led to their degradation.

Because of the growing biomass scarcity

and growing population density, inter-settlement tensions appear to be growing over the use of common property resources and there is now a strong desire to convert free access common property resources once again into community managed common property resources. The experiments of the 1980s bear testimony to the fact that village communities are desirous of taking control of the common resources.

The major obstacle to community management of grazing and forest lands is the prevalence of laws, inherited from the British, which put the control of grazing lands, forest lands and most of India's surface water systems into the hands of government agencies. The bureaucracy left behind by the British, which independent India's political leaders failed to change, has consistently fought against changes in these laws. But to promote community governance, these laws will have to be changed. Moreover, open and democratic village institutions will have to be created with adequate legal powers to manage the community resources. Though community lands can one day become a major source of revenue for village institutions, Indian village communities are today not in a financial position to invest in the restoration of the community lands. A regular flow of funds will, therefore, have to be provided to community institutions to help them plan and undertake ecological restoration within the village agroecosystem.

Numerous programmes have been launched by the Indian government to create employment opportunities in rural areas and ensure a basic subsistence wage to the country's rural poor. These programmes can be a major source of funds to mobilize labour for ecological restoration, most of which consists of labour-intensive activities. Current thinking in India is to enact a legally guaranteed Right to Work, which can then be used, amongst other things, to mobilize labour on a massive scale for undertaking ecological restoration activities like afforestation and soil and water conservation. It is estimated that such a programme would require about seven billion US dollars a year. If employment generation and ecological regeneration can be linked together, and village institutions created together with laws that strengthen village decision making, it should not be difficult to restore India's vast wastelands to their optimal productivity (for a detailed exposition of this thesis, see Agarwal and Narain 1989). In the process, the country will be able to reduce pressure on areas of rich biodiversity.

SOUTH-NORTH ASPECTS OF BIODIVERSITY CONSERVATION

It has been argued that biodiversity is a global resource. This argument is almost tantamount to colonial intervention. The success of the Western environmental movement over global environmental issues like the atmosphere, the ozone layer and biodiversity is forcing the world to think more in terms of global solidarity and the need to take globally beneficial decisions. But this movement is viewed with great suspicion in the Third World. The Western economies want to take global decisions which will have serious implication for Third World development but are not prepared to alleviate the crushing poverty and debt burden of the developing countries or improve financial flows so that poor nations have the financial space to take more appropriate development decisions. Global rules are being set, global discipline is being demanded, but global sharing is not on the global agenda.

The world's environmental problems can be divided into two categories: (1) those whose alleviation demands global management, and (2) those whose alleviation demands community management. If both these sets of problems are to be solved, the nation state, in the years to come, will have to give up certain aspects of governance to global systems and other aspects of governance of community systems. In other words, increasingly, the world will simultaneously have to develop concepts of global governance and global citizenship, on one hand, and of community governance and community citizenship, on the other.

The concept of citizenship embodies within itself not only certain societal obligations which have to be fulfilled but also a specific body of rights. In all democratic societies, there is a specific framework of rights within which citizens are expected to discharge their obligations. Unfortunately, as a global society we remain fragmented where a free-for-all prevails and each nation is left largely to its own wherewithal to meet its needs.

Morality and the need for equality apart, this global free-for-all is becoming extremely incongruous with the world economy which is becoming increasingly interdependent and integrated. With the bastions of Eastern Europe falling, this process of global economic integration will be even more rapid in the years to come. Biomass products like tea, timber, bananas, coffee, chocolate, pineapples and peanuts are produced on the land of the developing world, but to a large extent consumed in the rich economies of the North. Developing countries are not poor because they do not work or because they are lazy, but because the terms of trade for their products, especially biomass products, have been consistently and rapidly deteriorating. Western consumers have not yet shown the political will to pay the true cost of producing these products. The result is that poverty sustains itself within the developing world, pushing hundreds and thousands to **scrape the earth** to a point that it yields nothing, and then move on to the last frontier – the slums of towns and cities. Within such a global system, it would be impossible to eradicate poverty or reverse ecological degradation.

All over the world, there is growing consciousness about environmental economics and the need to incorporate ecological costs of production into national income and wealth accounts. But what is the point of doing this in a developing country if the rich and powerful consumers of the world are not prepared to pay the true cost of their consumption? At the moment biomass products are too cheaply valued. The farmers of the developing world neither get a fair remuneration for their labour nor does the land get an adequate rent which would pay for the ecological costs of growing crops.

Global environmental concerns cannot be chosen on an *ad hoc* basis by the rich and powerful political actors in the world. If issues like biodiversity, climatic change and depletion of the ozone layer are to be put on the global agenda, to benefit future generations, it is equally important to put on the global agenda the problems of land and water degradation which are discounting the future of both present and future generations in the Third World. A global action programme based on a set of global obligations and rights ought to be developed which will help us to move towards a better and more fair global management of our environment.

Just before she died, Barbara Ward, one of the founders of the Western environmental movement, suggested an international system of income tax so that wealth can be automatically transferred as a matter of right and not just as aid and charity. In all civilized and democratic countries of the world, progressive taxation to even out inequalities within the society is an accepted norm. In the international arena, she lamented, we are not even prepared to discuss the need for this form of civilized behaviour, and aid remains a matter of *ad hoc* choice and charity. This international income tax should not be seen as a matter of charity, but a statutory obligation to pay, if nothing else, the ecological costs of our consumption, and it should be imposed on all rich consumers of the world, both in developed and developing countries.

The funds so raised could then be used to financc an international Right to Work so that any one living anywhere in the world can be assured a basic minimum subsistence wage. Instead of being forced to scrape the earth, the labour of the world's poor can be mobilized to restore the productivity of the earth, giving them the financial space to wait for returns that will come five to ten years later. If this step, which would ensure minimum social justice in the world, could be accompanied with the establishment of community governance systems, then a system of basic grassroots democracy, so vital for good environmental management, would also have been created. Within such a framework, it would probably be fair to demand that Third World biodiversity be treated as a global resource.

CONCLUSIONS

The Indian subcontinent is rich in biodiversity. The region's animal and plant wealth has suffered erosion through human activities for a long time. Over the last few hundred years, there have been massive impacts because of the economic needs of the region's colonisers. Demands for economic development after independence have further increased these impacts. The large majority of the population, however, continues to live within a biomass based subsistence economy. Plants and animals provide a large part of the basic requirements for food, energy and shelter. The pressure on the remaining biomass stock is therefore high.

Programmes to protect areas rich in biodiversity have led to tensions with economic development proponents, on one hand, and the poor, on the other. Tensions are common on the fringes of protected areas. Conservation groups play an important role in supporting the government to deal with these pressures by building up public opinion in favour of protection. But as it is immoral to deprive the poor of their limited sources of survival, it is vital that local communities be involved in the management of the protected areas and derive benefits from the resources offered by them. Simultaneously, major efforts have to be made to regenerate the ecologically degraded lands so that local communities can meet their biomass needs from them and reduce exploitative pressure on protected areas. Experience shows that such regeneration programmes succeed only when village democracy is strengthened and local communities gain control over their natural resource base. International environmental programmes must be developed to support employment generation for ecological regeneration. Unless programmes are developed that aim to reduce the levels of poverty that today exist in the Third World, the efforts to conserve biodiversity in the Third World will be viewed with suspicion.

LITERATURE CITED

Agarwal, A. and B. Bhatt. 1985. Cooking energy: Firewood in the cities, pp. 267–276. In: A. Agarwal and S. Narain (eds) *The State of India's Environment 1984–85: The Second Citizen's Report*. Centre for Science and Environment, New Delhi.

Agarwal, A. and S. Narain. 1989. *Toward Green Villages: A Strategy for Environmentally Sound and Participatory Rural Development*. Centre for Science and Environment, New Delhi.

Agarwal, A. and S. Narain. 1990. Are dams the answer? The Hindustan Times, March 18, 1990. New Delhi, India.

Armitage, F. B., P. A. Joustra and B. Ben Salme. 1980. *Genetic Resources of Tree Species in Arid and Semi-Arid Areas*. FAO, Rome.

Bhat, P. N. 1981. Conservation of animal genetic resources in India, pp. 86–95. In: *Animal Genetic Resources: Conservation and Management*. FAO Animal Production and Health Paper 23. FAO, Rome.

Bhat, P. N., P. P. Bhat, B. U. Khan, O. B. Goswami and B. Singh. 1981. *Animal Genetic Resources in India*. Indian Veterinary Research Institute, Izatnagar.

Fernandes, W., J. C. Das and S. Rao. 1989. Displacement and rehabilitation: an estimate of extent and prospects, pp. 62–90. In: W. Fernandes and E. G. Thurkral (eds) *Development, Displacement and Rehabilitation*. Indian Social Institute, New Delhi.

Gadgil, M. 1982. Wildlife, pp. 165–173. In: A. Agarwal, R. Chopra and K. Sharma (eds) *The State of India's Environment 1982: A Citizen's Report*. Centre for Science and Environment, New Delhi.

Ghotge, S. and R. Chopra. 1985. Mining, pp. 19–26. In: A. Agarwal and S. Narain (eds) *The State of India's Environment 1984–85: The Second Citizen's Report*. Centre for Science and Environment, New Delhi.

IRRI/IBPGR. 1983. *1983 Rice Germplasm Conservation Workshop*. International Rice Research Institute, Los Banos.

Jain, S. K. and A. R. K. Sastry. 1980. *Threatened Plants of India: A State of the Art Report*. Department of Science and Technology, New Delhi.

Kothari, A., P. Pande, S. Singh and D. Variava. 1989. *Management of National Parks and Sanctuaries in India: A Status Report*. Indian Institute of Public Administration, New Delhi.

Mehra, K. L. and R. K. Arora. 1982. *Plant Genetic Resources of India: Their Diversity and Conservation*. National Bureau of Plant Genetic Resources, New Delhi.

Palmberg, C. 1981. A vital fuelwood gene pool is in danger. Unasylva 33: 133.

20 Sustainability! The integral approach

Arne Næss
Center for Development and Environment,
Division for Environmental Studies,
P. O. Box 1116, Blindern,
N-0317 Oslo 3,
Norway

ABSTRACT

Whatever our job we need to integrate life theory and life practice, clarify our value priorities, distinguish life quality from mere standard of life, and contribute in our own way to diminish unsustainability. In the rich countries this requires taking the ethical norm of universalizability serious: we should not live on a material level we cannot seriously wish others to reach. Ecological sustainability implies long range sustainability of the richness and diversity of life forms on Earth, not only conservation of resources for humans. The life forms have intrinsic value. Why? Don't bother giving many reasons, we must all start with something we trust, whether we are mathematicians or artists. Acceptable development requires ecological sustainability and economic progress in rich as well as poor countries, but not necessarily any economic growth measured by gross national product. Slow human population decrease will in the remote future serve human life – including their economy – and serve non-human life in every way.

RICH LIFE, SIMPLE IN MEANS!

The Greek philosopher Diogenes of Sinope, or, Diogenes in the Barrel, is well known because of his meeting with Alexander the Great. When Alexander, sitting in front of his barrel, or pitcher, asked him to express a wish, he is said simply to have answered, "Please get out of the sun!". Alexander could have granted him more. He could have received gifts produced through great expenditures of energy and natural resources. Diogenes was active in the rich cultural life of Athens, but that did not require any such gifts. Diogenes' solution to his housing problem is a non-verbal expression of globally, regionally and locally sustainable lifestyle, and his answer is a verbal suggestion of the same. Alexander was reminded of the ecologically innocent character of vital needs, and, not to forget the proverbial wit and joyful, spontaneous character of Diogenes, he got a lesson in "rich life, simple in means". It is sad that most followers of Diogenes through the centuries misunderstood him, taking him to

Conservation of Biodiversity for Sustainable Development.
Edited by O. T. Sandlund, K. Hindar and A. H. D. Brown. 1992. Scandinavian University Press, Oslo.

be a proponent of a simple, rather than a rich, life.

The reason I mention Diogenes of Sinope is to remind you that classical Greek philosophy as well as philosophy in the Middle and Far East combined the verbal and non-verbal. You were not a philosopher if you did not do so. Today we say about somebody that he or she is a philosopher and use it for such a combination. But the intended range of philosophies seems small. People have mostly what the professionals call "popular Stoicism" or "popular Epicureanism" in mind. The very special form of Western academic philosophy does not require a philosopher. No combination of theory and practice is called for. If it were, some of you would perhaps be called "sustainers", practising philosophy combining theory of biodiversity and sustainable development with a lifestyle which is, or at least intended to be, appropriate.

There is much talk about a "new" ethics of respect for life on Earth, but it cannot focus on the lifestyle of those who emphatically agree. One strong reason for this neglect is the plain fact that the complication and structural density of modern society make it practically impossible to combine the consistent use of simple means with full participation in social and political struggles.

Depressing? Yes, but we may allow ourselves to say that Diogenes in the Barrel went too far. An Indian practising philosopher perhaps also carried the concern for citizen's self-sufficiency and equi-mindedness too far, when he, old and satisfied with life ("enough is enough"), brought together some wood, started a fire, placed himself on top, and cremated himself (in conformity with a practice that unfortunately caused deforestation in certain localities). Yoga of some sort is part of most classical Indian philosophies. They integrated theory and practice. We may all try the same even if less spectacular.

SUSTAINABILITY OF RICHNESS AND DIVERSITY OF LIFE ON EARTH

Discussing various interpretations of the term "sustainable development" we may start with the following formulation: There is sustainable development if and only if it ensures that it meets the vital needs of the present human population without compromising the ability of future generations to meet their own vital needs (Bakke 1988). This formulation resembles those of the World Commission for Development and Environment (WCED) in the report *Our Common Future*, also known as the Brundtland Report (WCED 1987). However, there is a major difference: I have substituted "vital needs" for just "needs". Lists of needs made by decision makers of the rich countries include items which clearly are exorbitant and ununiversalizable. Even the **desire and demand** for more parking space is generally talked about in terms of satisfying a **need**. Those parking spaces are often placed in cities where children have had their precious playing grounds, or where people have lived for a long time but cannot afford a car, and among those who are incapable of resistance against destruction of their habitat.

There are fortunately possibilities to decrease unsustainability and ununiversalizability of the "average" lifestyles in economically rich countries without decrease of life quality (Næss 1989). To exemplify such a decrease is a great aim for someone who wish to be taken seriously in the present struggles.

Needs are accepted in the rich countries which, if met universally, exclude sustainable development in every acceptable sense. The term "vital" is vague and ambiguous, but it is a good starting point for a critical approach to the term "need" in its relation to "demand" used by market theorists in the rich countries of the world.

Postulating that we, as humans, have a special ethical obligation toward other human beings, the vital needs of humanity require some kind of priority. But this does not mean that we do not have serious obligations toward non-human beings, and the classes and systems of such beings, culminating with obligations relative to the Earth as a whole. However, these can be perceived as mainly negative obligations related to the destructive consequences of our exponentially increasing interferences in the ecosystems.

As to the satisfaction of human vital needs, there is at present a substantial number of humans living in a desperate state of poverty or oppression clearly preventing a minimum satisfaction of those needs. They live mostly in countries one may call poor, using an economic measuring rod, where the term "sustainable development" is referring basically to sustainable economic progress, not necessarily to economic growth. The implied task is gigantic considering that soon the population in the poor countries will double while the area of cultivable land will not. Under similar difficult circumstances long ago Europeans migrated to North and South America introducing new vast unsustainability and a lamentable decrease of cultural diversity. What if Columbus had landed in 1992, not in 1492?

AN ECOLOGIST'S WISH: MAY ALL COUNTRIES BE DEVELOPING COUNTRIES!

It marks a major victory for the global ecological movement that the WCED announces clearly that sustainable development unconditionally requires ecological sustainability. It is a necessary but not sufficient condition. The consequences of this admission are wide-reaching because ecological sustainability requires significant economic, technological, social, political and cultural changes in most or all countries. Here I shall first dwell on a terminological consequence: The term "developed country" implies automatically "ecologically developed country". We should now ask: "Are there any?"

If we retain the underdeveloped/developed terminology we should therefore class all ecologically unsustainable countries as underdeveloped without having to add the adverb "ecologically". It is implied by our terminology that these countries are underdeveloped. The richest industrial country is not a developed country if it is not in a process of ecologically sustainable development.

A revised terminology has already been introduced in Norway. We ask: "When will Norway rise from being underdeveloped to a developing country?" That is, when will the rate of unsustainability (globally measured) decrease in a stable way? My guess is that by the year 2100 this may well happen. As it is now, Norway, like the US, has not turned the tide of pollution, energy consumption, and other variables which must be taken into account. There are not even plans to do so. The population would vote against them.

Those who think this view is rather pessimistic are likely to neglect the necessary wide perspective: It is the global situation that counts, and we must be aware that in the years 2000, 2010, 2020,..., with even bigger population and economic growth measured in terms of GNP in the Second and the Third World, the projected level of Norwegian pollution, growing energy consumption, etc. is intolerable. Talking about 2100, I have assumed that political ability to take ecological problems seriously is much higher than today, but not in a revolutionary way. That would require a considerable increase of active interest within the Norwegian population at large. Many more sustainers!

One may hold that it is wiser to modify the old usage. My only point in that regard is that there is ample ground, when we encounter its old use in discussions, to ask what the terms are meant to express, and then, to strongly discourage any use that might support the old belief that the rich industrial nations are developed and somehow able to show the poor countries how to develop. I am afraid this belief is still not uncommon within the power elites of the Second World and Third World.

THE INTRINSIC VALUE OF LIFE AS A POLITICAL PREMISE FOR PROTECTION

In the 1980s, it was out of the question for a wide group of politicians to declare that life on Earth, or life in the universe as a whole, has any value in itself apart from the rather narrow serving of human needs. In the sense that we feel it obvious to care for children for their own sake, the richness and diversity of life is also worthy of being cared for for its own sake. Such thoughts were not incorporated in the report *Our Common Future* (WCED 1987). By now, perhaps only one country, New Zealand, can show a public document which affirms intrinsic value. If, within a few years, such affirmation gets to be commonplace, it will not automatically make much difference in human practice, but it will **add to the force of argumentation** in favor of generous, wide sustainability. "You **said** it has intrinsic value! Why don't you act accordingly?"

The formidable capacity of our brain makes it easy for us to "see ourselves in others", not only in other humans, but in every living being. Sometimes it is even easier "to identify" with certain animals than with certain humans. Compassion, aided by the brain, is something that encompasses not only everything capable of pain. The interest, in a wide, easily understandable sense, of a tiny plant to live and blossom is obvious, and under suitable circumstances, we act to serve this interest. The definition of sustainable development cannot ignore these facts. A wide and deep perspective is obviously needed. Human capacity to think and to feel, human dignity, requires it (Næss 1986).

Our contemplation of the development of life on Earth through countless millions of years, together with the development of the richness and diversity of life forms, almost inevitably make mature, informed humans adopt a wide perspective. In short, we demand that full ecological sustainability shall mean or include conservation of the richness and diversity of life forms. We cannot slavishly accept the narrow interpretations of the Brundtland Report, interpretations which most of the signatories, including Gro Harlem Brundtland herself, personally, I think, find too narrow.

Sustainable development

Let us look closely at the term "sustainable development" as it occurs on page 8 in *Our Common Future* (WCED 1987). It says: "Humanity has the ability to make development sustainable – to ensure that it meets the needs of the present without compromising the ability of future generations to meet their own needs". Plausibly, but narrowly interpreted, what must be considered sustainable is compatible with maximum destruction of life conditions on Earth, a maximum of extinction of life forms and habitats of life forms, a maximum of gross human interference with landscapes and ecosystems – **as long as** these maxima are believed nevertheless to permit satisfaction of human needs, as these are conceived at any definite time. And conceived by whom?

It is clear that many decision makers by "reforestation" do not imply to get back real forests. The artificial tree plantations with fast growing trees do not support the biodiversity of a forest. The number of species in the plantations may be one fourth, or less, of that of the original forest. If the goal is to distinguish the forest **itself** from the animal and plant life in it, the ignorance of ecosystem thinking is clear, and the road is open for a maximal destruction that only too late will be seen to be incompatible with the satisfaction of vital needs of additional billions of people born in the years to come.

In the present conflicts, the usual narrow interpretation of "sustainable development" is convenient for planners of gigantic destructive policies, because it is difficult to convince people that future generations will lack the ability to take care of themselves whatever we find it suitable to do. People read about technological inventions, even revolutions, which are said to push the limits of responsible growth indefinitely. They do not read about the lack of economic and political will to make global use of the inventions made the last 30 years.

Rejecting the narrow concepts of ecological sustainability, which at least one plausible interpretation of the Brundtland Report admits, the way is open for acceptance of wider interpretations. Preferable are such that are not completely implausible interpretations of the document. They should be in harmony with what at least some of the 22 people underwriting the document have had in mind. The criterion I am going to make use of in what follows do not explicitly refer to humans, but it is implied that human cultures are life forms: **There is ecological sustainability if and only if the richness and diversity of life forms are sustained.**

"Richness" does not only refer to abundance of specimens of species but to their wide distribution, provided that wide distribution is realized today, or was recently realized, and that it is practicable to restore the former situation.

The criterion is applicable to the Earth as a whole, to regions, nations, and societies, but only to some extent to localities. It is of little use to talk about ecological unsustainability in **small** areas of localities of monoculture due to city developments. More cities are ecologically needed to contain the growing populations. A region may be called sustainable even if there are plenty of such localities. Otherwise, sustainability would not be reached in the next century. On the other hand, there is not global sustainability if some (large) regions are not. (One may of course give the criterion more precise meanings, but I am not trying to do so here).

The term "development", however ambiguous, must be used because of its importance in policy documents, but I would like to present a terminology proposal: **A development is globally sustainable if and only if there is a long term trend that assures, or that may justifiably be considered to assure, global ecological sustainability, and also assures long range elimination of abject poverty.**

The special obligations we have as humans for our own species require us in the long run to assure a population size that is necessary to provide conditions for reaching the ultimate goals of humanity and to satisfy vital needs. Beyond that, our obligations to life in general and the Earth as a whole acquire priority.

Sustainability in the Third World

If some of the Third World countries reach ecological sustainability in the next century which one will be the first? Costa Rica? The educational level is high, standard of living is increasing, the government is interested, and a significant number of "parataxonomists" (raised in the country) and others help mapping out the fauna and flora which contributes to an increasing respect and joy of life. Furthermore, much is done to develop sustainable uses of tropical biodiversity, thus integrating concern for ecological sustainability in the society. The cooperation of researchers and local people is flourishing. The ecologist Daniel H. Janzen is the best known researcher working along "social ecological" lines and collecting millions of dollars in aid for projects in Costa Rica. These activities decrease the ecological unsustainability in certain ways, but large scale deforestations still go on. Some researchers guess that only about 10 % of the habitats will be saved from complete human domination including extensive regions of monocultures and asphalt. It is therefore a wide open question when sustainability will be reached, if at all.

One may wish that all other tropical countries developed in the auspicious way of Costa Rica. The corresponding amount of money that would have to be collected in the rich countries would not be in the range of millions, but billions. Benevolent bureaucracies would have to be available as well as an army of ecologists, with assistants, working closely together with the local populations. Their insight is indispensable.

EXISTENTIAL CONSEQUENCES FOR HUMAN BEHAVIOR

The great Danish philosopher, Søren Kierkegaard, the father of existentialism as a philosophical movement, insisted that humans are always in "deep water". Their decisions must be made on the basis of a total integrated view. This implies, in principle, to go back to ultimate premises and to a conception of the main goal of human life, whether pleasure, happiness or achievement of some sort. It implies also that if you consider a certain question to be of immediate relevance for actions in your life, your community of life on Earth, you must have an answer, whether expressed through deliberate words or through deeds. And we should remember that even if we do not answer deliberately, our actions express answers. In life one cannot say "leave **me** out!". Ignorance and incompetence furnish explanations, but not excuse – not automatically. The question is relevant: are we informed to an extent that should be expected of us?

Crudely expressed: If it is an important decision either to turn right or to turn left, to do neither also has important consequences. One answer may be "I am too tired to reflect, I'll go left!" This and similar kinds of decisions must of course be tolerated. The main thing is the awareness with equanimity, that a choice is made anyhow.

The practicing philosopher is one that feels obligated to answer, but who does not hereby pretend that it is worth while for others to listen. Perhaps all our answers are more or less imperfect.

Those who are seriously engaged somehow to contribute to decrease unsustainability locally, regionally or globally, may do it by a combination of specialized sort of job, like researchers, or as a generalist way of contribution, showing as much as explaining their choices in life. They are then to be classed as practicing philosophers whatever their degree of ignorance of academic philosophy. Sometimes this ignorance may be a plus.

In our very special kind of culture in the western rich countries, verbal articulations in the form of reasons are highly appreciated. Somebody may ask: "What is your reason for your value priority of A over B?" You may have a reason R_1, but then you are asked: "What is your reason for accepting R_1 as adequate?" Suppose you answer with an R_2 or you admit that you do not pretend to have a sort of reason R_2 such that if it were untenable you would give up the priority of A over B. That is, you stop the chain R_1, R_2,..., R_n, already after the first number of the series. Mostly such a behavior is wise, because it is likely that you sooner or later give misleading reasons, reasons which do not really fit your ultimate or complex motivation.

One of the many great achievements of Aristotle was a clear denial that you could prove everything you assert. You cannot even give reasons for everything you assert. You stop somewhere, normally outside science, and doing this, you may, if appropriate, quote Aristotle: Say that belief in the possibility of providing reasons for everything shows lack of education, and that you like to be considered educated. Some insights and intuitions you simply trust, but admit fallibility.

Most ardent supporters of a wide form of sustainability accept as evident many pronouncements that others find non-evident. They ask for reasons, for premisses from which the pronouncements can be derived logically, scientifically or by other forms of generally accepted reasoning. "Which are your **reasons**", they ask, and the answers, R_1-formulations, seem mostly unconvincing, for instance, begging the question. If R_2-formulations are offered, they may sometimes be unconvincing even for the formulator.

The important point here is the trivial onc that it cannot possibly be asked of you **never** to stop. My advice is to stop where you announce something you personally find intuitively obviously true or correct, or which you cannot imagine to give up except for reasons you never have heard of and cannot see convincing. This is not dogmatism. You are not worse off than mathematicians and logicians who again and again use the basic rules of inference which they do not try to prove or validate scientifically.

All this would be useless to say if we do not constantly see people who **unreasonably** feel guilty when they do not give reasons, or even scientists who hide and never announce their basic norms or evaluations because these lack "scientific" reasons. A great cultural loss.

Humans are never wholly "functionaries" – behaving wholly as functionaries – they are always, as specimens of adult, sane *Homo sapiens*, responsible as persons. Also the timid expressions "as I see it...", "in my personal opinion..." are misplaced if such ultimate views are at stake. You should not try to step outside yourself, try to be a **mere** witness to your own intuitions. But what about conflicting intuitions – are they not sometimes a cause of violence and war? Yes, but also sometimes cause of peace, progress and nonviolence of the most superb kind.

BIODIVERSITY!

Full global and regional biodiversity is necessary to reach full ecological sustainability. Full ecological sustainability is necessary to realize sustainable development. Biodiversity is required to satisfy the vital needs of humanity. This is now generally acknowledged. That the biodiversity of this planet should be protected also for its own sake was first internationally recognized through The United Nations Charter for Nature. The initiative to this charter was taken by a group of poor nations, and the General Assembly adopted it in 1982 by a vote of 111 to 1, USA casting the sole dissenting vote.

It serves the cause of biodiversity to maintain that it has a value in itself apart from narrowly conceived usefulness. But it also helps when people who maintain this testify to its profound implications through their lifestyle.

High level humanitarian norms justify ecologically negative policies. They are usually short range and could largely be avoided through cooperation by rich and poor nations on a greater scale than ever before.

Because of its touchy nature, I wish to end with a remark on the size of the human population as seen in a cultural philosophical perspective: A future long range, slow decrease of the human population would to some extent increase the chances of full biodiversity, sustainable development, deep cultural diversity, and the prospect of satisfying vital needs and reaching cultural and philosophical goals (Næss 1990).

ACKNOWLEDGMENTS

I am grateful to Pär Ingvar Haukeland for revising and editing the manuscript.

LITTERATURE CITED

Bakke, H. 1988. Opening address at the Research Policy Conference on Environment and Development, pp. 17–21. In: *One Earth – One World.* Report from a Research Policy Conference on Environment and Development. Oslo, Norway, 22–23 March 1988. NAVF, Oslo.

Næss, A. 1986. Intrinsic value: Will the defenders of nature please rise?, pp. 504–515. In: M. E. Soulé (ed) *Conservation Biology: The Science of Scarcity and Diversity.* Sinauer, Sunderland, MA.

Næss, A. 1989. *Ecology, Community and Lifestyle.* (Translated and edited by David Rothenberg). Cambridge University Press, Cambridge, UK.

Næss, A. 1990. Sustainable development and deep ecology, pp. 87–96. In: J. R. Engel and J. G. Engel (eds) *Ethics of Environment and Development.* Belhaven Press, London.

World Commission on Environment and Development. 1987. *Our Common Future.* WCED, Oxford University Press, Oxford.

SECTION 7

Recommendations: actions in biodiversity conservation

During the International Conference on Conservation of Genetic Resources for Sustainable Development, where the papers in this book were originally presented, the participants worked out three sets of recommendations for action in biodiversity conservation addressing policy makers, managers and researchers. In this section we include the executive summary of these recommendations. The complete version of the recommendations is available from the Norwegian Institute for Nature Research (NINA, Tungasletta 2, N-7005 Trondheim, Norway).

RECOMMENDATIONS

Biodiversity, the product of thousands of millions of years of biological evolution, has always provided humans with food, shelter, and cultural enrichment. Now we are destroying vast numbers of species and gene combinations and entire ecosystems at a time when human dependence on genetic resources and ecosystem services is increasing. The current erosion of biodiversity is incompatible with planetary integrity and human dignity and with material welfare.

It is clear that sustainable economic uses of these genetic and other natural resources will depend on fundamental changes in the ways humans interact with each other and with other species. Such changes will depend on a recasting of the political principles and contracts that govern international relations, production and consumption and the exchange of information and commodities between human groups and nations. Consumers must pay the real costs of resource utilization. It will not be possible to conserve biodiversity while global poverty and inequality are rampant, and these problems must be addressed and dealt with in order to get the full support of the people of the developing world. In addition, a strictly utilitarian conceptualization of species and genetic diversity is inconsistent with the long-term protection of biodiversity, and the maintenance of genetic resources for the benefit of future generations.

Certain conditions must be met before science and technology can further the objectives of economic, social, and biological well-being, benefitting humans and minimizing further biological and genetic degradation and impoverishment. In particular:

1. Current economic (e.g. production, trade, and finances) and demographic trends and policies that cause rampant habitat destruction and fragmentation and create new poverty and inequality must be altered. So must the unsustainable exploitation of biological resources, the careless introduction of invasive species, and the pollution of air and water that are causing massive changes in the environment.
2. Governments, agencies, and foundations must encourage the discovery of new biological and social principles and knowledge that will facilitate the description, analysis, protection, and sustainable utilization of genetic resources.
3. Local communities must assert and be granted their rights to enjoy, manage, and profit from the wise use of biodiversity, and should also share the responsibility with national governments and international agencies for its protection.

All human beings are responsible for sustainable use of the planet's resources. In particular:

POLICY MAKERS NEED TO:

- Support the negotiations for the proposed Convention on Biological Diversity and Sustainable Use of Biological Resources and urge that this Convention should be seen in a broader framework of other international conventions and agreements that deal with global poverty and inequality.
- Establish national, regional, and international biodiversity objectives and strategies, including appropriate payments for genetic resources to the countries of origin.
- Establish networks of protected areas, research centers for tropical ecology, centers for *ex situ* conservation, and educational programs.
- Support studies of traditional uses of germplasm and species.
- Support cleaner, less ecologically harmful, technologies.

- Cooperate on local, national and international levels in regulation, risk assessment and management of introductions of any organism (genetically engineered or not).
- Require multi-disciplinary and multisectorial teamwork in economic development projects and programs.
- Develop population and energy policies consistent with community interests and the maintenance of biodiversity.
- Create laws that allow local communities to decide, set policies, and manage.
- Integrate biodiversity conservation in policies for all relevant sectors, including agriculture, forestry, industrial development, and energy supply, and rationalize inter-agency (ministry) responsibilities for biodiversity matters.

ADMINISTRATORS AND MANAGERS NEED TO:

- Facilitate biodiversity surveys, inventories, and long-term monitoring programs.
- Involve local communities in decision making, policy and management, including harvesting and other uses.
- Sponsor the preservation and support of traditional agroecological practices.
- Support research that may foster the maintenance of genetic and species diversity.
- Consider how best to avoid the genetic introgression of natural and captive stocks.
- Develop guidelines for evaluation of the ecological and genetic consequences of releasing novel organisms, whether they are exotic or produced by conventional breeding or genetic engineering.
- Protect populations that may contain genes that provide resistance to environmental stress.
- Cooperate with other managers and agencies in local, regional and international consortia for the protection of genetic resources and biodiversity, both *in situ* and *ex situ* on an integrated basis.
- Document, value and publicize the benefits of local biodiversity and ecosystem services, and genetic diversity.
- Include economists, political, natural and social scientists in planning and management teams.

RESEARCHERS NEED TO:

- Develop standard procedures for the measurement of biodiversity.
- Implement biodiversity surveys, inventories, ethnobiological studies, and long-term monitoring programs.
- Develop better methods for hazard evaluation, analyzing the long-term impacts of development, and for population and ecosystem viability analyses.
- Develop methods for managing and controlling novel organisms, whether they are exotic or produced by conventional or *in vitro* genetic procedures.
- Initiate long-term studies on demographic, genetic and environmental variation.
- Elucidate the relationship between the standing genetic variability of populations and their long-term fitness, i.e. their capacity to adapt to environmental change.
- Document the history and analyze the long-term effects of selective harvesting and other uses of semi-natural ecosystems, including traditional methods of exploitation and management.
- Study methods of agricultural intensification that depend on traditional, agroecological practices.
- Analyze the role of particular (strongly interacting) species and mutualisms in maintaining diversity, and roles of ge-

netic and demographic variation and structure in the persistence of such species.

- Develop better methods for the harvest and/or storage of recalcitrant seeds, tissues, gametes, embryos, for *in vitro* fertilization, and criteria for captive propagation and reintroduction of endangered species.
- Initiate research programs at the species, ecosystem and landscape levels to facilitate the design of protected area systems; this includes restoration, patch dynamics, conservation corridors, metapopulation dynamics, area and edge effects, and human uses of protected areas.
- Develop and analyze models integrating ecological and socioeconomic processes.

Index